W9-BYV-841

SEEDS
Physiology of
Development and
Germination

SEEDS

Physiology of Development and Germination

J. Derek Bewley

Plant Physiology Research Group
Department of Biology
University of Calgary
Calgary, Alberta, Canada

and

Michael Black

Department of Biology
Queen Elizabeth College
University of London
London, England

Plenum Press • New York and London

Library of Congress Cataloging in Publication Data

Bewley, J. Derek, 1943–

Seeds: physiology of development and germination.

Bibliography: p.
Includes index.
1. Seeds — Physiology. 2. Germination. I. Black, Michael. II. Title.
QK661.B49   1984   582′.0467   84-13444
ISBN 0-306-41687-5

J. Derek Bewley
Plant Physiology Research Group
Department of Biology
University of Calgary
Calgary, Alberta T2N 1N4
Canada

Michael Black
Department of Biology
Queen Elizabeth College
Campden Hill
London W8 7AH
England

© 1985 Plenum Press, New York
A Division of Plenum Publishing Corporation
233 Spring Street, New York, N.Y. 10013

Printed in the United States of America

Preface

Since the publication of our monograph on seed physiology and biochemistry (*The Physiology and Biochemistry of Seeds in Relation to Germination,* Springer-Verlag, 1978, 1982), it has been suggested to us that a text covering the same subject area would be appropriate. This book is our response. Unlike the previous volumes, however, this text is not intended to be either a critical or a comprehensive account. Instead it is a more generalized consideration of the essential aspects of seed physiology and biochemistry as we see them. It also includes a substantial amount of new and different material. In a work of this sort it is inevitable that some simplifications must be made, but we hope, nevertheless, that we have presented the most reasonable conspectus of areas of controversy and uncertainty. In this respect, literature citations have been kept to a minimum and do not interrupt the text; they are placed at the end of each chapter and are intended to be used as a source for further references.

We hope that this book will be of value to students and teachers in universities, colleges, and other institutes of higher learning whose courses include plant biology. Although it is particularly appropriate for studies of seed biology, it should also find broader applications in general plant physiology, agriculture, and horticulture. We have assumed in parts of our discussions that the reader has some familiarity with the basics of plant physiology and biochemistry, but we have included details of some fundamental biochemical processes where necessary for clarity of discussion.

Our thanks are offered to those who have contributed in various ways to the production of this book—in the provision of illustrations, photographs, and unpublished material, and in discussions. M.B. gratefully acknowledges financial assistance from the British Council for travel to Calgary. J.D.B. thanks the University of Calgary for the award of a Killam Resident Fellowship for part of the time during which this book was written. Our debt to Erin Smith of the Department of Biology, University of Calgary, for typing the manuscript so expertly and expeditiously is inestimable. We deeply appreciate her skillful and cheerful assistance.

J. D. Bewley
M. Black

Contents

Chapter 3

Storage, Imbibition, and Germination

Chapter 4

Cellular Events during Germination and Seedling Growth

Chapter 6

Some Ecophysiological Aspects of Germination

Chapter 7

Mobilization of Stored Seed Reserves

Chapter 8

Control of the Mobilization of Stored Reserves

Chapter 9

Seeds and Germination: Some Agricultural and Industrial Aspects

Chapter 1

Seeds
Germination, Structure, and Composition

1.1. INTRODUCTION

The new plant formed by sexual reproduction starts as an embryo within the developing seed, which arises from the ovule. When mature, the seed is the means by which the new individual is dispersed, though frequently the ovary wall or even extrafloral organs remain in close association to form a more complex dispersal unit as in grasses and cereals. The seed therefore occupies a critical position in the life history of the higher plant. The success with which the new individual is established—the time, the place, and the vigor of the young seedling—is largely determined by the physiological and biochemical features of the seed. Of key importance to this success are the seed's responses to the environment and, on a biochemical level, its food reserves, which are available to sustain the young plant in the early stages of growth before it has become an independent, autotrophic organism, able to use light energy. Man also depends on these activities for almost all of his utilization of plants. Cultivation of most crop species depends on seed germination, though, of course, there are exceptions when propagation is carried out vegetatively. Moreover, seeds such as the cereals and legumes are themselves major food sources whose importance lies in the storage reserves of protein, starch, and oil laid down during development and maturation.

The biological and economic importance of seeds is evident, therefore. In this book we will give an account of processes involved in their development, in germination and its control, and in the utilization of the seed reserves during the early stages of seedling growth.

1.2. SEED GERMINATION—SOME GENERAL FEATURES

In the scientific literature the term *germination* is often used loosely and sometimes incorrectly, and so it is important to clarify its meaning. Germination begins with water uptake by the seed (imbibition) and ends with the start of elongation by the embryonic axis, usually the radicle. It therefore includes

1

numerous events, e.g., protein hydration, subcellular structural changes, respiration, macromolecular syntheses, and cell elongation, none of which is itself unique to germination. But their combined effect is to transform a dehydrated, resting embryo with a barely detectable metabolism into one that has a vigorous metabolism culminating in growth. Germination *sensu stricto* therefore does not include seedling growth, which commences when germination finishes. Hence, it is incorrect, for example, to equate germination with seedling emergence from the soil since germination will have ended sometime before the seedling is visible. Seed testers often refer to germination in this sense because their interests lie in monitoring the establishment of a vigorous plant of agronomic value. Although, as physiologists, we do not encourage such a definition of the term *germination,* we acknowledge its widespread use by seed technologists. Processes occurring in the nascent seedling, such as mobilization of the major storage reserves, are also not part of germination: they are postgermination events.

A seed in which none of the germination processes is taking place is said to be *quiescent.* Quiescent seeds are resting organs, generally at a low hydration level, with metabolic activity almost at a standstill. A remarkable property of seeds is that they are able to survive in this state, often for many years, and subsequently resume a normal, high level of metabolism. For germination to occur, quiescent seeds generally need only to be hydrated under conditions that encourage metabolism, for example, a suitable temperature and in the presence of oxygen.

Components of the germination process may, however, occur in a seed that does not achieve radicle emergence. Even when conditions are apparently favorable for germination, so that imbibition, respiration, synthesis of nucleic acids and proteins, and a host of other metabolic events all proceed, culmination in cell elongation does not occur, for reasons that are still poorly understood: such a seed expresses *dormancy.* Seeds that are dispersed from the mother plant already containing a block to the completion of germination show *primary dormancy.* Sometimes, a block(s) to germination develops in hydrated, mature seeds when they experience certain environmental conditions, and such seeds show *induced* or *secondary* dormancy. Dormant seeds are converted into germinable seeds (i.e., dormancy is broken) by certain "priming" treatments such as a light stimulus or a period at low or alternating temperature which nullify the block to germination but which themselves are not needed for the duration of the germination process. The relationships between these properties of seeds are shown in Fig. 5.1.

1.3. MEASUREMENT OF GERMINATION

The extent to which germination has progressed can be determined roughly, say by measuring water uptake or respiration, but these measure-

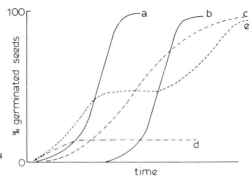

Figure 1.1. Generalized time courses of germination. See text for details.

ments give us only a very broad indication of what stage of the germination process has been reached. The only stage of germination that we can time fairly precisely is its termination! Emergence of the axis (usually the radicle) from the seed normally enables us to recognize when germination has gone to completion, though in those cases where the axis may grow before it penetrates through the surrounding tissues, the completion of germination can be determined as the time when a sustained rise in fresh weight begins.

We are generally interested in following the germination behavior of large numbers of seeds, for example, all the seeds produced by one plant or inflorescence, or all those collected in a soil sample, or all those subjected to a certain experimental treatment. The degree to which germination has been completed in a population is usually expressed as a percentage, normally determined at time intervals over the course of the germination period. Figures 1.1 and 1.2 show some examples of germination curves, about which some general points should be made. Germination curves are usually sigmoid—a minority of the seeds in the population germinates early, then the germination percentage increases more or less rapidly, and finally the relatively few late

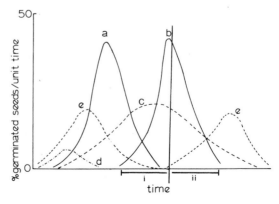

Figure 1.2. Time distributions of germination, derived from Fig. 1.1. The vertical line on curve (b) indicates the midpoint of the germination period. Note that the curve is positively skewed; i.e., more seeds germinate in the first half (i) of the period than in the second half (ii). See text for details.

germinators emerge. The curves are often positively skewed because a greater percentage germinates in the first half of the germination period than in the second (see the discussion on uniformity). But although the curves have the same general shape, important differences in behavior between populations are evident. For example, curve (d) in Fig. 1.1 flattens off when only a low percentage of the seeds has germinated, showing that this population has a low *germination capacity;* i.e., the proportion of seeds capable of completing germination is low. Assuming that these seeds are viable, the behavior of the population could be due to dormancy or to environmental conditions, such as temperature or light, which do not favor germination of most of the seeds.

The shape of the curves also depends on the *uniformity* of the population, i.e., the degree of simultaneity or synchrony of germination. It is especially clear that the seed population represented by curve (e) in Fig. 1.1 is not uniform. Here, a limited percentage of seeds succeeds in germinating fairly early, but the remainder begin to do so only after a delay. The population therefore seems to consist of two discrete groups—the quick and the slow germinators [see Fig. 1.2, curve (e)]. This example also illustrates the point that populations with the same germination capacity [e.g., (a) and (e) in Fig. 1.1] can differ in other respects. The population depicted in curve (c) in Fig. 1.1 also lacks uniformity as individual seeds complete their germination in very different periods of time [see also Fig. 1.2, curve (c)]. On the other hand, highly uniform behavior is displayed by seeds represented by curves (a) and (b) in Fig. 1.1, the steepness being due to the fact that the majority of seeds complete the germination process over a relatively short time period [see also Fig. 1.2, curves (a) and (b)]. It is important to note that distribution curves of the type shown in Fig. 1.2 are often positively skewed because more seeds germinate in the first than in the second half of the germination period; i.e., there is frequently not a normal distribution of germination in time.

But although seed samples may be alike as far as both germination capacity and uniformity are concerned [e.g., (a) and (b) in Figs. 1.1 and 1.2], they may be very different in one respect—their *rate* of germination. The rate of germination can be defined as the reciprocal of the time taken for the process to be completed, starting from the time of sowing. This can be determined for an individual seed, but it is generally expressed for a population. Here, the mean time to complete germination ($\bar{t}$) is equal to $\Sigma(t \cdot n)/\Sigma n$, where t is the time in days, starting from day 0, the day of sowing, and n is the number of seeds completing germination on day t. The mean germination rate (R) therefore equals $\Sigma n/\Sigma(t \cdot n)$. A value sometimes used is the coefficient of the rate of germination (CRG), which equals $R \times 100$. Alternatively, a measure of the rate of germination can be based on the time required by an arbitrary percentage of seeds, generally 50%, to complete the germination process. At this point it is worthwhile returning briefly to the matter of uniformity. A seed population which is highly uniform is one in which individual germination rates are close to the mean rate of germination for the population as a whole. Uniformity can therefore be expressed as the variance of individual times around the mean time. If we assume a normal distribution of the time to complete germination (often this is not strictly the case, however—see the previous discussion), then the coefficient of uniformity of germination (CUG) is given by: $CUG = \Sigma n / \Sigma[(\bar{t} - t)^2 \cdot n]$, and the higher the value, the greater is the uniformity.

It should now be clear that the populations represented by curves (a) and (b) in Fig. 1.1 have very different germination rates and that the parallelism of the curves indicates only that they have similar uniformity. Of course, populations with similar germination rates may differ markedly in uniformity or in other respects, such as germination capacity.

The behavior of a seed population with respect to germination, therefore, has several quantitative aspects that must be considered, and quantification obviously should not be limited to one parameter—say maximum germination percentage (germination capacity) or germination rate. Several attempts have been made to incorporate two parameters, rate and capacity, in a mathematical expression, so that population behavior can be described by a single statistic, but some of these are not entirely satisfactory, for several reasons. Mathematical treatments which take account of both the rates of germination and total germination, and which may be adequate for dealing with germination data in some circumstances, include use of the normal distribution function and of polynomial regression methods of curve fitting. Probit analysis can also be of great value in handling germination data. It is beyond the scope of this book to deal further with these mathematical methods, but details can be found in works listed at the end of this chapter.

1.4. SEED STRUCTURE

The seed develops from the fertilized ovule; this process is discussed in more detail in Chapter 2. At some stage during its development the angiosperm seed is usually comprised of (1) the embryo, the result of the fertilization of the egg cell in the embryo sac and one of the male pollen tube nuclei; (2) the endosperm, which arises from the fusion of two polar nuclei in the embryo sac with the other pollen tube nucleus; (3) the perisperm, a development of the nucellus; and (4) the testa or seed coat, formed from one or both of the integuments around the ovule. Although all mature seeds contain an embryo (sometimes poorly developed), and many are surrounded by a distinguishable seed coat, the extent to which the endosperm or perisperm persists varies between species. Sometimes the testa exists in a rudimentary form only, the prominent

Table 1.1. Is the Dispersal Unit a Seed or a Fruit?—Some Examples

Seed	Fruit (and type)
Legumes (e.g., peas, beans)	Cereals (caryopsis)
Cotton	Lettuce, sunflower, and other Compositae (cypsela)
Rape	Ash and elm (samara)
Castor bean	Hazel and oak (nut)
Tomato	Buttercup, anemone, and avens (a collection of
Squashes (e.g., cucumber, marrow)	achenes)
Coffee bean	

outermost structure being the pericarp or fruit coat derived from the ovary wall; in these cases, the dispersal unit is not a seed, but a fruit (Table 1.1). In gymnosperm seeds there is no fusion between the male and polar nuclei leading to the formation of a triploid endosperm—in these the storage tissue in the mature seed (which is functionally similar to the true endosperm) is haploid and is the modified megagametophyte (Figs. 1.3 and 2.1).

More rarely, seeds are produced by nonsexual processes, such as by apomixis (i.e., from diploid cells in the ovule)—such seeds are often indistinguishable from those of the same species resulting from sexual reproduction. The dandelion, and some other composites, have come to rely exclusively, or almost exclusively, on apomictic reproduction.

Let us now consider briefly each of the seed components.

1.4.1. Embryo

The embryo is comprised of the embryonic axis and one or more cotyledons. The axis incorporates the embryonic root (radicle), the hypocotyl to which the cotyledons are attached, and the shoot apex with the first true leaves (plumule). These parts are usually easy to discern in the dicot embryo (Figs. 1.3 and 2.2), but in monocots—particulary the Gramineae—identifying these parts is considerably more difficult. Here the single cotyledon is much reduced and modified to form the scutellum (Figs. 1.3 and 7.1), the basal sheath of the cotyledon is elongated to form a coleoptile covering the first leaves, and in some species (e.g., maize, Fig. 3.22) the hypocotyl is modified to form a mesocotyl. The coleorhiza is regarded as the base of the hypocotyl sheathing the radicle.

The shapes of embryos and their sizes in relation to other structures within the seed are variable. In monocot and dicot species with a well-developed endosperm in the mature seed, the embryo occupies less of the seed than in nonendospermic seeds (Figs. 1.3 and 2.24). Cotyledons of endospermic seeds are often thin and flattened since they do not store much in the way of reserves

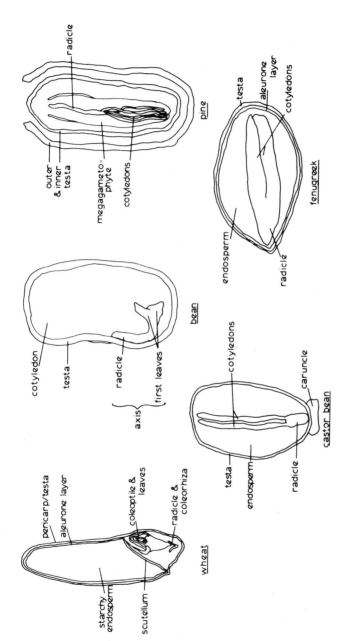

Figure 1.3. The location of tissues found in monocot, dicot, and gymnosperm seeds. Not drawn to scale.

(e.g., castor bean); in nonendospermic seeds, such as many of the legumes, the cotyledons are the site of reserve storage and account for almost all of the seed mass (Fig. 1.3). Cotyledons of nonendospermic, epigeal (Fig. 3.21) species (such as some members of the squash family) which are borne above the ground after germination and become photosynthetic are relatively not as large, nor do they contain as much stored reserves as the subterranean hypogeal type. The cotyledons are absent from seeds of many parasitic species; in contrast, the embryos of many coniferous species contain several cotyledons (polycotyledonous) (Figs. 1.3 and 2.1).

Polyembryony, i.e., more than one embryo in a seed, occurs in some species, e.g., *Poa alpina* and *Citrus* and *Opuntia* spp. This can arise because of cleavage of the fertilized egg cell to form several zygote initials, development of one or more synergids (accessory cells in the embryo sac), the existence of several embryo sacs per nucellus, and various forms of apomixis and adventitious embryony (from diploid cells of the nucellus). In *Linum usitatissimum* and other species some of the embryos formed by polyembryony are haploid.

Not all seeds contain mature embryos when liberated from the mother plant. Orchid seeds contain minute and poorly formed embryos and no endosperm. The final developmental stages of the embryo in other species occur after the seed has been dispersed, e.g., *Fraxinus* (ash) species and *Heracleum sphondylium*.

1.4.2. Nonembryonic Storage Tissues

In most species the perisperm, derived entirely from the maternal nucellar tissue of the ovule, fails to develop and is quickly absorbed as the embryo becomes established. In a few species, of which coffee and *Yucca* are the most noted examples, the perisperm is the major store of the seed's food reserves. In these seeds the endosperm is absent, although in others it may be developed to a greater (*Acorus* spp.) or lesser (*Piper* spp.) extent than the perisperm. In beet seeds both the perisperm and cotyledons of the embryo contain substantial reserves—there is no endosperm.

Seeds can be categorized as endospermic or nonendospermic in relation to the presence or absence in the mature seed of a well-formed endosperm. Some seeds are not generally regarded as being endospermic, even though an endosperm is present—in these cases the endosperm may be only a remnant of that broken down during seed development (e.g., soybean and peanut), or it may be only one to a few cell layers thick (e.g., lettuce). Here another structure, usually the cotyledon, is the principal storage organ. Some endosperms are relatively massive and are the major source of stored reserves within the mature seed, e.g., in cereals, castor bean, date palm, and the endospermic legumes such as fenugreek, carob, and honey locust. In the cereals and some endospermic legumes (e.g., fenugreek) the majority of cells in the endosperm are nonliving at maturity, the cytoplasmic contents having been occluded by the stored reserves during development. But on the outside of the endosperm remains a living tissue, the aleurone layer, which does not store reserves, but rather may be ultimately responsible for the release of enzymes for their mobilization (Figs. 1.3 and 7.6). Endosperms with a high water retention capacity may have a dual role: to provide reserves for the germinated embryo, and to regulate the water balance of the embryo during germination (e.g., in fenugreek seeds). An unusual endosperm is that of the coconut, in that part of it remains acellular and liquid.

1.4.3. Seed Coat—Testa

Variability in the anatomy of the testa is considerable, and it has been used taxonomically to distinguish between different genera and species. Hence a discussion of the range of seed coat structures is well beyond the scope of all but specialist monographs. The testa is of considerable importance to the seed because it is often the only protective barrier between the embryo and the external environment (in some species the fruit coat, and even the endosperm, support or provide a substitute for this role). The protective nature of the seed coat can be ascribed to the presence of an outer and inner cuticle, often impregnated with waxes and fats, and one or more layers of thick-walled protective cells (Fig. 5.5). Coats may contain mucilaginous cells which burst upon contact with water, providing a water-retaining barrier around the seeds. Such barriers may also restrict oxygen uptake, as will the presence of phenolics, and there are other structural features in some coats that restrict exchange of gases between the embryo and environment. Some coats are largely impermeable to water and consequently can restrict the metabolism and growth of inner tissues.

The coloring and texture of seed coats are distinguishing features of many seeds but sometimes cannot be used taxonomically because they may change due to environmental and genetic influences during development (such polymorphism in seeds is discussed in Section 5.2). Upon detachment from the mother plant, the seed coat bears a scar, called the hilum, marking the point at which it was joined to the funiculus. At one end of the hilum of many seed coats can be seen a small hole, the micropyle. Rarely, hairs or wings develop on the testa to aid in seed dispersal (e.g., in willow, lily, *Epilobium* spp.); more usually the dispersal structures are a modification of the enclosing fruit coat. Outgrowths of the hilum region may be seen—the strophiole, which restricts water movement into and out of some seeds, and the aril, which often contains chemicals; in some species these attract animals important in dispersal of the seeds. In castor bean the aril is associated with the micropyle and is called the caruncle. Arils are variable in shape, forming knobs, bands, ridges, or cupules, and are often brightly colored. The aril of the nutmeg is used as a source of the spice mace; the seed coat contains different chemicals and is ground up for the spice named after the seed itself.

1.5. SEED STORAGE RESERVES

About 70% of all food for human consumption comes directly from seeds (mostly those of cereals and legumes), and a large proportion of the remainder is derived from animals that are fed on seeds. It is not surprising, therefore, that there is a wealth of literature concerned with the chemical, structural, and nutritional composition of seeds. Unfortunately, we will be able to do it only the briefest of justice. Most of our knowledge of the chemical composition of seeds is for cultivated species since they comprise such a large share of our food source and also provide a great many raw materials for industry. Information on seeds of wild species and wild progenitors of our cultivated crops is relatively scarce. But with increasing interest in new food sources and in improved genetic diversity within domesticated lines, the seeds of wild plants are now receiving more attention.

In addition to the normal chemical constituents found in all plant tissues, seeds contain extra amounts of substances stored as a source of food reserves to support early seedling growth. These are principally carbohydrates, fats and oils, and proteins. Seeds contain other minor, but nevertheless important reserves (e.g., phytin); of these, a number are recognized as being nutritionally undesirable or even toxic (e.g., alkaloids, lectins, proteinase inhibitors, phytin, and raffinose oligosaccharides).

The chemical composition of seeds is determined ultimately by genetic factors and hence varies widely among species and their varieties and cultivars. Some modifications of composition may result from agronomic practices (e.g., nitrogen fertilizer application, planting dates) or may be imposed by environmental conditions prevalent during seed development and maturation, but such changes are usually relatively minor. Through crossing and selection, plant breeders have been able to manipulate the composition of many seed crops to improve their usefulness and yield. Modern cultivars of many cereals and legumes store significantly higher quantities of food material than either earlier cultivars or their wild progenitors. Even so, some nutritional deficiencies remain to be rectified; e.g., the composition of the storage proteins of legumes and cereals is such that they do not provide all of the amino acids required by simple-stomached (monogastric) animals such as humans, pigs, and poultry

Table 1.2. The Food Reserves of Some Important Crop Species[a]

	Average percent composition			Major storage organ
	Protein	Fat	Carbohydrate[b]	
Cereals				
Barley	12	3[c]	76	Endosperm
Dent corn (maize)	10	5	80	Endosperm
Oats	13	8	66	Endosperm
Rye	12	2	76	Endosperm
Wheat	12	2	75	Endosperm
Legumes				
Broad bean	23	1	56	Cotyledons
Garden pea	25	6	52	Cotyledons
Peanut	31	48	12	Cotyledons
Soybean	37	17	26	Cotyledons
Other				
Castor bean	18	64	Negligible	Endosperm
Oil palm	9	49	28	Endosperm
Pine	35	48	6	Megagametophyte
Rape	21	48	19	Cotyledons

[a]After Crocker and Barton (1957) and Winton and Winton (1932).
[b]Mainly starch.
[c]In cereals, fats are stored within the scutellum, an embryonic tissue.

(see Sections 1.5.3 and 2.3.4). Some indication of the variability in the composition of stored reserves in seeds is to be found in Table 1.2: the important storage tissue within each seed is indicated.

As indicated in Table 1.2, the major storage reserves may be deposited within the embryo—usually the cotyledons, although in the Brazil nut they occur within the radicle/hypocotyl—or within extraembryonic tissues such as the endosperm, the megagametophyte (in gymnosperms), or, rarely, the perisperm (e.g., coffee, *Yucca*). In many seeds the stored reserves may occur within both embryonic and extraembryonic tissues, but in different proportions, e.g., in maize (Table 1.3). Different reserves may even be located within dif-

Table 1.3. Percent Composition of Stored Reserves in Different Parts of a Maize (cv. Iowa 939) Kernel[a]

Reserve	Whole grain	Endosperm (starchy and aleurone layer)	Embryo (including scutellum)
Starch	74	88	9
Lipid	4	< 1	31
Protein	8	7	19

[a]After Earle *et al.* (1956).

ferent storage tissues; in fenugreek seeds, for example, the endosperm is the exclusive source of carbohydrate (as cell wall galactomannan), but the cotyledons contain the protein and fats. The reserves may be distributed unevenly within any one storage tissue; returning to the maize grain, we find that there are protein-rich regions (horny endosperm) and starch-rich regions (floury endosperm) within the endosperm (see Fig. 2.24). Chemical differences can also exist within a species in relation to the distribution of any particular reserve—in rape, the fat in the cotyledons and hypocotyl contains different proportions of erucic and palmitic acids.

We shall now discuss each of the major reserves in more detail and briefly elaborate on the chemistry of some of the minor reserves. Chapter 2 deals with the deposition of stored reserves, and therein, of necessity, will be found some further discussion of their chemical composition and localization.

1.5.1. Carbohydrates

Carbohydrates are the major storage reserve of most seeds cultivated as a food source (Table 1.2). Starch is the carbohydrate most commonly found in seeds, although "hemicelluloses," "amyloids," and the raffinose series oligosaccharides may be present and may sometimes be the major carbohydrate reserve. Other carbohydrates that occur in nonstorage forms are cellulose, pectins, and mucilages.

Starch is stored in seeds in two related forms, amylose and amylopectin; both are polymers of glucose. Amylose is a straight-chain polymer some 300–400 glucose units in length—adjoining glucose molecules are linked by α-1,4 glucosidic linkages (Fig. 1.4, A). Amylopectin is much larger (10^2–10^3 times); it consists of many amylose chains linked via α-1,6 bonds to produce a multiple-branched molecule (Fig. 1.4, B). Starch is laid down in discrete subcellular bodies called starch grains. Most grains are composed of about 50–75% amylopectin and 20–25% amylose. Certain mutants of cereals have a higher- or lower-than-average content of amylose (Table 2.2), and starch grains of waxy mutants of maize are devoid of this polymer. In wrinkled peas, amylose accounts for two thirds or more of the starch compared with about one third in smooth peas and other legumes.

Starch grains may have an appearance characteristic for the individual species, and they may be predominantly spherical (barley), angular (maize), or elliptical (runner bean). Grain shape is to a large extent determined by amylose content—the higher the amount of amylose, the rounder the grain. Grain size is very variable, from 2–100 μm in diameter, even in the same seed. In the rye endosperm, for example, there are large oval starch grains which are up to 40 μm in diameter; there are also numerous smaller grains less than 10 μm in

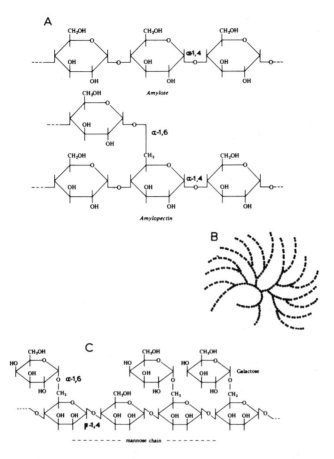

Figure 1.4. (A) The chemical composition of starch and (B) the structure of amylopectin. (C) The chemical structure of galactomannan. The chemical links between the molecules are noted.

diameter embedded in a fine network of cytoplasmic protein (Fig. 1.5A). In barley, the spherical starch grains are separable into two groups, large ones and small ones. Although the latter account for about 90% of the total number of grains, they comprise only 10% of the total starch by weight.

Hemicelluloses are the major form of stored carbohydrates in some seeds, particularly in certain endospermic legumes. Usually starch is absent from tissues where hemicelluloses are present in appreciable quantities. The endosperm or perisperm of the ivory nut, date, and coffee is extremely hard because of hemicelluloses laid down as very thick cell walls; there are no hemicellulose storage bodies. Many of the hemicelluloses are mannans, i.e., long-chain poly-

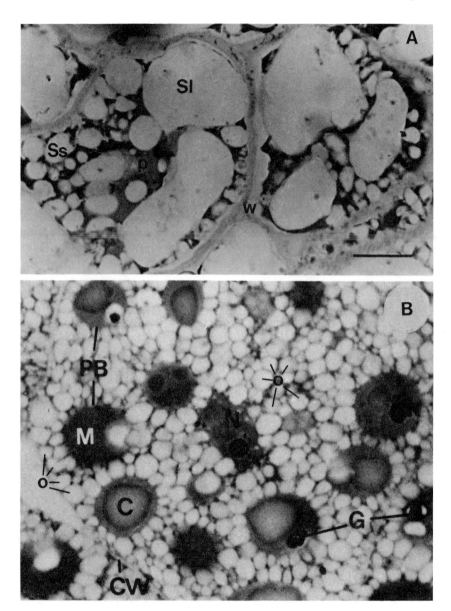

Figure 1.5. (A) Large (Sl) and small (Ss) starch grains embedded in the protein matrix (p) of the starchy endosperm cells of rye. w, cell wall. Bar = 10 μm. Courtesy of M. L. Parker (1981). (B) Endosperm of castor bean showing extensive packing with oil (fat) bodies (o). PB, protein body; C, crystalloid; G, globoid; M, matrix; N, nucleus; CW, cell wall. Courtesy of J. S. Greenwood.

mers of mannose (linked by $\beta 1,4$ bonds) with variable but small quantities of sugar present as side chains [e.g., galactose (linked by $\alpha 1,6$ bonds) in galactomannans: Fig. 1.4, C]. The number of galactose side chains present affects the consistency of the galactomannan; it varies from being very hard in almost pure mannans (coffee mannans contain 2% galactose) to being mucilaginous at much higher galactose contents. Within the Leguminosae the mannose:galactose ratio may have a taxonomic significance as well. Glucomannans, in which a large proportion of the mannose residues is replaced by glucose, are restricted to the endosperm of certain monocots, specifically members of the Liliaceae and Iridaceae. They, too, are components of thickened cell walls. Other hemicelluloses, the xyloglucans, which are substituted cellulose-type molecules, i.e., a linear β-1,4-linked glucose backbone with short xylose (and galactose) side chains, are found in cell walls of certain dicot embryos and endosperms, e.g., members of the Caesalpinoideae, a subfamily of the Leguminosae. Xyloglucans are also called "amyloids" because they react positively to a stain for starch.

Free sugars are rarely the main storage carbohydrate, but in sugar maple, which lacks starch, they may account for nearly 11% of the dry weight of the mature seed. Disaccharides (sucrose) and oligosaccharides (raffinose series oligosaccharides, Section 4.1.2) are commonly found as minor reserves in the embryo and reserve tissues. There is increasing evidence that they are an important source of sugars for respiration during germination and early seedling growth.

1.5.2. Fats and Oils (Lipids)

Most storage lipids of seeds are neutral fats, or oils if they are liquid above about 20°C; some seeds may contain appreciable quantities of phospholipids, glycolipids, and sterols too. Lipids are insoluble in water but soluble in a variety of organic solvents including ether, chloroform, and benzene. They are esters of glycerol and fatty acids

$$CH_2O \cdot OC \cdot R^1$$
$$|$$
$$R^2 \cdot CO \cdot OCH$$
$$|$$
$$CH_2O \cdot OC \cdot R^3$$

and also are called triglycerides or triacylglycerols. The number of carbon atoms in the fatty acid chains, denoted as R^1, R^2, and R^3, may be identical but usually is not.

Fatty acids are identified according to the number of carbon atoms and double bonds in their chain. Saturated fatty acids contain an even number of carbon atoms and no double bonds; e.g., palmitic acid (16:0—16 carbon atoms:no double bonds) is the most common saturated fatty acid in seed oils. But the predominant fatty acids in seeds are the unsaturated ones, and of these oleic (18:1) and linoleic (18:2) account for more than 60% by weight of all oils in oil seed crops. Less common fatty acids are erucic acid (22:1), a component of some rapeseed oils (although it has now almost been bred out of most commercial cultivars because it is toxic) and of the oil of *Crambe abyssinica,* and ricinoleic acid (12-hydroxy 18:1), the major component of castor (bean) oil. The storage lipids of jojoba are unusual in that they are wax esters of long-chain fatty acids and alcohols, and they are liquid. They are an important source for the manufacture of high-pressure lubricants (see Section 2.3.3).

The fatty acid composition of some important crop species is shown in Table 1.4. Both maize and sunflower oil are widely used as cooking oils and in margarine. Among their desirable properties for this purpose is their high linoleic acid content. Catalytic hydrogenation of this oil to form polyunsaturates changes its melting point: the degree of unsaturation left after hydrogenation determines whether the fat or oil is a solid at room temperature (required for margarine) or a liquid (cooking oils). Variable hydrogenation of peanut oil yields peanut butter of different consistencies. Another desirable property of maize and sunflower oils is their low linolenic acid content, for this triunsaturated fatty acid oxidizes readily during food storage to produce off-flavors. The advantage of using plant oils over animal fats in the diet is that because of their high polyunsaturate level after hydrogenation they are purported to reduce atherosclerosis, a claim not wholly substantiated in medical tests. Fats such as lard (Table 1.4) from animal sources are high in oleic acid and cannot

Table 1.4. The Major Fatty Acid Composition of Commercial Oils and Fats from Various Plant Sources[a]

Species	Palmitic (16:0)	Stearic (18:0)	Oleic (18:1)	Linoleic (18:2)	Linolenic (18:3)
Sunflower	6	4	26	64	0
Maize	12	2	24	61	< 1
Soybean	11	3	22	54	8
Cotton	27	3	17	52	0
Peanut	12	2	50	31	0
Oil palm	49	4	36	10	0
Linseed (flax)	—	—	—	77	17
Animal fat (lard)	29	13	43	10	0.5

[a]After Weber (1980) and Miller (1931).

easily be hydrogenated to form polyunsaturates; this process causes rapid saturation of the one double bond per molecule. Drying oils, such as those used in paints and lacquers, are also polyunsaturated after hydrogenation (e.g., linseed oil from flax and rapeseed oil) but contain a higher linolenic acid content. Upon exposure to air a free-radical polymerization reaction occurs, initiated by oxygen, which cross-links the oils to form a tough film.

The triglyceride reserves in seeds are laid down in discrete subcellular organelles—oil bodies (fat bodies or wax bodies, depending on the consistency of the stored product). They range in size from 0.2–6 μm in diameter, according to species. In high-fat-storing seeds the oil bodies occupy a substantial volume of the cell, as in the castor bean endosperm (Fig. 1.5, B). The ontogeny of the lipid body is discussed toward the end of Section 2.3.3.

1.5.3. Proteins

According to Osborne's classification, seed proteins can be divided into four classes in relation to their solubility: (1) *albumins*—soluble in water and dilute buffers at neutral pHs; (2) *globulins*—soluble in salt solutions but insoluble in water; (3) *glutelins*—soluble in dilute acid or alkali solutions; (4) *prolamins*—soluble in aqueous alcohols (70–90%). This classification was proposed at about the turn of this century, and although it is far from ideal, since modern workers now use different solvents and extraction procedures, a better one has still to be devised.

Most investigations on these various classes of proteins have been conducted on edible seeds of crop species. Hence we will concentrate here on the seeds of cereals and legumes whose protein composition is important in both human and animal nutrition. The percentage of protein in cereals is debatably sufficient to satisfy the requirements of human beings, when adequate calories are obtained, but it is insufficient to support the growth of farm animals. Moreover, the amino acid composition of both cereal and legume proteins is not ideally suited for the diet of human beings or animals.

The approximate proportions of the main protein classes in cereals are shown in Table 1.5, along with the commonly used names of some proteins. The major storage protein in maize, barley, and sorghum is of the prolamin type; in wheat it is glutelin, whereas in oats globulins predominate. The amino acid composition of the total protein fraction present in cereal grains (Table 1.6) is strongly influenced, not surprisingly, by the nature of the major storage protein. Albumins and globulins are not seriously deficient in specific amino acids, and hence, from a nutritional point of view, oats are a good source of dietary protein, particularly for breeding stock. In normal barley and maize grains, however, where prolamins are high, lysine is seriously limiting; trypto-

**Table 1.5. The Approximate Percent Protein Composition of Some
Cereals and the Common Names of Some Storage Proteins[a]**

Cereal	Albumin	Globulin	Prolamin	Glutelin
Wheat	9	5	40 (gliadin)	46 (glutenin)
Maize	4	2	55 (zein)	39
Barley	13	12	52 (hordein)	23 (hordenin)
Oats	11	56	9 (avenin)	23
Rice	5	10	5 (oryzin)	80 (oryzenin)
Sorghum	6	10	46 (kafirin)	38

[a]Based on Payne and Rhodes (1982).

phan is low and threonine levels are nutritionally inadequate. Prolamins are
high in proline and glutamic acid/glutamine. Attempts have been made to
modify the amino acid composition of the prolamin-rich cereals such as maize
and barley, and in particular to increase their lysine content (Section 2.3.4).
High-lysine mutants of maize (*opaque*-2) and barley *(hiproly)* exist; these also

**Table 1.6. Percentages of Amino Acids in Grain Protein of Oats,
Normal and *Opaque*-2 Maize, and Normal and *Hiproly* Barley[a]**

Amino acid	Maize		Barley		
	Normal	*Opaque*-2	Normal	*Hiproly*	Oats
Alanine	10.0	7.2	3.8	4.3	5.0
Arginine	3.4	5.2	4.6	4.9	6.9
Aspartic acid	7.0	10.8	6.0	6.8	8.9
Cystine[b]	1.8	1.8	1.1	0.9	1.6
Glutamic acid	26.0	19.8	26.8	23.9	23.9
Glycine	3.0	4.7	3.6	3.9	4.9
Histidine	2.9	3.2	2.1	2.2	2.2
Isoleucine[b]	4.5	3.9	3.7	3.9	3.9
Leucine[b]	18.8	11.6	6.7	7.1	7.4
Lysine[b]	1.6	3.7	3.4	4.2	4.2
Methionine[b]	2.0	1.8	1.2	1.5	2.5
Phenylalanine[b]	6.5	4.9	5.9	5.9	5.3
Proline	8.6	8.6	12.6	11.3	4.7
Serine	5.6	4.8	4.3	4.4	4.2
Threonine[b]	3.5	3.7	3.4	3.6	3.3
Tryptophan[b]	0.3	0.7	—	—	—
Tyrosine[b]	5.3	3.9	2.8	2.8	3.1
Valine	5.4	5.3	4.8	5.3	5.3

[a]Based on Frey (1977).
[b]Essential amino acids which must be provided in the animal diet, since the animal itself cannot synthesize them.

also contain less prolamin storage protein and more lysine-rich glutelin (Table 2.3), resulting in an altered overall amino acid composition (Table 1.6).

Characteristically, storage (holo-) proteins are oligomeric; that is, they are made up of two or more subunits that can be separated, after extraction, using mildly dissociating conditions. These subunits in turn may be made up of a number of polypeptide chains, which may vary slightly in amino acid composition between "homologous" subunits. This variation results in considerable heterogeneity of the native protein. For example, the prolamin fraction of maize (zein) consists of major subunits of mol. wt. 23,000 and 21,000 (23 and 21 kD) and a minor subunit of mol. wt. 13,500; separation of these subunits on the basis of their molecular charge (i.e., by isoelectric focusing) shows that together they are comprised of nearly 30 polypeptides. Similar complexity is shown by the wheat prolamin gliadin, which is separable into four major holoproteins (α, β, γ, and ω) made up totally of at least 46 discrete polypeptides. The wheat glutelins are even more complex, being polymeric molecules made up of 15 component proteins ranging from $11-133 \times 10^2$ kD. Hence most storage proteins should not be thought of as a single protein, but rather as a complex of individual proteins bound together by a combination of intermolecular disulfide groups, hydrogen bonding, ionic bonding, and hydrophobic bonding.

Legume seeds are the second most important protein source, on a world basis, after cereals. Nutritionally, they are generally deficient in the sulfur amino acids (cysteine and methionine), but unlike cereal grains, their lysine content is adequate (Table 1.7). The major storage proteins are globulins, which account for up to 70% of the total seed nitrogen. The globulins consist of two major families of proteins which differ in mol. wt. and which sediment during ultracentrifugation with sedimentation coefficients (S values) of approximately 7 (average 7–8) and 11 (average 11–13). These are the 7 S (called the vicilin group) and 11 S (the legumin group) proteins; both are holoproteins composed of regularly assembled subunits. The holoproteins form a series of related polymers of the vicilin or legumin type, although each type may differ quantitatively and qualitatively in subunit composition. For example, different pea *(Pisum sativum)* cultivars contain two or three different vicilins (av. 186 kD), each comprised of one or more major subunits near 18, 30, 50, or 75 kD, and minor amounts of 12-, 14-, and 24-kD subunits. Four distinct forms of pea legumin (av. 360 kD) have been isolated also, composed mainly of 20- and 40-kD subunits, with variable minor amounts of 18-, 25-, 27-, and 37-kD subunits. All four pea legumins are comprised of two different acidic subunits of 40 kD and six different basic subunits of 20 kD joined by disulfide bridges: the different forms of legumin are determined by their minor subunit composition.

The major storage protein in soybean cotyledons is the legumin glycinin, which is probably comprised of four acidic and four basic subunits making up

Table 1.7. The Amino Acid Composition of the
11 S and 7 S Globulin Storage Proteins of
Soybean Seeds (Expressed as mol. %)[a]

Amino acid	11 S	7 S
Alanine	6.2	3.7
Arginine	5.6	8.8
Aspartic acid	11.7	14.1
Cystine	0.6	0.3
Glutamic acid	21.4	20.5
Glycine	7.5	2.9
Histidine	1.7	1.7
Isoleucine	4.1	6.4
Lysine	3.9	7.0
Methionine	1.3	0.3
Phenylalanine	4.6	7.4
Proline	6.5	4.3
Serine	6.0	6.8
Threonine	3.8	2.8
Tryptophan	0.8	0.3
Tyrosine	2.7	3.6
Valine	5.2	5.1

[a]After Derbyshire et al. (1976).

the 12.3 S holoprotein of 330 kD. The 7 S storage fraction, β- and γ-congly-cinin, also contains a heterogenous group of proteins. β-Conglycinin, which is the predominant vicilin form, has an average mol. wt. of 160,000 and is made up of three subunits: α and α' of 57 kD and β of 42 kD. These may combine into six different isomeric forms made up of different subunit compositions (i.e., $\alpha'\beta$, $\alpha\beta$, $\alpha\alpha'\beta$, $\alpha\beta$, $\alpha\alpha$, and α), each varying slightly in amino acid and carbo-hydrate composition (the carbohydrate is present usually as glucose, mannose, or glucosamine, from 0.5–1.5%, because many subunits of 7 S storage proteins are glycosylated during their deposition—see Section 2.3.4). Soybean storage protein also contains a 2 S form, α-conglycinin, which is heterogeneous and includes several enzymes as well as trypsin inhibitor activity (Section 7.5.5). Storage proteins of other legumes show equal complexity, and although it is not necessary to deal with this in greater depth here, some examples are given in Table 1.8.

Not all seed storage proteins have nutritionally desirable properties. Pro-teinase inhibitors (Section 7.5.5) including trypsin inhibitor may reduce the effectiveness of proteolytic enzymes in the digestive tracts of animals. Lectins, in contrast to proteinase inhibitors, are usually glycoproteins and have the capacity to bind to animal cell surfaces, sometimes causing agglutination (par-ticularly of erythrocytes), e.g., concanavalin A in jack bean (Canavalia ensi-

formis) seeds—hence lectins are sometimes called phytohemagglutinins. This property is probably superfluous as far as the seed is concerned, and many lectins are innocuous from a nutritional standpoint. Some lectins are highly toxic, however, including ricin D from castor bean *(Ricinus communis)* and abrin from *Abrus precatorius,* both of which are mixtures of nonagglutinating toxins and nontoxic agglutinins. Any toxic effect of lectins can be generally eliminated by proper heat treatment. For example, castor bean meal may be fed as a protein supplement to cattle after extraction of castor oil, and heating.

Seed storage proteins are usually deposited within special cellular organelles called protein bodies. These range in diameter from 0.1–25 μm and are surrounded, at least during development, by a single membrane. In mature, dry storage tissues of some seeds, e.g., certain cereal grains (Section 2.3.4), the membrane is incomplete, or absent, leaving the protein dispersed in the cytoplasm. This is relatively uncommon, however. Some protein bodies are simple in that they consist of a protein matrix surrounded by a limiting membrane. Inclusions frequently occur, however, particularly crystalloids and globoids and, more rarely, druse (calcium oxalate) crystals (Fig. 1.6). The crystalloids are insoluble (in water or buffers) proteinaceous inclusions embedded in the soluble protein matrix; e.g., in castor bean (Fig. 1.7, A) the crystalloid is an insoluble 11 S protein, and the matrix is made up of 2 S and 7 S albumins, including lectins. Globoids are noncrystalline, globular structures and are the most commonly occurring inclusion in protein bodies (Fig. 1.7, A and B) although they vary in both size and number. In some species the globoids are found in protein bodies of one region of a seed but not in another (e.g., the aleurone-layer protein bodies—aleurone grains—of cereals usually contain

Table 1.8. Subunit Composition of the Globulin Storage Proteins of Some Legumes[a]

Species	Approximate sedimentation coefficients (S)	Average mol. wt. (kD)	Name of holoprotein	Subunits mol. wt. (kD)
Pisum sativum	7–8	186	Vicilin	12, 14, 18, 24, 30, 50, 75
(garden pea)	12–13	360	Legumin	18, 20, 25, 27, 37, 40
Vicia faba	7	150	Vicilin	31, 33, 46, 56
(broad bean)	11–14	328	Legumin	20, 37
Phaseolus vulgaris	6.5–7.5	150	Glycoprotein I	43, 47, 53
(French bean)	11	340	Glycoprotein II	30, 32, 34
Phaseolus aureus-				
Vigna radiata	8		Vicilin	24, 30, 50, 63
(mung bean)	11		Legumin	16, 44, 56
Glycine max	7–8	160	β-Conglycinin	42, 57
(soybean)	12	330	Glycinin	19, 37, 42

[a]Based on Miflin and Shewry (1981) and Larkins (1981).

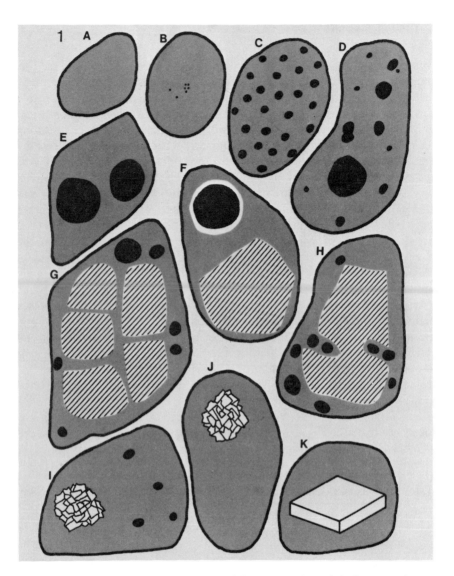

Figure 1.6. Diagrammatic representation of the different types of protein bodies present in storage tissues of mature dry seeds. (A) Protein body containing only matrix protein, e.g., in cotyledon mesophyll cells of some legumes. (B)–(E) Protein bodies with globoids of different sizes (black) embedded in the protein matrix. B, broad bean, soybean; C, some Compositae and Cruciferae; D, *Eucalyptus* and scutellum of wheat; E, cotyledon mesophyll of *Cassia*. (F) Globoid (black) surrounded by a soft globoid (white halo) and insoluble crystalloid protein (stripes) in the soluble protein matrix, e..g, cucumber cotyledons and castor bean endosperm. (G) and (H) as (F), but with more and scattered globoids and crystalloids in the matrix. G, cotyledon mesophyll of *Juglans regia;* H, radicle of Brazil nut. (I) and (J) Druse crystals of calcium oxalate with or without globoids in the protein matrix. I, hazel cotyledons; J, carrot endosperm. (K) Large single calcium oxalate crystal embedded in a protein matrix, e.g., carrot endosperm. From Lott (1981).

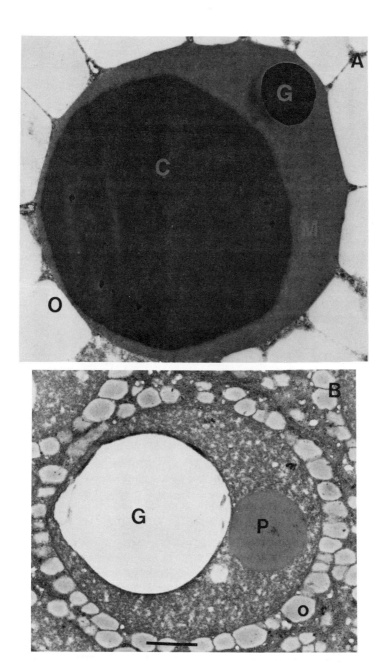

Figure 1.7. (A) Protein body of castor bean endosperm showing globoid (G) and crystalloid protein (C) embedded in a protein matrix (M). The protein body is surrounded by oil bodies (O). Courtesy of J. S. Greenwood. (B) Aleurone cell of barley with aleurone grain (protein body) containing globoid (G) and protein carbohydrate body (P) embedded in a protein matrix. The protein body is surrounded by oil bodies (O). Bar = 1 μm. From Jacobsen *et al.* (1971).

globoids, but protein bodies of the starchy endosperm never do). Globoids are the sites of deposition of phytin—the potassium, magnesium, and calcium salts of phytic acid (see Section 1.5.4); some globoids are surrounded by a soft-globoid region of unknown chemical composition. Barley aleurone-layer protein bodies also contain carbohydrate but in a unit distinct from the globoid, known as the protein–carbohydrate body (Fig. 1.7, B). Various enzymes may occur within the protein body, and during reserve mobilization other enzymes may be added so that it eventually becomes an autolytic vacuole (Section 7.5).

Protein bodies may contain only one type of storage protein. In certain legumes, for example, some protein bodies contain only albumin, or vicilin, or legumin, although most seem to contain both vicilin and legumin. In maize, the small protein bodies in the starchy endosperm contain proportionately more zein than do the larger bodies; protein bodies in the aleurone layer of this and other cereals do not contain any of the major reserve proteins.

1.5.4. Phytin

Phytin is the insoluble mixed potassium, magnesium, and calcium salt of *myo*-inositol hexaphosphoric acid (phytic acid), and although present in relatively minor quantities compared to the aforementioned reserves, it is an important source to the seed of phosphate and mineral elements. Additionally, iron, manganese, copper, and, more rarely, sodium are to be found in some phytin sources (Table 1.9).

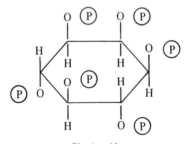

Phytic acid

The phytin is located exclusively within the globoid but, as mentioned in Section 1.5.3, not in all the protein bodies of the seed, nor is the mineral element composition of phytin the same in all cells; e.g., calcium is in highest amounts in the globoids of the radicle and hypocotyl regions of the embryo, but little, if any, of this element is found in the globoids of the cotyledons.

Phytic acid and its conjugates are generally regarded as being nutritionally undesirable since they can bind essential dietary minerals (e.g., zinc, cal-

Table 1.9. The Content of the Main Elements of Phytin in Various Species of Seeds, Expressed on a Percent Dry Weight Basis[a]

Species	Mg	Ca	K	P	Fe	Mn	Cu
Oat	0.4	0.19	1.1	0.96	0.035	0.008	0.005
Hazel	0.19	0.1	0.74	0.4			
Soybean	0.22	0.13	2.18	0.71			
Cotton	0.4	0.3	1.9	1.79	0.059	0.003	0.005
Barley	0.16	0.03	0.56	0.43			
Broad bean	0.11	0.05	1.13	0.51			
Sunflower	0.4	0.2	1	1.01			

[a]Taken from Weber and Neumann (1980).

cium, and iron), thus making them wholly or partially unavailable for absorption. The extent that this is a problem in the Western world is not fully understood, but the processing and preparation of many foods from seed sources for human consumption may remove most of the phytin. In third-world countries, where food is often less refined, the problem may be more acute.

1.5.5. Other Constituents

There is a wealth of minor constituents present within seeds that cannot be regarded strictly as storage components since the seed does not use them during germination or subsequent growth. Some, nevertheless, are worth mentioning briefly. Certain alkaloids, which are nonprotein nitrogenous substances, are important commercial sources of stimulants and drugs, e.g., theobromine from the cacao bean, caffeine from coffee and cocoa, strychnine and brucine from *Strychnos nux-vomica* (present as 2.5% of seed dry weight), and morphine from certain types of poppy. In nature, such compounds can prevent insects and animals from using a seed for food. Phytosterols such as sitosterols and stigmasterols are present in some seeds, e.g., in soybean. The latter is important pharmaceutically because it can be converted to the animal steroid hormone progesterone. Certain nonprotein amino acids may be found in considerable quantities in some seeds; e.g., canavanine accounts for 8.3% of the dry weight of *Dioclea megacarpa* seeds, hydroxytryptophan accounts for 14% of the dry weight of *Griffonia,* and dihydroxyphenylalanine (L-dopa) makes up 6% of dry *Mucuna* seeds. These amino acids are probably mobilized by the seed as a nitrogen source after germination. Glucosides are bitter-tasting components of some seeds, e.g., amygdalin from almonds, peaches, and plums and aesculin from horse chestnut; some, such as saponin from tung seeds, may be deadly to man and animals.

Phenolic compounds such as coumarin and chlorogenic acid, and their derivatives, and ferulic, caffeic, and sinapic acids occur in the coats of many seeds. These may inhibit germination of the seed that contains them or, being leached out into the soil, may inhibit neighboring seeds. Another germination inhibitor present in seeds is abscisic acid, although germination promoters and growth substances such as gibberellins, cytokinins, and auxins may also be present (for more details see Section 2.5).

USEFUL LITERATURE REFERENCES

SECTIONS 1.1 AND 1.2. SOME ADVANCED LITERATURE ON SEEDS AND GERMINATION

Bewley, J. D., and Black, M., 1978, 1982, *Physiology and Biochemistry of Seeds,* Volumes 1 and 2, Springer-Verlag, Berlin, Heidelberg, New York (covers all aspects of viability, germination, dormancy, and environmental control).

Heydecker, W. (ed.), 1973, *Seed Ecology,* Butterworths, London (multiauthor contributions to a scientific conference; covers selected aspects of germination, dormancy together with some agricultural and horticultural considerations).

Khan, A. A. (ed.), 1977, *The Physiology and Biochemistry of Seed Dormancy and Germination,* North-Holland, Amsterdam (multiauthor contributions).

Khan, A. A. (ed.), 1982, *The Physiology and Biochemistry of Seed Development, Dormancy and Germination,* Elsevier, Amsterdam (multiauthor contributions).

Kozlowski, T. T. (ed.), 1972, *Seed Biology,* Volumes I, II, III, Academic Press, New York (multiauthor contributions to many aspects of seed structure, germination, pathology, and ecology).

Roberts, E. H. (ed.), 1972, *Viability of Seeds,* Chapman and Hall, London (mostly viability but includes some physiology).

SECTION 1.3

Goodchild, N. A., and Walker, M. G., 1971, *Ann. Bot.* **35:**615–621 (measurement of germination).

Hewlett, P. S., and Plackett, R. L., 1979, *An Introduction to the Interpretation of Quantal Responses in Biology,* Edward Arnold, London (methods for mathematical analysis).

Janssen, J. G. M., 1973, *Ann. Bot.* **37:**705–708 (recording germination curves).

Richter, D. D., and Switzer, G. L., 1982, *Ann. Bot.* **50:**459–463 (quantitative expressions of dormancy in seeds).

SECTION 1.4

Corner, E. J. H., 1976, *The Seeds of Dicotyledons,* Cambridge University Press, Cambridge, U.K. (a comprehensive two-volume work).

Forest Service, U.S. Dept. Agriculture, 1974, *Seeds of Woody Plants in the United States.* U.S.D.A., Washington, D.C.

Rost, T. L., and Lersten, N. R., 1973, *Iowa State J. Research* **48**:47–87 (grass caryopsis anatomy).

Vaughan, J. G., 1970, *The Structure and Utilization of Oil Seeds,* Chapman and Hall, London.

SECTION 1.5

Crocker, W., and Barton, L. V., 1957, *Physiology of Seeds,* Chronica Botanica, Waltham, Massachusetts (seed constituent composition).

Derbyshire, E., Wright, D. J., and Boulter, D., 1976, *Phytochemistry* **15**:3–24 (seed proteins).

Duffus, C. M., and Slaughter, J. C., 1980, *Seeds and Their Uses,* Wiley, Chichester (economically important seeds).

Earle, F. R., Curtice, J. J., and Hubbard, J. E., 1956, *Cereal Chem.* **23**:507–515 (composition of corn kernel regions).

Frey, K. J., 1977, *Z. Pflanzenzuchtg.* **78**:185–215 (amino acids in cereal proteins).

Galliard, T., and Mercer, E. I. (eds.), 1975, *Recent Advances in the Chemistry and Biochemistry of Plant Lipids,* Academic Press, London (reviews on plant lipids).

Jacobsen, J. V., Knox, R. B., and Pyliotis, N. A., 1971, *Planta* **101**:189–209 (protein bodies in barley aleurone layers).

Larkins, B. A., 1981, in: *The Biochemistry of Plants. Proteins and Nucleic Acids,* Volume 6 (A. Marcus, ed.), Academic Press, New York, pp. 449–489 (seed storage proteins: review).

Lott, J. N. A., 1981, *Nordic J. Bot. 1,* 421–432 (protein bodies: review).

Meier, H., and Reid, J. S. G., 1982, in: *Encyclopaedia of Plant Physiol,* New Series, Berlin, Heidelberg, Volume 13A, pp. 418–471 (storage hemicelluloses).

Miege, M.-N., 1982, in: *Encyclopaedia of Plant Physiol,* New Series, Berlin, Heidelberg, Volume 14A, pp. 291–345 (protein types and distribution).

Miflin, B. J., and Shewry, P. R., 1981, in: *Nitrogen and Carbon Metabolism* (J. D. Bewley, ed.), Martinus Nijhoff/Dr. W. Junk, The Hague, pp. 195–248 (seed storage proteins: review).

Miller, E. C., 1931, *Plant Physiology,* McGraw-Hill, New York (seed constituent composition).

Parker, M. L., 1981, *Ann. Bot.* **47**:181–186 (rye endosperm structure).

Payne, P. I., and Rhodes, A. P., 1982, in: *Encyclopaedia of Plant Physiol,* New Series, Berlin, Heidelberg, Volume 14A, pp. 346–369 (cereal storage proteins: review).

Weber, E., and Neumann, D., 1980, *Biochem. Physiol. Pflanzen.* **175**:279–306 (protein bodies and phytin).

Weber, E. J., 1980, in: *The Resource Potential in Phytochemistry. Recent Advances in Phytochemistry,* Volume 14, Plenum Press, New York, pp. 97–137 (composition of corn kernels).

Winton, A. L., and Winton, K. B., 1932, *The Structure and Composition of Foods,* Volume 1, Wiley, New York (review).

Chapter 2

Seed Development and Maturation

2.1. EMBRYOGENY AND STORAGE TISSUE FORMATION

Before we can consider the specific physiological and biochemical processes intimately involved in seed development, it is necessary to review briefly the morphological and anatomical aspects of embryo and storage tissue formation. The variations in patterns of development are numerous, so we will present here only a generalized picture of what occurs in a gymnosperm, a dicot, and a monocot angiosperm.

In gymnosperms, the egg nucleus lies within the female gametophyte and is fertilized by a motile gamete released from the pollen tube. The resultant zygote then divides to produce several free nuclei, around which cell walls are laid down to form the proembryo (Fig. 2.1, a and b). Subsequent cell divisions give rise to embryonal and suspensor cells (Fig. 2.1, c and d). These may develop to form a single embryo and an elongated suspensor, but often in the conifers there is cell separation to form four embryos (polyembryony) (Fig. 2.1, d). Only one of these develops further and the others degenerate. Within the seed the central portion of the megagametophyte (the haploid female gametophyte) breaks down to form a cavity into which the embryo expands (Fig. 2.1, e). Storage reserves—lipid, starch, and protein—are deposited within the persistent parts of the megagametophyte (Fig. 1.3), to be used after germination.

The unique feature of fertilization in angiosperms is the participation of two male nuclei. One nonmotile nucleus released from the pollen tube fuses with the egg nucleus to form the diploid zygote, and the other fuses with two polar nuclei to produce a triploid nucleus. Unlike in the gymnosperms, there is no free nuclear stage; the first division results in an axial (distal or apical) cell and a basal cell. In dicot seeds, the suspensor is formed from the basal cell, which may also contribute to the embryo; most of the embryo is derived from the axial cell. In monocots, the basal cell does not divide but forms the terminal (haustorial) cell of the suspensor; the embryo and the other few suspensor cells are produced from the axial cell. By definition, the embryos of the mature seeds of the dicots possess two cotyledons, whereas there is only one in monocots.

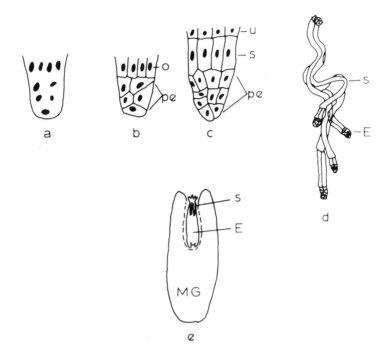

Figure 2.1. Embryo development in conifers. (a) Free nuclear stage. (b) Proembryo (pe) showing internal division of cells, with open tier (o) cells above. (c) Three-tiered mature proembryo
(pe) beneath the suspensor tier (s), which elongates to form the primary suspensor, and the upper
tier (u), which degenerates. (d) Polyembryonic stage. Elongated suspensor(s) bearing the
embryonal masses of cells (E). (e) The single maturing embryo (E) elongates into a cavity in the
megagametophyte (MG); the suspensors (s) degenerate.

The single cotyledon of the Gramineae is reduced to the absorptive scutellum
(Fig. 1.3). Mature gramineous embryos also include a specialized thin tissue
which covers the radicle (coleorhiza) and one which is around the plumule and
covers the first foliage leaf (coleoptile) (Fig. 1.3). The pattern of development
of the embryo of a nonendospermic legume, the pea, is shown in Fig. 2.2 and
of a gramineous monocot in Fig. 2.3 (see legends for details).

The true endosperm, the cells of which are triploid, is found only in the
angiosperms; it is derived from the triple fusion nucleus. Two types of endosperm development have been noted: (1) nuclear, where the endosperm undergoes several free nuclear divisions prior to cell wall formation (e.g., apple,
wheat, squash), and (2) cellular, where there is no free nuclear phase (e.g.,
Magnolia, Lobelia). Both these types occur in monocots and dicots. A less common form of development, found only in some monocots, is the helobial type,

where free nuclear division is preceded and followed by cellularization. Nutrients are drawn from adjacent tissues into the endosperm during its development, and new products are also laid down therein. Thus, the growing embryo becomes enveloped in, or intimately associated with, an available food source upon which it can draw during its maturation and subsequent germination/growth stages. In grasses, e.g., cereals, there is limited utilization of the endosperm during embryo maturation, but the part that is depleted forms the intermediate layer which lies between the starchy endosperm and the scutellum

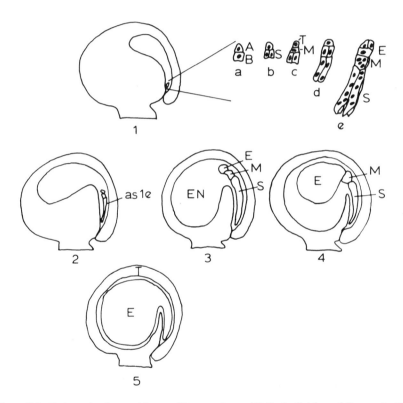

Figure 2.2. Embryo development in pea, *Pisum sativum*. (1) Early divisions of the zygote. The first division (a) is to form an apical (A) and a basal (B) cell. The basal cell divides transversely (b) to form two suspensor (S) cells, and the apical cell divides (c) to form a middle (M) cell and a terminal (T) cell. The suspensor cells (S) divide to become four, elongate (d, e), and become multinucleate; the middle cells undergo free nuclear division, and the embryonic cells (E) commence cell division. (2)–(5) The suspensor elongates and then embryo growth occurs into the embryo sac, to eventually utilize the reserves and to occlude the endosperm (EN) and the suspensor/middle cell (S, M). The mature embryo (E) occupies the whole of the inner region of the seed and is surrounded by the testa (T). Based on Marinos (1970) and Maheshwari (1950).

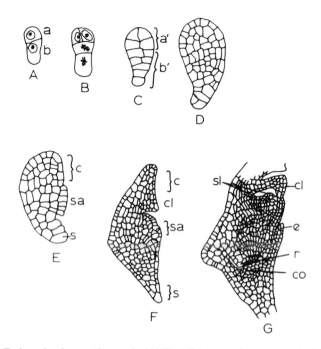

Figure 2.3. Embryo development in cereals. (A) First division produces an apical (a) and a basal (b) cell, both of which undergo divisions (B and C) to form derivatives, a′ and b′, respectively. After further division [(D)–(F)], regions that give rise to the short suspensor (s), cotyledon (c), stem apex (sa), and coleoptile (cl) are delimited. In the nearly mature embryo (G), these regions plus the coleorhiza (co), epiblast (e), stem apex and first leaf (sl), and root initials (r) are in evidence. After Johansen (1950).

of the mature embryo (Fig. 7.1). In nonendospermic dicot seeds, the endosperm reserves are depleted and occluded by the developing embryo (Fig. 2.2) and then reorganized in the cotyledons, which act as the source of stored reserves for the germinated embryos. The endosperm is retained as a permanent storage tissue in the endospermic dicot seeds (e.g., castor bean, Fig. 1.3), and in endospermic legumes (e.g., fenugreek) this storage tissue may be surrounded by an aleurone layer. The nucellus develops into the perisperm in a few nonendospermic species (e.g., *Coffea, Yucca*), but this is rare.

The outer structures of the ovule, the integuments, undergo marked reorganization during seed maturation to form the seed coat (testa). Seed coat structure and thickness are extremely variable between different species, and sometimes even in the seeds of the same species grown under different environmental conditions. Seed coats are reviewed briefly in Chapter 1, but space available here precludes a more extensive discussion of their development and morphology.

2.2. SOURCE OF ASSIMILATES FOR GRAIN AND SEED FILLING

2.2.1. Cereals

Carbohydrates such as free sugars, starch, and other polysaccharides reach a maximum concentration in the vegetative parts of the mother plant around the time of anthesis (dehiscence of the anthers), after which they start to decrease. Some of this stored carbohydrate is translocated to the growing grain, and although estimates are variable, it may provide up to 15–20% of the final dry weight of the grain. The contribution to grain filling by current photosynthesis in different plant parts is related to their potential photosynthetic activity, their longevity during grain ripening, and to the light environment in the crop canopy.

The photosynthetic rates exhibited by different parts of the plant are quite variable from crop to crop. For wheat (Fig. 2.4, A) and barley, net photosynthesis in the flag leaf and ear is relatively high, and these regions provide the major nutrient source to the grain, although each may make quantitatively different contributions in different crops and cultivars (Table 2.1). In oats and rice, the flag leaf and penultimate leaf appear to be of equal importance in supplying assimilates for grain filling. In corn (Fig. 2.4, B), sugars produced by the leaves above the ear are translocated efficiently into the kernel, but translocation from leaves below the ears is poor. In general, then, the upper, or uppermost, leaves direct their assimilates mostly to the grains and the stem, and the lower leaves to the roots and tillers. But even within the ear itself the distribution of assimilates may vary. In wheat, for example, assimilates from the flag leaf are distributed preferentially to the lower central spikelets (the first to anthese), and the setting, and hence yield, of the grains in the upper

Table 2.1. The Photosynthetic Contribution by Various Vegetative Plant Parts during Postanthesis to the Total Carbon Content of Mature Grains[a]

	Barley	Awnless wheat
Ear		
Gross photosynthesis	79	24
Day respiration	−24	−28
Night respiration	−10	−11
Net photosynthesis	45	−15
Leaf		
Net photosynthesis	55	115
	100	100

[a]Based on data of Dr. G. N. Thorne, quoted in Milthorpe and Mooby (1979).

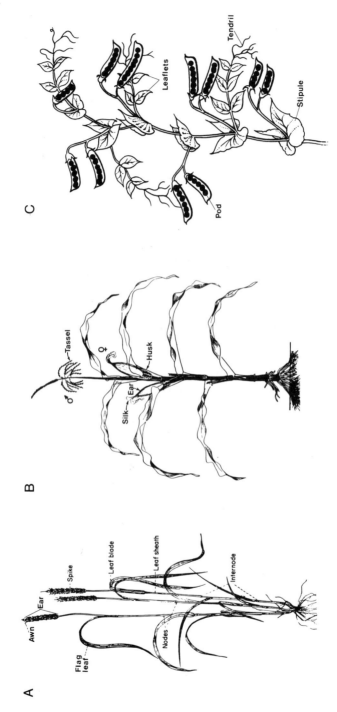

Figure 2.4. Mature (A) wheat and (B) maize (corn) and (C) pea plant at fruiting stage.

florets may be reduced by the more rapid development of grains from the first flowers to reach anthesis. On the other hand, photosynthates produced by various parts of the ear, e.g., the awns, tend to supply the grains in closest proximity.

The longevity of the green tissue in the various plant parts during the ripening period may be a factor in grain yield. Rice leaves remain green almost until grain maturity, whereas the ear turns yellow fairly early during ripening. In wheat, however, the leaves become yellow (i.e., senesce) before the ear does, and the latter may be the more important in providing photosynthates at the late stages of maturity. At this time, however, respiration by the grain may exceed the net import of sugars and other substrates, resulting in a small loss in grain weight.

The light environment impinging on the crop canopy plays a major role in determining the real photosynthetic activity of a given plant part. The developing ears of barley and wheat are fully exposed to sunlight and can realize their full photosynthetic potential. In contrast, the ears of certain improved rice cultivars have the tendency to bend and are positioned below the flag leaf; hence they are heavily shaded and make an insignificant contribution to the total assimilates utilized in grain filling. It is notoriously difficult to estimate precisely the photosynthetic contribution of different parts in a crop canopy under field conditions, but factors such as leaf area index, leaf age and longevity, angle of incidence to the sun, shading and ear structure (awns or awnless) all play a role, as well as environmental variables like temperature and nutrient and water availability.

A considerable proportion of the nitrogen utilized in protein synthesis in the developing grains is obtained from remobilization of proteins in the senescing vegetative organs. In wheat, the flag leaf appears to be the largest contributor of grain nitrogen (Fig. 2.5). There is evidence that in several cereals the efficiency of nitrogen redistribution varies with genotype, the high grain-N cultivars typically having the highest efficiency for nitrogen redistribution. Such differences, in turn, may be related to the activity levels of proteolytic enzymes in the senescing organs of the mother plant.

2.2.2. Legumes

The pea pod, which is derived from the ovary wall, more or less completes development before swelling of the enclosed seeds commences, i.e., prior to deposition of the major storage reserves within the cotyledons. Although green, the seeds themselves photosynthesize only weakly and lose as much CO_2 by respiration as they gain by photosynthesis. A major source of carbon for the

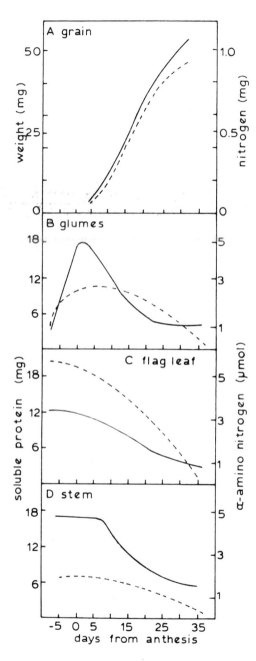

Figure 2.5. (A) Dry weight (——) and total nitrogen content (----) of grains of wheat during development. (B)–(D) Soluble protein content (----) and α-amino nitrogen (——) in the glumes, flag leaf, and stem. The decline after anthesis is due to remobilization to the grain. After Waters *et al.* (1980).

developing seed is photosynthesis by the pod, which recycles the respiratory carbon released by the seed into the inner cavity; the pod uses negligible quantities of external CO_2. An equally important contribution to the carbon status of the seed is provided by photosynthesis of the leaflets of the subtending leaf (i.e., at the same node) (Fig. 2.4, C). The photosynthetic activity of the stipules is only a minor contributor of carbon. Most legumes are unable to mobilize their assimilates stored prior to fruiting; so storage reserve formation in the seed depends heavily on assimilates formed during fruiting itself (shown in Fig. 2.6 for the grain legume *Lupinus angustifolius*), and seed yield is likely to be extremely sensitive to adverse environmental influences during filling.

As with the cereals, the increased protein content of the developing seed is largely at the expense of nitrogen derived from the senescing mother plant (including the pod), although nitrogen fixation by root nodules during reserve deposition may augment this supply.

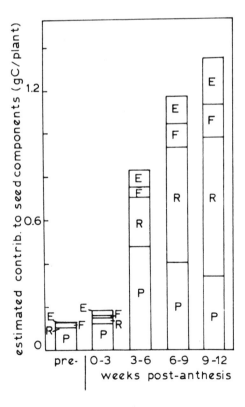

Figure 2.6. Contribution to seeds of *Lupinus angustifolius* cv. Unicrop by photosynthate fixed at different periods during development. Total time needed for seeds to ripen: 15 wk after anthesis. There is little contribution of prefruiting photosynthate (pre-) to any fraction, except protein. F, fat; P, protein; R, residue (includes polysaccharides); E, ethanol-soluble components (includes sugar and amino acids). After Pate *et al.* (1980).

2.2.3. Translocation of Assimilates into the Developing Seed

Sucrose is the translocated form of sugar from the sites of photosynthesis to the developing seeds. The major pathway for its long-distance transport from the vegetative parts is via the phloem. Developing cereal seeds have no direct vascular connections with the mother plant, and a short-distance transport mechanism operates to move the assimilates from the vascular tissues to the region of reserve deposition, the endosperm. The nature of this short-distance assimilate pathway varies between cereals. In wheat and barley all nutrients are supplied via the vascular tissue running in the crease and must first pass through the funiculus–chalazal region, then through the nucellar projection, and finally through the aleurone layer before entering the starchy endosperm (Fig. 2.7, A). In corn, which does not have a crease, there are specialized cells at the base of the developing endosperm which facilitate movement from the termination points of the phloem in the pedicel into the starchy endosperm and embryo (Fig. 2.7, B).

As the sucrose passes from the vascular tissue into the cereal endosperm, it is cleaved by invertase to glucose and fructose. Sucrose is then reformed after uptake into the endosperm cells. Although this hydrolysis and resynthesis may appear to be an energetically wasteful process, it nevertheless is an integral step in the transfer of sugar from the vascular tissue to the grain. Moreover, it is possible that this short-distance assimilate pathway between the phloem and the endosperm can operate to regulate the rate of sucrose transport into the grain. This, in turn, could act to limit the accumulation of starch and the growth of the grain itself, although in some cereals the capacity actually to utilize the assimilate might be a more important regulatory factor. The uptake of amino acids into developing cereal grains has received less attention, but it is generally accepted that asparagine and glutamine are the important translocated forms from the mother plant. Formation of other amino acids must occur within the developing grains themselves. The amino group of the amide-containing amino acids (i.e., glutamine and asparagine) provides the nitrogenous component for newly synthesized amino acids, and carbon skeletons are furnished by the translocated carbohydrates.

During early stages of development the (nonendospermic) legume seed obtains its nutrients from the endosperm surrounding the embryo in the embryo sac (Fig. 2.2). But the assimilates required later for reserve deposition in the cotyledons are translocated from the mother plant. This is facilitated by a vascular strand which branches from the vascular tissue running through the pod and then passes through the funiculus, into the integuments (incipient seed coat) (Fig. 2.7, C). Passage of assimilates through the funiculus, and from the seed coat into the cotyledons, by diffusion, is aided by the presence of transfer cells. These are specialized cells with ingrowths of the cell wall to increase the

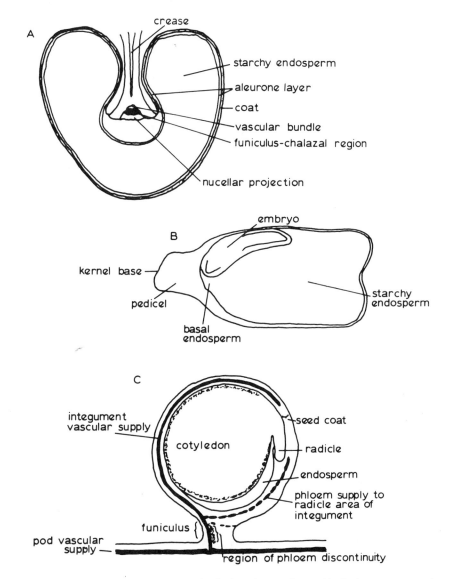

Figure 2.7. (A) Cross section through a developing wheat grain at midpoint between apex and base to show the relationship between the vascular tissue and the starchy endosperm, between which the short-distance transport system operates. (B) Longitudinal section through a developing maize grain. Here the short-distance transport system is through the kernel base (fruit stalk) and pedicel (maternal tissue in which the phloem ends) to the basal part of the starchy endosperm. (C) Mid-sagittal section of a developing seed of the garden pea to show the vascular supply to the developing ovule and in the integuments (seed coat). Heavy dashed line indicates the relative position of the phloem supply to the radicle area. Stippled areas are the probable positions of transfer cells in the funiculus and cotyledonary epidermis. Adapted from (A) Sakri and Shannon (1975); (B) Shannon (1972); and (C) Hardham (1976).

surface area for the absorption or export of assimilates. In some legumes, e.g., in some cultivars of soybean and field and garden pea, the translocated sucrose produced by photosynthesis in the leaves and pods may be stored temporarily as starch in the pod prior to remobilization and transfer to the developing seed. In species that are cultivated for their edible pods, e.g., runner bean *(Phaseolus vulgaris)*, the fleshy pod serves both as a photosynthetic capsule for recycling CO_2 respired by the seeds and as a permanent depository for ostensibly seed-bound sugars.

The importance of the phloem transport system for the supply of assimilates to the developing fruit is emphasized by data obtained from studies on *Lupinus albus*. Here, the phloem is suggested to supply an average 98% of the carbon, 89% of the nitrogen, and 40% of the water entering the fruit from the parent plant. The remaining nitrogen and water are presumed to be supplied via the xylem. Sucrose accounts for 90% of the carbon in the phloem, and asparagine and glutamine carry 75–85% of the nitrogen of xylem and phloem; hence these assimilates provide the major source of raw materials for the anabolic processes associated with reserve deposition.

In recent years, it has been suggested that hormones, particularly gibberellins and abscisic acid, play some role in the regulation of reserve deposition, either by modifying the activity of synthetic enzymes or by regulating the flow of assimilates into the developing seed. This is discussed briefly in Section 2.5.

2.3. DEPOSITION OF RESERVES WITHIN STORAGE TISSUES

Most mature seeds contain at least two or three stored reserves in appreciable quantities (Table 1.2), and to a large extent they are generally synthesized concomitantly during seed development. But for the sake of clarity and convenience we have chosen to discuss singly the synthesis and deposition of the major reserves—carbohydrate, fat, and protein—and the important minor reserve—phytin. Later, however, we will briefly consider the chronology of the various synthetic events in storage organs from their early development to maturity.

2.3.1. Starch Synthesis

Sucrose, the sugar translocated from the mother plant to the seed, is the substrate for starch formation. First it is converted to fructose and UDPGlc (uridine diphosphoglucose) by sucrose synthetase (sucrose-UDP glucosyl transferase) as follows:

$$\text{Sucrose} + \text{UDP} \rightleftharpoons \text{Fructose} + \text{UDPGlc}$$

Both fructose and UDPGlc are converted to Glc-1-P (glucose-1-phosphate), the former via fructose-6-phosphate and glucose-6-phosphate, and the latter by activity of the enzyme UDPGlc pyrophosphorylase as follows:

$$UDPGlc + PPi \rightleftharpoons Glc\text{-}1\text{-}P + UTP$$

The Glc-1-P is now converted to ADPGlc, another sugar nucleotide, by ADPGlc pyrophosphorylase as follows:

$$Glc\text{-}1\text{-}P + ATP \rightleftharpoons ADPGlc + PPi$$

The ADPGlc donates its glucose to a small glucose primer, thus increasing its chain length by one unit. The process is repeated until the starch molecule is completed, the enzyme involved being ADPGlc-starch synthetase (ADPGlc-starch glucosyltransferase), as follows:

$$ADPGlc + Glc_{(n)} \rightleftharpoons Glc_{(n+1)} + ADP$$

There is some debate as to whether there exists an alternative synthetic pathway involving phosphorylase enzymes, as follows:

$$Glc\text{-}1\text{-}P + Glc_{(n)} \rightleftharpoons Glc_{(n+1)} + Pi$$

It is unlikely that starch is synthesized in this manner, although phosphorylases might play a role in the formation, from maltose, of the small glucose primer molecules required for ADPGlc starch synthetase activity.

As indicated in Chapter 1, starch exists in two forms, the straight-chain amylose (comprised of α-1,4-linked glucose residues), and the branched-chain amylopectin (linear amylose chains with branches linked by α-1,6-glucosidic bonds). Therefore, the presence of a branching enzyme (Q enzyme) to forge the α-1,6 links is also required. This enzyme introduces short chains of α-1,4-linked glucose residues to amylose; these chains then have more glucose units added on by the synthetase enzyme.

Our understanding of the role of enzymes in starch synthesis has been greatly aided by the discovery that certain mutants, particularly of cereals, exhibit variations or defects in their starch–synthetic pathways. Some of these are listed for maize in Table 2.2.

Starch is stored in organelles called amyloplasts, which are derived from proplastids. The amyloplasts surround the starch grains (or granules) with a double membrane and possibly incorporate some of the enzymes of starch synthesis. The number of starch grains per amyloplast varies with the species: in barley, maize, rye, and wheat, for example, there is only one grain present,

Table 2.2. Some Mutants of Maize, Characteristics of Their Sugars and Starch, and Possible Relationship to Synthetic Enzyme Activities[a]

Mutant (gene)	Response	Cause	Water-soluble polysaccharides (%)	Starch (%)
Amylose extender (ae)	Fewer branch points in amylopectin	Absence of one of the branching enzymes (IIa)	—	—
Sugary (su)	Smaller, more highly branched, and soluble amylopectin (phytoglycogen)	Altered activity of a branching enzyme (I)	36	30
Shrunken-2 (sh₂)	Reduced starch deposition	Reduced ADPGlc pyrophosphorylase activity	2	25
Waxy (waxy)	Little amylose production, mostly amylopectin	No granule bound starch synthetase: no protection of amylose from branching enzymes	—	—

[a]Normal maize contains 1.3% water soluble polysaccharides and 65% starch.

whereas in rice and oats there are many. Starch synthesis in cereals (Fig. 2.9, A) is initiated in the proplastids with the formation of one (Fig. 2.8, A, rye, wheat, and maize) or more (rice, oats) small starch granules, which at first occupy only a small volume of this organelle. There is then a gradual increase in size of the granule (or granules, Fig. 2.8, B) until the mature amyloplast is completely filled. Characteristically, the mature starch grains appear to be composed of concentric rings or shells. Each ring probably represents one day's accumulation of starch in the grain. Some cereals, e.g., barley, contain both large and small amyloplasts in the mature endosperm. It appears that the large ones are initiated early after anthesis and the smaller ones only later (after about 2 wk). The presence of both large and small amyloplasts serves effectively to pack the mature cells of the starchy endosperm and thus, along with the protein bodies, occlude the living contents (Fig. 1.5, A).

In legumes, starch is deposited in the cells of the cotyledons. It has been observed that in the garden pea small, single starch grains become visible in the stroma of some chloroplasts within 2–3 wk after anthesis, although it is the translocated products of photosynthesis by the pod and leaves that provide the sugars for starch grain growth. This single grain increases in size, causing deformation (Fig. 2.8, C) and eventual loss of the internal chloroplast structure. A few small chloroplasts do not form starch, and in remaining intact they account for the pale (pea) green color of the mature seed.

A distinctly different pattern of starch accumulation occurs in another legume, soybean. Here there is a substantial increase in starch content in the cotyledons during development, followed by a decline, so that in the mature seed little starch remains (Fig. 2.9, B). This is because the starch is remobilized in the latter stages of development to provide carbon skeletons for fat and protein synthesis. Presumably the products of starch hydrolysis supplement the restricted flow of assimilates coming into the cotyledons from the mother plant.

2.3.2. Deposition of Polymeric Carbohydrates Other than Starch

Starch is not the only carbohydrate stored in cereals and legumes, and even where it is present, it may not be the major form of stored polymeric sugar. In cereals, the cell walls of the starchy endosperm are comprised of considerable quantities of hemicelluloses and of glucans containing $\beta 1,3$ and $\beta 1,4$ links, although in minor amounts in relation to the amount of starch stored in the cytoplasm. In the garden pea, some 35–45% of the seed dry weight is attributable to starch, but in the field pea hemicelluloses in the cell walls eventually make up about 40% of the seed dry matter and starch only some 25%. Hemicellulose synthesis continues after starch synthesis has ceased (Fig. 2.9, C). Unfortunately, at the present time, little is known about the synthesis of these cell wall components in developing seeds.

In certain seeds, including some of the endospermic legumes, the major storage product is deposited in the cell walls as hemicellulose (Section 1.5.1). In species such as fenugreek *(Trigonella foenum-graecum)*, ivory nut *(Phytelephas macrocarpa)*, and date *(Phoenix dactylifera)*, the hemicelluloses— largely galactomannans—may be deposited in such large quantities that in the cells of the storage organs the living contents are occluded by the inward thickening of the cell wall. Galactomannan starts to be synthesized at an early stage of development in fenugreek and continues to be laid down in the cell walls almost until maturity. The formation of galactomannan might occur in vesicles of the rough endoplasmic reticulum (RER), the polymer then being secreted through the plasmalemma and into the surrounding cell wall; the biochemical mechanism of polymer synthesis and secretion is unknown. The mannose itself is probably first produced from phosphorylated sucrose and then converted to the nucleotide sugar, as follows:

$$\text{Sucrose} \longrightarrow \text{Fructose-6-P} \underset{①}{\rightleftharpoons} \text{Mannose-6-P} \underset{②}{\rightleftharpoons} \text{Mannose-1-P}$$

$$\text{Galactose} \longrightarrow \text{(Galacto) mannan} \underset{④}{\longleftarrow} \text{GDP-Mannose} \underset{③}{\nearrow}$$

① Phosphomannoisomerase
② Phosphomannomutase

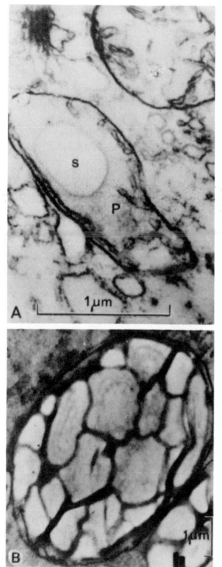

Figure 2.8. (A) Proplastid (P) in young barley endosperm cell with small starch granule(s) from which the starch grain will grow. (B) Maturing starch grains in a developing amyloplast from a young oat endosperm. (C) Developing starch grain (SG) in the chloroplast of a garden pea endosperm cell with the grana (G) distorted and pushed against the plastid membrane (PLM). (A) and (B) from Buttrose (1960) and (C) from Bain and Mercer (1966).

③ GTP:Man-1-P guanylyl transferase
④ GDP mannose:galactomannan-mannosyltransferase

There is no change in the proportion of galactose to mannose units in galactomannan during development, which suggests that both of these sugars are incorporated into the polymer at the same time.

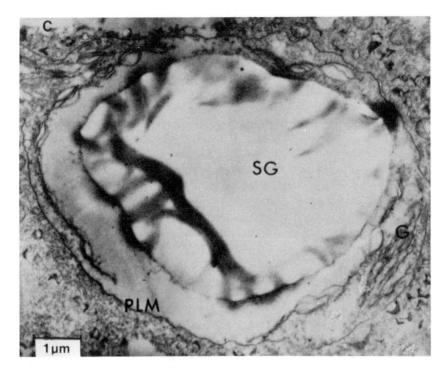

Figure 2.8. (*Continued*)

2.3.3. Fat (Oil) Synthesis

There is considerable variation in the major fatty acid constituents of stored fats in seeds. Some examples are presented in Table 1.4, although the diversity is much greater than outlined there. For the sake of brevity and clarity we will not dwell on this diversity but will restrict ourselves to highlighting some of the important facets of fat biosynthesis in developing seeds.

A generalized scheme for triglyceride biosynthesis in oil-rich seeds and the probable localization of various steps within the cell are shown in Fig. 2.10. Sucrose is transported into the developing seed and converted via hexose to 6-phosphogluconate in the cytosol. Glucose-6-phosphate dehydrogenase, the enzyme responsible for the production of the gluconate from Glc-6-P, is present in the cytosol, whereas the enzyme responsible for the next step in the pentose phosphate pathway, 6-phosphogluconate dehydrogenase, is located largely within the proplastid. Completion of the pentose phosphate pathway to produce fructose-6-phosphate occurs within this organelle too. This product is converted by glycolysis to acetyl CoA, which is then carboxylated to yield malonyl CoA. Both the pentose phosphate pathway and glycolysis produce reducing power,

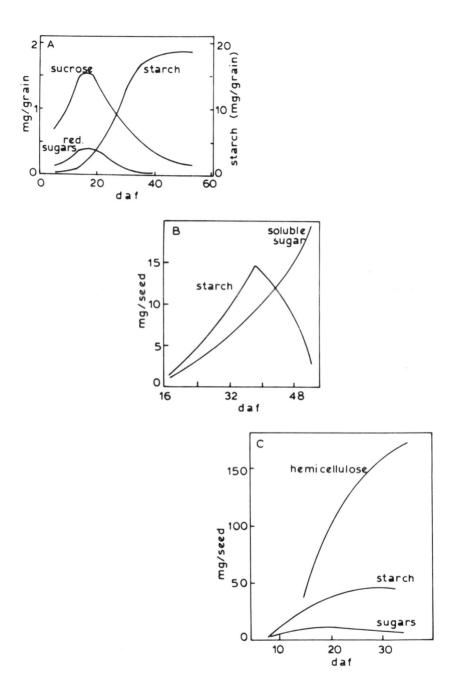

Figure 2.9. Time courses of the synthesis of polymeric carbohydrates and changes to the soluble sugars in developing (A) wheat, (B) soybean, and (C) field pea with increasing days after flowering (daf). After (A) Turner (1969); (B) Adams *et al.* (1980); and (C) Pate and Flinn (1977).

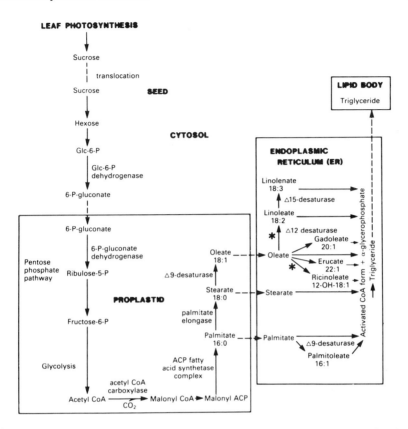

Figure 2.10. Scheme for the synthesis of fatty acids in the cells of lipid-storing organs. Synthesis involves three compartments of one cell, the cytosol, proplastid, and endoplasmic reticulum (ER), the latter becoming modified to form the lipid body (see Fig. 2.13). The conversion of oleate to more unsaturated fatty acids (asterisked) probably requires first the incorporation of olelyl–CoA into phosphatidyl choline, and then desaturation to linoleoyl–phosphatidylcholine before release of the free fatty acid. Hydroxylation, an essential step in ricinoleic acid formation, may also occur on phosphatidyl choline. Mammalian tissues are not able to desaturate oleic acid to yield linoleic acid. Hence this fatty acid, which is an essential part of the diet, must be obtained from plant sources. Based on Simcox *et al.* (1977) and Stumpf (1977).

in the form of NADPH and NADH, respectively, and adenosine triphosphate (ATP) (glycolysis only), which will be utilized in the synthesis of the long-chain fatty acids. The malonyl CoA is accepted by an acyl carrier protein (ACP), which is part of a multienzyme complex—the ACP fatty acid synthetase complex. After condensation of malonyl CoA with a further acetyl CoA molecule to yield acetoacetyl (4C) CoA and CO_2, more acetyl (2C) residues are sequentially condensed with acetoacetyl CoA on the ACP complex until a palmitic acid (16:0)–ACP complex is formed. One more acetyl residue is added by the elongation process to yield a stearic acid (18:0)–ACP complex, from which the fatty acid is released. This saturated fatty acid is aerobically desaturated to produce oleic acid (18:1), which has one double bond. The oleic acid is then transported to the ER for further modification; e.g., other fatty acids with greater amounts of desaturation are produced (e.g., linoleic—18:2 and linolenate—18:3), as are certain fatty acids found in only a few species, such as ricinoleic acid (12-OH–18:1) in castor bean and erucic acid (22:1) in some cultivars of rape and *Crambe abyssinica*. Desaturation occurs while the fatty acids are in the CoA form. Although it has still to be confirmed for many species, the free fatty acids are probably activated to form the appropriate fatty acyl CoAs while associated with the ER; these are condensed with (phosphorylated) glycerol to eventually form the triglyceride (Fig. 2.11).

Work on *Crambe,* a species of Cruciferae rich in erucic acid, illustrates many of the principles of seed oil accumulation (Fig. 2.12, A and B). During the initial or lag phase of seed development little or no storage triglyceride is present, owing to the absence of the appropriate enzymes. Some galactosylglycerides and phosphoglycerides, rich in linolenic acid, 18:3, can be detected at this time; these are components of the membranes of the cotyledon chloroplasts. In the second phase, some 8–30 days after flowering, there is a gradual accumulation of neutral lipid, accompanied initially by a transient rise in linoleic (18:2) and oleic (18:1) acids and then a steep increase in the amount and proportion of erucic acid, 22:1. In the final stage of seed development, later than 30 days from flowering, fat accumulation ceases, and presumably the enzymes involved in the synthetic processes are destroyed.

There are several variations to this generalized pattern of fat synthesis. In safflower, for example, there is a temporary accumulation of oleic acid (18:1) at the time when fat synthesis commences (Fig. 2.12, C), presumably because the rate of its synthesis initially exceeds the capacity of the desaturase enzymes to produce linoleic acid (18:2). Later the proportion of oleic acid falls as the major storage fat is formed. Some attention has been paid to the synthesis of the unique wax esters of jojoba *(Simmondsia chinensis)* seeds. The stored products are liquid waxes comprised of oxygen esters of long-chain fatty acids and alcohols, and they are of great importance because they are difficult to synthesize commercially; the only other natural source of such waxes is the endangered sperm whale. Because jojoba wax is liquid it is useful for the manufacture of extremely high-pressure lubricants such as those now used in the power transmissions of heavy-duty vehicles. Unlike many of the other oil-bearing plants of agronomic importance, the jojoba plant is a low shrub and only yields seed after taking several years to reach maturity. Studies on the seed show that wax synthesis reaches its maximum potential when the seed is about half its mature weight, and that the desirable composition of waxes is reached about 20 days before the completion of development.

The characteristic fatty acid composition of plant lipids is largely genetically determined, although environmental factors may result in modifications during seed development. Seed oils of plants grown in cool climates generally tend to be more unsaturated than those grown in warm climates. The major influence is usually on the predominant fatty acid of the seed. In flaxseed oil there is a marked decline in the proportion of linoleic acid at $10°C$ compared to that at $30°C$ and a corresponding increase in its precursor, oleic acid. The

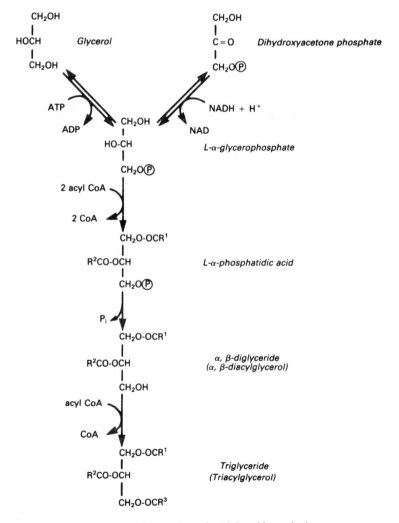

Figure 2.11. The pathway for triglyceride synthesis.

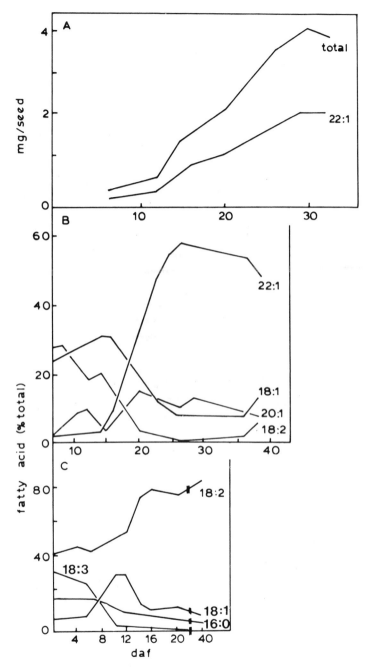

Figure 2.12. (A) Accumulation of total fat and of erucic acid (22:1) in developing seeds of *Crambe abyssinica*. (B) Changes in the composition of some fatty acids in *C. abyssinica* seed storage fats during development. (C) Changes in the fatty acid composition of safflower *(Carthamus tinctorius)* seed storage fats during development. daf, days after flowering. After (A) Gurr *et al.* (1974); (B) Appleby *et al.* (1974); and (C) Ichihara and Noda (1980).

precise nature of the climatic influence on fatty acid composition remains to be elucidated.

The reserve fats and oils found within the cells of storage organs are confined to organelles called lipid (fat or oil) bodies. The origin and development of these bodies have been the subject of considerable controversy over the years, but recent evidence points clearly to their origin in the ER. Newly formed fat accumulates between the two layers of the ER double membrane, leading to its swelling (Fig. 2.13). When the fat-filled vesicle reaches a critical size, it may bud off completely (Fig. 2.13, type A), bud off with a little ER still

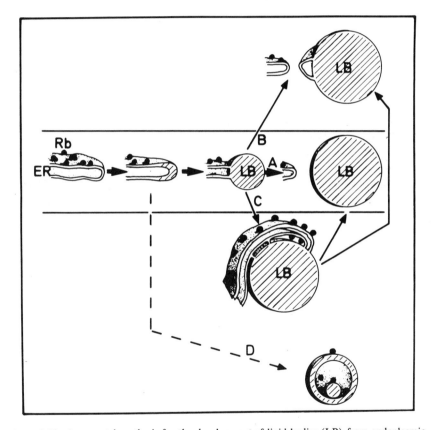

Figure 2.13. A current hypothesis for the development of lipid bodies (LB) from endoplasmic reticulum (ER). Several alternatives exist: (A) lipid bodies pinch off the ER without a remaining piece of membrane, e.g., broad bean *(Vicia faba)* and pea *(Pisum sativum);* (B) a small piece of membrane remains attached to the lipid body, e.g., watermelon *(Citrullus vulgaris);* (C) ER strand maintains contact with the nascent lipid body, e.g., pumpkin *(Cucurbita pepo)* and flax *(Linum usitatissimum);* (D) vesicles of lipid-rich membranes develop from ER vacuoles by continued lipid incorporation into the membrane middle layer, e.g., watermelon. Rb, ribosomes. After Wanner *et al.* (1981).

attached (type B), retain multiple contacts with the ER system (type C), or pinch off as an independent "microsome" (type D)—in the membranes of which are deposited fats. Since the storage fats accumulate in the membrane between the phospholipid layers, the outer structure surrounding the mature lipid body is a half-unit membrane.

2.3.4. Storage Protein Synthesis

In general, the syntheses of the various storage proteins of a particular seed are initiated at approximately the same time during development, and proteins are accumulated at about the same rate. There are some exceptions, however, and in some species the appearance of new proteins during late development (even during the early stages of drying) has been noted. Moreover, the subunit composition of some proteins synthesized throughout development may change during the later stages; e.g., in soybean the α and α' subunits of conglycinin appear 15–17 days after flowering, but the β subunit appears only 5–7 days later (Fig. 2.14, E). The final amount of stored protein varies markedly, and usually there is a characteristic major reserve protein within any one species (Table 1.5). The quantitative and qualitative changes occurring during protein deposition within a cereal grain, a legume, and an oil-storing dicot are illustrated in Fig. 2.14, A, C, and D. Although the basic mechanism of protein synthesis is the same in cereals and dicots, the mode of deposition within the protein bodies shows some considerable variations; hence these two groups of seeds will now be considered separately.

2.3.4.1. Cereals

In barley, wheat, and maize the major (prolamin) storage proteins are synthesized on polysomes closely associated with the ER (Fig. 2.15). The newly synthesized proteins pass through the ER membrane into the lumen where, because of their hydrophobic nature, they aggregate into small particles. These are brought together by cytoplasmic streaming to form larger aggregates, which eventually become protein bodies. In wheat (and perhaps barley), it is

--→

Figure 2.14. (A) and (B) Changes in the endosperm protein fractions during kernel development of normal Bomi barley (A) and the high-lysine barley mutant Risø 1508 (B). —▲—, Albumins (plus free amino acids); —○—, globulins; —X—, hordeins (prolamins); —●—, glutelins (C) The accumulation of vicilin (○), legumin (●), and albumins (x) in developing cotyledons of broad bean *(Vicia faba)*. (D) Deposition of the 11 S crystalloid protein (○) and of matrix and noncrystalloid proteins (2 S and 4–6 S) (●), including lectins, in developing endosperms of castor bean *(Ricinus communis)*. (E) Accumulation of α, α', and β subunits of the 7 S storage protein β-conglycinin in developing soybean seeds. (A) and (B) After Brandt (1976); (C) after Manteuffel *et al.* (1976); (D) based on Gifford *et al.* (1982); and (E) after Gayler and Sykes (1981).

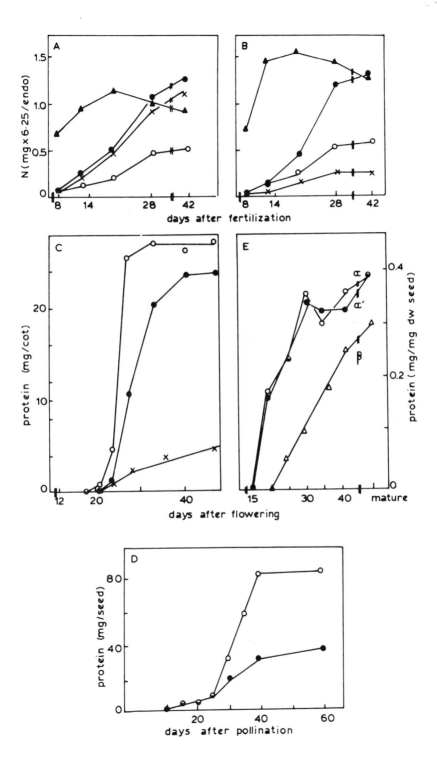

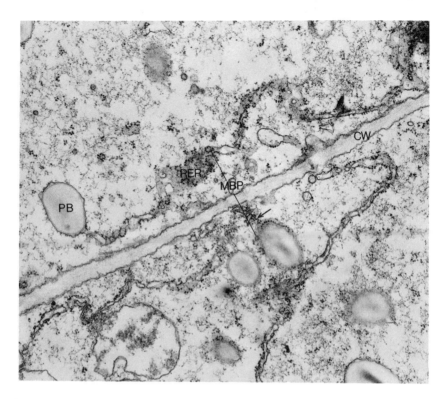

Figure 2.15. Electron micrograph of a developing cell of the starchy endosperm of maize to illustrate the formation of protein bodies (PB) forming from rough endoplasmic reticulum (RER) and the proximity of membrane-bound polyribosomes (MBP). Note the continuity between the protein body membranes and extended RER cisternae (arrow). CW, cell wall. By Larkins and Hurkman (1978).

likely that the aggregates inside of the lumen of the ER put a strain on the membrane and cause it to rupture. The membrane may re-form free of the protein aggregate, and the protein "body" itself is obviously not bounded by an outer membrane. In other cereals, such as millet, rice, maize, and sorghum, where the protein bodies remain as distinct membrane-bound entities even in the mature seed, presumably the strain on the ER is less severe such that it distends but does not rupture. Alternatively, the ER itself may be more elastic in nature.

Isolation of messenger RNAs for storage proteins from the polysomes associated with the ER (rough endoplasmic reticulum, RER), and their subsequent translation *in vitro,* have been achieved for developing maize, wheat, and barley grains. The *in vitro* synthesized products strongly resemble the prolamin storage proteins characteristic of the species, but they have a slightly higher molecular weight (about 2 kD) than the native protein. This is because within the cells of the developing endosperm itself (i.e., *in vivo*) the newly synthesized protein is modified (shortened) during its passage through the ER membrane into the lumen. This translocation across the membrane is facilitated by the presence of a special "signal peptide" at the amino end of the nascent protein chain. This signal peptide allows for the passage of the protein across the membrane, but it is removed by proteolytic activity at the same time (Fig. 2.16). This cotranslational modification or shortening of the storage protein occurs *in vivo,* but not *in vitro* in the absence of a membrane preparation. However, as shown by Larkins and co-workers (1979), the cotranslation processing of storage proteins can be achieved even when the messenger RNA for zein is introduced into an animal cell. They injected this message into oocytes of *Xenopus* and found that not only were all the zein components completely

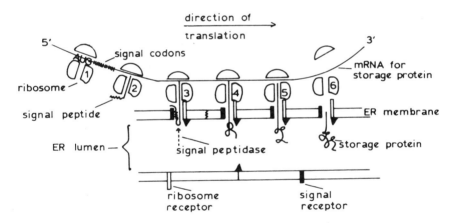

Figure 2.16. A schematic representation of the signal hypothesis for the translocation of storage proteins across the membrane of the rough endoplasmic reticulum (RER). Several stages of the translation of a messenger RNA for a storage protein on a membrane-associated polysome are shown. The polysome contains six ribosomes. The two near the start (5′) end of the mRNA are not yet bound to the membrane. The nascent protein chain "grows" in a passage through the large (lower) ribosomal subunit. The signal peptide is indicated as a jagged line at the amino terminus of the nascent protein (see second and third ribosome), or as being cleaved (i.e., cotranslationally) from the protein as translation proceeds, and located in the membrane (see membrane between ribosomes 3 and 4). The site of the peptidase that cleaves the signal peptide is indicated by an upward-pointing arrow connected to a dashed line. The ribosome receptor and signal receptor activity are represented by two different integral membrane proteins, although they could be two parts of one protein. Based on Blobel *et al.* (1979).

Table 2.3. Distribution of Soluble Protein Fractions in Endosperms of
Normal and *Opaque*-2 Maize[a]

	Percentage of total protein	
Protein fraction	Normal	*Opaque*-2
Albumins (plus free amino acids)	3.7	14.7
Globulins	1.7	4.4
Prolamins (including zein)	54.2	25.4
Glutelins	40.4	55.5
% Lysine in total protein	1.6	3.7

[a]After Jiminez (1966), quoted in Weber (1980).

synthesized, but also that they were processed correctly and secreted into membrane vesicles!

Cereal proteins are deficient in certain amino acids, in particular lysine
and tryptophan. Certain high-lysine mutants have been developed, and studies
have been made to determine the nature of protein deposition in these lines. In
the high-lysine mutant Risø 1508 of barley (Fig. 2.14, B) and the *opaque*-2
mutant of maize (Table 2.3) there is a reduction in the deposition of lysine-
poor prolamin (hordein and zein, respectively) in the developing endosperm
and an increase in albumins, globulins, and glutelins, all of which have a higher
lysine content than zein or hordein. The *opaque*-2 mutant of maize does not
produce mRNAs for one of the zein proteins, and there is suppression of a
select group of genes for this storage component. Other metabolic changes may
play a role also, however; for example, the catabolism of lysine in the endosperm of normal maize appears to be greater than in the high-lysine mutant.

In rice and sorghum, glutelins occur as storage proteins within protein
bodies, but in other cereals they form an insoluble matrix in which the starch
grains and protein bodies lie. In maize, barley, and wheat the glutelins appear
to be synthesized as soluble storage proteins that congeal when the grain undergoes its maturation–desiccation. Although the glutelins contribute considerably to the protein content of the starchy endosperm, little is known about their
origin and complexity.

2.3.4.2. Dicots

The formation of protein bodies in the cotyledons and persistent endosperms of the dicots studied to date follows a distinctly different pattern from
that in the cereal endosperm. This has been illustrated particularly well for the
developing cotyledons of the garden pea *(Pisum sativum)* (Fig. 2.17). Early
during development, around 12 days after flowering, the cotyledon cells contain

one or two large vacuoles, which occupy the majority of the volume of the cell. Some protein deposition within the vacuole has commenced at this time (Fig. 2.17, A). By day 15 many of the vacuoles are considerably smaller and more spherical in shape (Fig. 2.17, B and legend), and by day 20 these have become filled with storage protein and form the distinct and discrete protein bodies (Fig. 2.17, C). The pattern of development, then, is gradual fragmentation over a few days of a highly convoluted central vacuole, and this leads to the formation of the smaller protein bodies. This sequence of events probably occurs in the cells of other developing legume cotyledons; it also takes place in the developing endosperm of the oil-storing castor bean seed, except that vacuolar subdivision occurs before any protein deposition occurs. An apparent exception is the mustard seed *(Sinapis alba),* where it is claimed that in some cells, at least, the large central vacuole becomes filled with protein and does not undergo such extensive fragmentation.

Although deposition of proteins is within the vacuoles/protein bodies, their synthesis occurs on the rough endoplasmic reticulum, and there is vectorial transport to the storage organelles. This has been demonstrated in cotyledons of legumes, such as the pea. Here, the major reserve proteins of the pea are the globulins legumin and vicilin, which constitute about 70% of the total stored protein in the mature seed. Legumin is an oligomeric 12 S protein, consisting of six acidic (mol. wt. 40,000) and six basic (mol. wt. 20,000) subunits, each acidic subunit being joined by disulfide bridges to a basic peptide. The subunits are initially synthesized as a group of polypeptides on membrane-bound polysomes as mol. wt. 60,000 precursors, which are sequestered within the ER. Analysis of the messenger RNA for legumin has shown that the coding sequence (gene) for the acidic legumin subunit is contiguous with that of the basic legumin subunit, and that the acidic subunit gene is translated immediately before the basic subunit gene. Hence, the newly synthesized subunits are joined together—by a short amino acid chain—forming the high-molecular-weight precursor. This legumin precursor may be stable for several hours, but eventually it is processed, posttranscriptionally, by hydrolytic cleavage to form an acidic and basic subunit (Fig. 2.18). These subunits then arrange themselves and become linked by disulfide bridges. At the present time there is no report of a signal peptide on the newly synthesized legumin precursor, although it is reasonable to expect that one exists. The glycinin storage protein subunits of soybean are translated similarly on adjoining genic regions and then cleaved posttranslationally. Pea vicilin is also synthesized in precursor form on the RER, but with a more complex range of molecular sizes than legumin—at least two of the vicilin subunits have signal peptides when made, and these are removed on the ER.

The mode of transport from the ER to the protein body and the site of posttranslational modification of the legumin and vicilin precursors have only

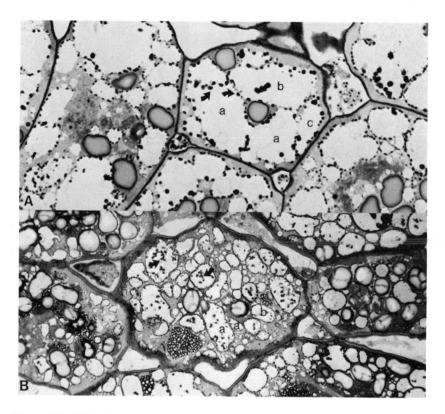

Figure 2.17. Light micrograph of cells of pea cotyledons (A) 12, (B) 15, and (C) 20 days after flowering. (A) A large vacuole is present in the cells. Serial sections through the cell (see original reference) show that (c) is an extension of vacuole (a), which is also continuous with (b). Arrows, proteins deposited in vacuole. (B) Several discrete vacuoles are in evidence, which are irregular in shape; e.g., serial sections through the cell show that (a) is a single vacuole, but it is distinct from (b). Arrows, protein deposits. (C) Discrete, spherical protein bodies are in evidence; (a), (b), and (c) are not connected to each other. An indication of the vacuole or protein body diameter, in relation to the diameter of the developing cotyledon cell, can be obtained from the following data:

Days after flowering	Cell diameter (μm)	Vacuole/protein body diameter (μm)
10	42	39
15	81	11
15	84	3
20	99	1

Note the large reduction in vacuole size as they fragment to form protein bodies. Original data in Craig *et al.* (1979, 1980).

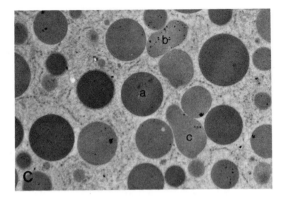

Figure 2.17. (*Continued*)

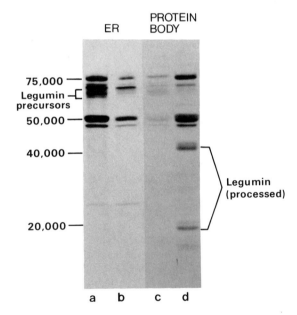

Figure 2.18. Fluorograph of proteins separated by polyacrylamide gel electrophoresis showing the disappearance of radioactively labeled legumin precursor from the endoplasmic reticulum (ER) and the appearance of processed legumin in the protein body during development of pea cotyledons. Cotyledons at 23 days after flowering were incubated in radioactive amino acids, during which time the legumin precursors became radioactive owing to their incorporation. The cotyledons were then removed from the radioactivity, and the fate of the labeled proteins in the ER and protein body fractions was studied by extraction and separation after (a) and (c) 15 min or (b) and (d) 135 min. From Chrispeels *et al.* (1982).

just recently started to be studied. Transport of proteins from the RER to the protein bodies takes some 20–30 min. This could be mediated by the budding- or "blebbing"-off of protein-filled ER-derived vesicles, or by transport of proteins through the ER lumen to the Golgi apparatus, where they are packed into transport vesicles. Either type of vesicle could fuse with the protein bodies and deposit their contents within. The least likely alternative is that reserve proteins are released into the cytosol itself and then taken up directly by the protein bodies. Posttranslational modification of legumin occurs after the precursor has left the ER (note the absence of processed legumin from the ER fraction in Fig. 2.18, a and b): whether this processing and assembly into oligomers occurs largely during transport to the protein body—the Golgi has been mooted as the appropriate site—or in the protein body itself is a matter for debate. One current theory contends that newly synthesized precursors of legumin (and vicilin) are assembled into oligomers in the ER and then transported to the protein bodies for processing. Although the precursor forms of legumin do not accumulate within the developing protein bodies (owing to its rapid processing), the processing of vicilin is much slower, and both precursor and processed vicilin subunits have been found to remain for many hours within the protein bodies themselves. A brief summary of this sequence of events is presented in Table 2.4.

Vicilin, unlike legumin, but like a variety of storage proteins in other dicots, is a glycoprotein; i.e., it is glycosylated during its synthesis. The carbohydrate moiety of many of the glycosylated storage globulins contains mannose and glucosamine, the latter attached by an N-glycosidic bond between N-ace-

Table 2.4. The Sequence of Events Involved in the Synthesis of Storage Globulins in Developing Pea Cotyledons[a]

1. Translation of mRNAs for storage globulins on RER
2. Translocation into RER lumen of nascent polypeptides—probably facilitated by the presence of a signal peptide
3. Some vicilin subunits glycosylated on RER (during synthesis?)
4. Large-molecular-weight precursors of legumin and vicilin assembled into oligomers for transport to protein bodies
5. Oligomeric proteins packed into ER vesicles (or pass through ER lumen to Golgi for packing into vesicles), which bleb off and fuse with protein bodies.
6. Fragmentation of large central vacuole to form smaller vacuoles, which become protein bodies (occurs as storage protein synthesis is commencing)
7. Proteins released into protein bodies. Posttranslational processing of legumin completed: 60-kD precursor cleaved to release acidic (40-kD) and basic (20-kD) subunits, which join by disulfide bridges. Six acidic and six basic chains assemble to form native protein. Vicilin processing occurs over a longer time period—percursors of 75, 70, 50, and 49 kD are posttranslationally modified to produce subunits of 34, 30, 35, 18, 14, 13, and 12 kD

[a]See text for details. Based on data in Croy et al. (1982) and Chrispeels et al. (1982) and references therein.

tyl-D-glucosamine (GlcNAc) and an asparagine residue on the storage protei
as follows:

$$\begin{array}{c} \text{polypeptide} \\ | \\ (\alpha\text{-Mannose})_n\text{-}\beta\text{-mannose}(1\text{-}4)\text{-}\beta\text{-GlcNAc}(1\text{-}4)\beta\text{-GlcNAc-asparagine} \\ | \\ \text{polypeptide} \end{array}$$

The formation of this linkage (which can occur almost as frequently as there are asparagine residues in the protein) involves the transfer of UDP-N-acetyl-D-glucosamine and GDP-mannose to lipid intermediates (dolichol phosphates) on the RER (and Golgi), which in turn transfer the oligosaccharide to the protein. Glycosylation may occur cotranslationally or posttranslationally. Its importance is unclear, but it is interesting that prevention of the synthesis of the carbohydrate side chains of vicilin affects neither the synthesis of the storage protein, nor its assembly into oligomers, nor its transport into protein bodies.

The synthesis of storage protein precursors and their mode of processing show considerable variation between species, and only a few examples will be presented in order to make this point. In cotton, the principal storage proteins appear to be synthesized initially as 69-kD and 60-kD precursors (Fig. 2.19, A). The larger-molecular-weight precursor (a) undergoes only a small modification, initially, perhaps by cotranslational removal of the signal peptide as it passes through a membrane (perhaps the ER, but this has not been demonstrated). The molecular weight of the smaller precursor increases to about 70,000 owing to glycosylation—a signal peptide might be lost too. Both precursors are then cleaved further, slowly, and perhaps within the protein bodies, to yield the mature 52-kD and 48-kD storage proteins (Fig. 2.19, A). The peptides that are cleaved off to produce the proteins from their precursor are very large (15–20 kD), much larger than any signal peptide, and they may also accumulate within the protein bodies. Comparable posttranslational changes in molecular size during the processing of the 7 S α subunit of the soybean storage protein are shown in Fig. 2.19, B.

In developing endosperms of castor bean, the main storage protein, the 11 S crystalloid complex, is formed by posttranslational cleavage of a 50-kD precursor into polypeptides of mol. wt. 30,000 and 20,000, which become linked by disulfide bridging (Fig. 2.19, C and D). Formation of the matrix protein agglutinins in the castor bean is considerably more complex, however. Agglutinin type I is comprised of two polypeptides of 37 kD and 31 kD in the mature state, but they are encoded as 59-kD and 35-kD precursors. Cotranslational processing cleaves 1.5 to 2.0-kD segments (signal peptides), and there

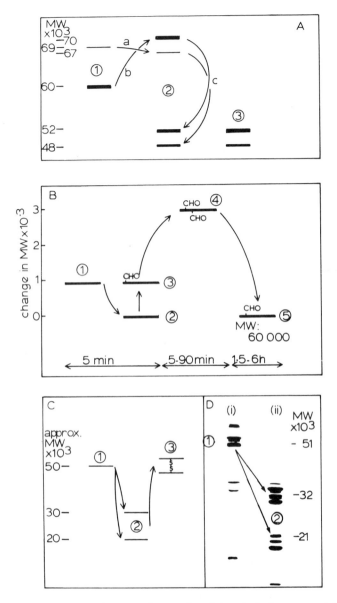

Figure 2.19. (A) The proposed pattern of processing of the principal storage protein in developing cotton *(Gossypium hirsutum)* cotyledons. (1) Precursors. Events (a), cotranslational loss of signal peptide; (b), cotranslational loss of signal peptide and increase in molecular weight owing to glycosylation. (2) Processed products. Event (c), *in situ* cleavage. (3) Molecular weights of abundant storage proteins in the mature seed. (B) Schematic outline of the sequence of events involved in the synthesis of the 7 S α subunit of soybean storage protein. (1) is the primary translation product of membrane-bound polyribosomes. This protein is cotranslationally cleaved to form (2) and cotranslationally glycosylated to form (3). After posttranslational glycosylation to product (4), this is subjected to proteolysis and/or deglycosylation to form the α subumit (5). Time from initial synthesis to completion of processing is approximately 6 h. The final 7 S α

is also cotranslational modification of the large-molecular-weight precursor as it increases to 66–69 kD owing to glycosylation. Posttranslational cleavage yields the authentic glycosylated 37-kD agglutinin and a 34-kD protein. The precursor of the smaller agglutinin chain is reduced in mol. wt. to about 32 kD by cotranslational removal of the signal peptide; subsequently, posttranslational "trimming," i.e., removal of small segments of peptides, yields the 31-kD protein.

Considerable interest is being expressed in the synthesis of messenger RNAs for storage proteins in relation to the timing of synthesis of the proteins themselves. We can see from Fig. 2.20 that in pea the bulk of the synthesis of the storage proteins in the cotyledons occurs mostly during the cell expansion phase of growth, following cell division, and it ceases as the seed completes its maturation drying. The content of both DNA and RNA in the cotyledons increases during the cell expansion phase. Synthesis of RNA, the bulk of which will be ribosomal RNA, during the first half of this phase is to be expected because the expanding cells are commencing the large-scale synthesis of storage and metabolic proteins. The continuing synthesis of DNA in cells that are no longer dividing is less easy to explain. It has been suggested that polyploidization makes available extra copies of the genes for storage proteins. But the weight of evidence is against this possibility; in the dicots and cereals studied so far there is no evidence for any selective amplification of the genes for storage proteins during seed development. Most storage protein genes are either single-copy genes, or there are no more than four or five copies per haploid genome. Thus the extra DNA production may merely have a storage function as a reserve of deoxynucleotides, to be utilized following germination to support the extensive cell divisions in the growing embryo.

The amount of mRNA for storage proteins increases during development, to reach a peak at about the time of maximum protein deposition (e.g., in soybean, Fig. 2.21 and Table 2.5). Then, as the seed begins to mature and dry out, the amount of mRNA declines; storage protein synthesis declines likewise. This pattern of events suggests a simple transcriptional control of protein synthesis—essentially, the amount of storage protein produced is determined by the level (and stability) of its mRNA present. This probably holds true for

---

subunit has an approximate mol. wt. 60 kD. (C) Proposed scheme for the processing of the major 11 S crystalloid storage protein in developing castor bean endosperms. (1) is a large precursor of approximate mol. wt. 50 kD which is posttranslationally cleaved to form two polypeptides (2). These are then linked together by disulfide bridges (3) to yield a storage protein of approximate mol. wt. 50 kD. This is shown on polyacrylamide gels in (D) as (i) the radioactively labeled approximate mol. wt. 50-kD precursor (1) processed in time into the (ii) low-molecular-weight (approximately 30-kD and 20-kD) polypeptides (2). The two polypeptides are prevented from forming the final product (C, 3) by using reducing conditions during the extraction and gel electrophoresis. (A) After Dure and Galau (1981). (B) After Sengupta *et al.* (1981). (C) and (D) by D. J. Gifford.

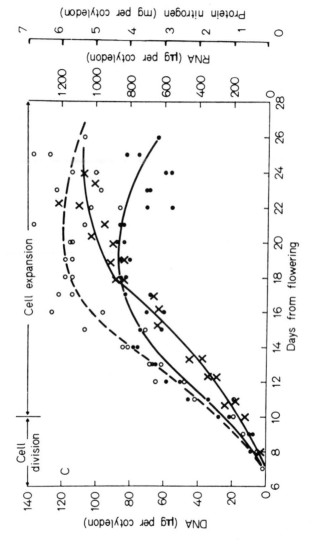

Figure 2.20. Changes in DNA (o), RNA (●), and protein (X) content during pea cotyledon development. After Millerd and Spencer (1974).

Figure 2.21. Changes in messenger RNA content in soybean cotyledons during development and after germination. △, mRNA for α and α' subunit of 7 S storage protein (conglycinin); ○, mRNA for 11 S storage protein (glycinin); ●, mRNA for trypsin inhibitor. Developmental changes: c, cotyledon; em, early maturation; mm, midmaturation; lm, late maturation; q, quiescent; pg, postgermination. After Goldberg *et al.* (1981).

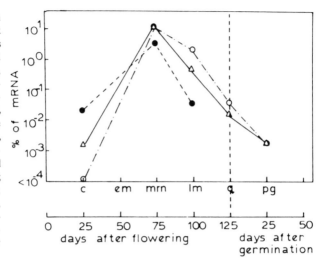

most proteins, but there may be exceptions. In soybean and some other legumes, for instance, the subunit composition for the 7 S storage protein (conglycinin) may change during development. Also, in soybean, the β subunit of the 7 S protein is only synthesized several days after two of the α subunits (Table 2.5), yet the mRNAs for all the α and β subunits of the soybean 7 S proteins are present at the earliest time when only the α subunits are being synthesized (i.e., at 18–20 days). This suggests that there is in operation a translational control mechanism which, early during development (before 28

Table 2.5. Summary of Developmental Events in the Accumulation of Storage Proteins and Their Messenger RNAs in the Cotyledons and Axes of Soybean
(Glycine max)[a]

Days after anthesis	Event
14–18	mRNAs for 7 S and 11 S storage proteins first detectable
18–20	Two subunits (α and α') of the 7 S protein begin to accumulate in axes and cotyledons. Increased levels of mRNAs for all 7 S protein subunits
19–21	Acidic and basic subunits of 11 S protein begin to accumulate in cotyledons. Increased levels of mRNAs for 11 S proteins
> 28	β Subunit of 7 S protein and A-4 acidic subunit of 11 S protein begin to accumulate in cotyledons. α° Subunit of 7 S protein begins to accumulate in axes
> 50	mRNAs for 7 S and 11 S proteins no longer detectable
60	Mature dry seed

[a]Based on information in Meinke *et al.* (1981).

days), prevents the use of the mRNAs for synthesis of the β subunit and also of the mRNAs for the changed α subunit ($\alpha°$) produced only at later stages of development.

Over the past 5 years there has been an enormous increase in interest in the nature and synthesis of storage proteins in developing seeds, particularly with a view to improving their nutritional quality. A vast proportion of all human food is derived directly or indirectly (by feeding to livestock) from seeds. Yet, as noted earlier, cereal seeds are deficient in certain essential amino acids (lysine, threonine, and tryptophan), and so are legumes (in the S-containing amino acids, methionine and cysteine). Monogastric animals, such as human beings, pigs, and poultry, do not possess the anabolic processes necessary to synthesize these essential amino acids from other C and N sources. Strategies have been developed, or are being researched, to overcome these deficiencies.

In cereals the prolamins are the major storage proteins; they are high in proline and amide and are low in lysine (Section 1.5.3). Spontaneous mutants of some of the cereals have been isolated, e.g., the *opaque*-2, *brittle,* and *shrunken* mutants of maize, which produce less of the prolamin, zein (Table 2.3), and more of the usually minor storage proteins which have a higher lysine content. The increased nutritional value of these high-lysine mutants to human beings and to experimental animals is well recognized, but, unfortunately, there are several negative aspects associated with the mutant seeds. As a consequence of the reduced prolamin content the kernels are more brittle than the normal type and tend to shatter easily during storage; also, the baking properties of the mutant maize meal are poor. The most serious deficiency, however, is that the mutations lead to a severe disruption in carbohydrate metabolism (i.e., starch synthesis), and yield may be decreased by 10–60% (Table 2.2). So far, then, attempts to breed low-prolamin cultivars of maize have been unsuccessful, and little more success has been achieved with other cereals. Natural and induced (by using radiation) mutants of barley have been produced, but none are available yet as commercial lines. A promising sorghum mutant (P721 *opaque*) has been produced by mutation breeding, which leads to a decline in yield by only 10–15%. Other types of mutants have been sought also, e.g., those with an enhanced ability to synthesize free amino acids, owing to changes in the regulatory mechanisms operating the biosynthetic pathways. There is some evidence that when the lysine (in cereals) or methionine (in legumes) content of seeds is increased, these amino acids appear in proteins with greater frequency. Certainly, supplementing soybean cotyledons developed in culture with methionine can enrich the storage protein fraction in this amino acid by more than 20%. This has led to attempts to increase the content of S-containing amino acids in legume seed proteins by increased application of sulfate to the mother plant during seed development.

But it is the advent of genetic engineering and the possibility of isolating and cloning genes that have created the greatest excitement and interest in relation to improving protein quality. Considerable progress has already been made in isolating and identifying the genes for storage proteins. Messenger RNAs for storage proteins have been isolated from membrane-bound polyribosomes; their complementary DNA (cDNA) has been made using reverse transcriptase, made double stranded with DNA polymerase I, and introduced into bacteria via plasmid vectors. There the gene is expressed, and storage proteins are synthesized. This shows, then, that the genes can be transferred from one organism to another. One hope is to improve the quality of storage proteins by isolating and modifying their genes using such techniques, or to transfer genes for nondeficient proteins from one seed to the genome of another. For example, it would be advantageous, theoretically, to introduce the gene for the storage globulin from legumes into the cereal genome, for this would eliminate the amino acid deficiencies created by either the prolamin or the globulin alone. There is a considerable number of problems still to be surmounted, however; e.g., (1) a suitable vector for transferring genes from one plant to another has to be found; (2) genes are very complex, and many have essential regulatory sequences which flank the coding sequences (exons); these must be transferred also, as must any intervening sequences (introns) which are transcribed into immature mRNA but are spliced out during its processing to the translatable form for the storage protein; (3) since seed storage proteins are not a single protein, but rather are comprised of a family of subunits or polypeptides, they must be encoded for by a family of genes. Hence the problem is not simply to transfer one gene, but to transfer a critical number of genes so that all the essential components of the holoprotein can be made. Moreover, (4) we need to know more about the multifaceted process of reserve synthesis itself, for the introduction of foreign genes might upset its regulation, resulting in a seed with an undesirable and/or unimproved protein content. Such problems are daunting, and many of them may be impossible to overcome, although research is continuing apace with guarded optimism, and with considerable financial investment by both the public and private sector.

2.3.5. Phytin Deposition

Despite the importance of phytin as a reserve compound in seeds, the pathways for, and the intracellular sites of, phytic acid biosynthesis are not clearly understood. It has been proposed that in developing seeds of sunflower and rice the phytic acid is synthesized within the protein bodies, where phytin is sequestered (Section 1.5.4). But supportive evidence for these organelles being the site of synthesis is poor. Recent evidence obtained from studies on

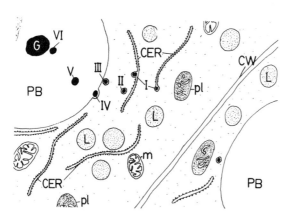

Figure 2.22. A diagrammatic representation showing the site of phytin synthesis and mechanism of deposition of phytin in protein bodies of developing endosperm cells of castor bean. Roman numerals I–VI show phytin at various stages of migration from the CER to the protein body globoid. CER, cisternal endoplasmic reticulum; CW, cell wall; G, globoid; L, lipid body; m, mitochondrion; PB, protein body; pl, plastid. After Greenwood and Bewley (1984).

the developing endosperm of castor bean is consistent with the hypothesis that phytin is formed initially within the ER (Fig. 2.22, step I), prior to being transported to the protein bodies (steps II and III). Fusion of the vesicular and protein body membranes (step IV) causes the phytin particles to be released into the lumen of the protein body (step V). The phytin particles condense therein to form the globoid (step VI).

2.3.6. Deposition of Several Reserves—Correlative Patterns

Although, as we discussed in the previous sections, considerable attention has been paid to the deposition of individual reserves during development, few studies have been carried out to elucidate the gross pattern of deposition of all the reserves in any particular seed. A couple of examples are given here.

Figure 2.23, A illustrates the temporal correlation between starch and protein accumulation during development of a cereal, wheat. Here, protein bodies arise during kernel development at about the same time as starch grains. In the legume, garden pea, again there is an almost identical temporal pattern of deposition of protein and starch (Fig. 2.23, B).

The increase in total content of reserves does not necessarily occur in all

——→

Figure 2.23. (A) Changes in endosperm starch (●) and protein (○) in developing grains of wheat. Change in fresh weight (X) also. Based on data in Jennings and Morton (1963). (B) Changes in fresh weight (○), moisture content (X), and dry weight, protein, and starch (all represented as ● since the changes are similar for all three) in developing garden pea embryos. Various stages of development are delineated. Adapted from Bain and Mercer (1966).

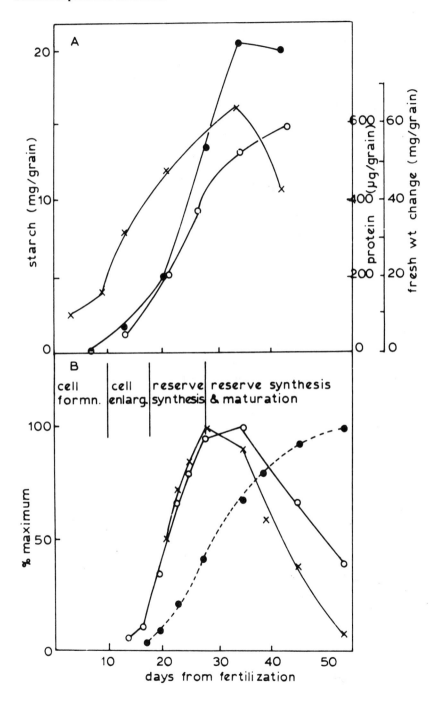

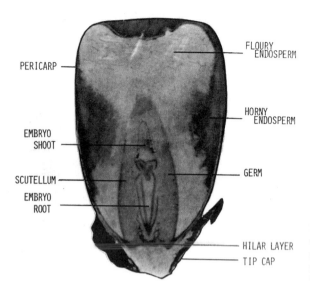

PERICARP

FLOURY
ENDOSPERM

HORNY
ENDOSPERM

EMBRYO
SHOOT

SCUTELLUM

GERM

EMBRYO
ROOT

HILAR LAYER

TIP CAP

Figure 2.24. Cross section of a dent maize kernel to show various regions of reserve deposition. From Weber (1980).

storage cells of the seed simultaneously. In the endosperm of castor bean, for example, deposition of proteins first occurs in some discrete regions prior to that in others, although at maturity the protein is located fairly evenly throughout the endosperm. Such even distribution is not found in all seeds, however. In mature maize grains there are recognizably protein-rich and starch-rich areas within the starchy endosperm; these are the horny endosperm and floury endosperm, respectively (Fig. 2.24). The proportion of each type of endosperm depends on the type and variety of corn. In popcorn, for example, the horny endosperm forms a thick shell around a small central core of floury endosperm. The starch grains of the horny endosperm are embedded in tough elastic colloidal material which resists steam pressure generated within the grains upon heating, until explosive forces are reached. The popping ability is therefore conditioned by the proportion of horny endosperm.

2.4. MATURATION DRYING—THE EFFECTS OF WATER LOSS ON DEVELOPMENTAL EVENTS

Maturation drying is an integral part of the development of most seeds; in fact, development is considered to be completed when the seed has dried. Subsequent hydration of the mature, nondormant seed leads to germination; hence it is possible that desiccation plays some role in the switch from developmental processes to those essential for germination. Many seeds will not germinate unless they are subjected to drying—immature seeds of many legumes picked fresh from the pod or grains of cereals removed from the developing ears fail to germinate when placed in water.

Although natural drying occurs relatively late during development (e.g., in pea, Fig. 2.23, B, and wheat), many seeds can withstand extensive water loss much earlier. Hence a considerable proportion of the time taken to complete development does not appear to be essential for the production of viable seed but is the time required to increase the seed mass, especially through deposition of reserves. Seed development can be divided into two phases in relation to the ability to withstand air drying: initially there is a desiccation-intolerant phase, when drying is fatal, followed by a tolerant phase, when subsequent rehydration leads to germination (Table 2.6). In fact, acquisition of desiccation tolerance is generally accompanied by development of the capacity of seeds to germinate, and these attributes may be acquired over a very brief period of development—only a few days (e.g., 22-day-developed axes of *Phaseolus vulgaris* are intolerant of desiccation, but those at 26 days are tolerant and are induced to germinate by the drying treatment, Table 2.6). The changes that occur in the seed during the change from intolerance to tolerance of desiccation are completely unknown.

Desiccation of axes of *P. vulgaris* after 32 days of development, which is in the desiccation-tolerant phase, causes them to germinate when subsequently rehydrated. This change in direction from development to germination is also mirrored in the pattern of protein synthesis exhibited upon rehydration, in that

Table 2.6. Germination of Dried or Fresh Intact Seeds of *Phaseolus vulgaris* Harvested at Various Times after Anthesis[a]

Days of development after anthesis	Fresh or dried	Germination
22	F	—
	D	—
26	F	—
	D	+
30	F	—
	D	+
35	F	Poor
	D	+
40	F	+
	D	+
Mature dry seed	F	+

[a]Seeds were removed from the pod at various times of maturation and placed in water. Populations of seeds that germinated are scored +, and those that did not germinate are scored —. During the normal course of development seeds begin to dry naturally after 36 days. Based on data in Long *et al.* (1981) and Dasgupta *et al.* (1982).

some protein synthesis uniquely associated with development ceases (particularly the synthesis of storage proteins), and other syntheses associated exclusively with germination commence. Presumably, then, desiccation causes the loss from the developing axis of mRNAs for developmental proteins and, moreover, permanently suppresses the genome so that no more of these mRNAs are made. At the same time, the genome for the production of germination-related mRNAs is activated—how this occurs is not understood. Production of enzymes exclusively associated with postgermination events is also induced by drying of the developing seed. For example, fresh, immature grains of barley and wheat do not produce α-amylase in response to gibberellic acid (Chapter 8), but do so after drying.

The fate of mRNAs during drying has aroused considerable interest over the past several years. Dure, using cotton cotyledons as a model, has proposed that there are several mRNA subsets involved in protein synthesis during germination. These are discussed in more detail in Section 4.4.2, but it is relevant briefly to mention them again here:

1. Residual mRNAs. These are mRNAs that are produced during seed development and are not destroyed during late maturation and desiccation. They are not important for germination and may be degraded early during imbibition. Any mRNAs for storage proteins not lost during drying itself may fall into this category.

2. Stored or conserved mRNAs. These are translated into proteins that are an integral part of germination or of postgermination events such as reserve mobilization. Many of the mRNAs for proteins/enzymes essential for the metabolic integrity of the developing and germinating seed are probably conserved, although there is no evidence for the presence in dry seeds of mRNAs for proteins unique to germination. There is a long-held claim that the mRNA for carboxypeptidase C (an enzyme involved in the mobilization of protein reserves in cotyledons of germinated cotton) is transcribed during development and conserved in the dry seed for postgermination translation. But the evidence for this claim is not strong, and considerable doubt has been expressed that any mRNAs for enzymes required for postgerminative events are conserved in the mature dry seed.

It is possible for developing embryos to germinate without a requirement for a drying process. This occurs naturally in some species (e.g., in mangrove and certain viviparous mutants of cereals, Chapter 9), but it occurs in others too when immature embryos are removed from the developing seed and placed in tissue (embryo) culture conditions. Isolated embryos may even continue to grow and accumulate storage proteins for several days in culture, as they would in the seed, before becoming briefly quiescent and then germinating. Presumably, the environment of the developing embryo on the mother plant is normally such that it encourages development and prevents germination—removal from this environment causes a hiatus such that development ceases. The controlling factor(s) provided by the mother plant are not known, although application of abscisic acid (ABA) to isolated developing embryos of rape in culture prevents their germination and encourages the continuation of developmental processes, including the synthesis of storage proteins (Fig. 2.25). In cotton embryos treated with ABA, some embryogenic protein synthesis continues, but this does not include the major storage proteins. Precocious germination of isolated wheat embryos in culture is prevented by ABA; at the same time a lectin (wheat germ agglutinin), produced exclusively during development, is synthesized. Hence, in this system, ABA appears to suppress germination and prolong a specific developmental event. It has been surmised that a steady supply of ABA (or some other translocatable factor that ABA wholly or partially mimics) from the mother plant to the embryo maintains its metab-

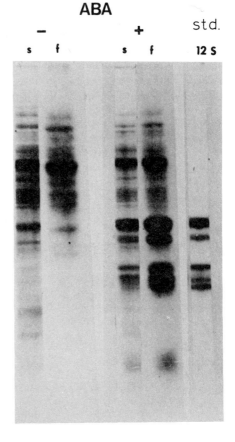

ABA

−   +   std.

s   f   s   f   12 S

Figure 2.25. Separation by polyacrylamide gel electrophoresis of proteins extracted from embryos of *Brassica napus* (rape) cultured on basal medium with (+) or without (−) the addition of 10^{-5} M ABA. Embryos were incubated on radioactive amino acid for 12 h after being cultured in the presence or absence of ABA. Proteins were extracted and separated by electrophoresis, and the radioactive (i.e., newly synthesized) proteins were detected by fluorography (f). Proteins detected by staining alone are on the s gels. The 12 S marker standards (std.) on the right are nonradioactive markers of where the 12 S storage proteins migrate to on the gels. Note the absence of newly synthesized 12 S storage proteins in embryos cultured on basal medium without ABA and their presence in embryos on basal medium containing ABA. This suggests that ABA maintains isolated rape embryos in a state of development and encourages synthesis of storage proteins—a developmental event. After Crouch and Sussex (1981).

olism in a developmental, i.e., largely anabolic, mode. Reduction in ABA reaching the embryo during the late stages of maturation could release seeds from the constraints of development and allow germination to occur. Desiccation may bring about a reduction in ABA levels or may reduce the sensitivity of seeds to ABA (or some other regulatory agent); hence premature drying removes the developmental constraints and the seeds can germinate upon subsequent rehydration. Changes in hormone levels, including ABA, during development are discussed in Section 2.5.

2.5. HORMONES IN THE DEVELOPING SEED

While the growing seed is accumulating its major storage reserves, changes are also occurring in its content of other important chemical substances—the growth regulators or hormones, auxin, gibberellins, cytokinins, and ABA (Figs. 2.26 and 5.4). These substances are thought to play important roles in the regulation of certain aspects of seed growth and development, as well as being involved in fruit growth and certain other physiological phenomena. Immature seeds or grains were the first higher-plant sources of most of the known plant hormones, and they continue to attract the attention of research workers studying biosynthesis, metabolism, and chemistry of these regulators. We will first briefly discuss the kinds of hormones present in devel-

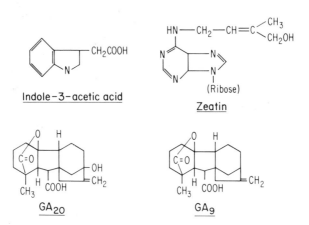

Figure 2.26. Growth regulators (hormones) in immature seeds. In ribosyl zeatin, ribose is present as shown. In the ribotide, the ribose is phosphorylated. Two gibberellins commonly found in developing seeds are shown. (See also Fig. 5.4 for structure of ABA.)

oping seeds, including aspects of their biosynthesis and metabolism, and then go on to consider their possible functions.

2.5.1. Composition and Location

2.5.1.1. Auxins

The major auxin in developing seeds is indoleacetic acid (IAA) (Table 2.7). This is formed in the seed tissues following the normal biosynthetic route from tryptophan and is not derived from the mother plant. Free IAA is widely found, and, in addition, various forms of so-called bound auxin are abundant. In immature kernels of maize, for example, there are IAA arabinoside, IAA *myo*-inositol, and IAA *myo*-inositol arabinoside. These compounds are considered to be precursors of IAA which, following germination, is liberated from them enzymatically and is probably transported to the coleoptile tip of the

Table 2.7. Growth Regulators (Hormones) of Some Immature Seeds[a]

Species	Auxin	GA	CK	ABA
Prunus cerasus	IAA	—	Zeatin	Present
Malus spp. (apple)	IAA	GA_4, GA_7, GA_9 GA_{12}, GA_{15}, GA_{17} GA_{20}, GA_{24}	Zeatin Ribosyl zeatin Zeatin ribotide	Present
Pisum sativum	IAA	GA_9, GA_{17}, GA_{19} GA_{20}, GA_{29}, GA_{38} GA_{44}, GA_{51}	—	Present
Zea mays	IAA "Bound" IAA	—	Zeatin Ribosyl zeatin Zeatin glucoside Zeatin ribotide Ribosyl zeatin glucoside Dihydrozeatin riboside	Present
Triticum aestivum	IAA	GA_{15}, GA_{17}, GA_{19} GA_{20}, GA_{24}, GA_{25} GA_{44}, GA_{54}, GA_{55} GA_{57}	Ribosyl zeatin	Present
Gossypium hirsutum	—	GA_1, GA_3, GA_4 GA_7, GA_9, GA_{13}	Isopentenyl adenine Isopentenyl adenosine	Present

[a]Absence of an entry against a species does not mean that no hormones of that kind exist in the seed but that chemical characteristics have not been recorded.

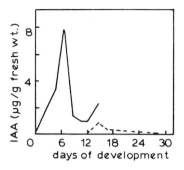

Figure 2.27. Auxin (IAA) in developing pea seeds (cv. Alaska). ----, IAA in embryo; ——, IAA in endosperm. After Euwens and Schwabe (1975).

growing seedling. In maize and other cereals, free IAA and its bound forms are extractable from the endosperm. In dicots too, such as the pea, auxin is first found in the endosperm and only later is it detectable in the embryo itself, after the endosperm has been resorbed (Fig. 2.27). The pattern of free IAA content in the pea is typical of that in developing seeds, in that the concentration rises to a peak value and then diminishes, with relatively little remaining at maturity. Variations on this theme are common: in apple, for example, there are two peaks in auxin concentration, the first coinciding with the change in the endosperm from a coenocytic to a cellular structure, and the second with the formation of new endosperm cells. The final disappearance of free IAA is due to metabolic conversion to bound forms and other products.

2.5.1.2. Gibberellins

By the end of 1982 64 gibberellins (GAs) were known, more than half of which had been found in developing seeds (see Fig. 2.26 and Table 2.7 for some examples). In addition, many GA conjugates have been identified, such as the polar, water-soluble glucopyranosides and glucopyranosyl esters. Because developing seeds are so rich in gibberellins they often feature as subjects for studies of gibberellin biosynthesis and metabolism. For example, work on cell-free systems from immature seeds of *Cucurbita maxima, Marah macrocarpus,* and *Pisum sativum* has shown that gibberellin synthesis follows the route: acetyl coenzyme A → mevalonate → isopentenyl pyrophosphate → chain elongation and ring closure → *ent*-kaurene → kaurenoic acid → GA_{12} aldehyde → gibberellins. Developing seeds, or cell-free embryo extracts, very actively interconvert GAs, which is why any one species contains several different kinds. In dwarf peas, for example, two conversion pathways have been identified: (1) $GA_{12} \to \to \to GA_{20} \to GA_{29} \to GA_{29}$ catabolite, and (2) $GA_{12} \to \to \to GA_9 \to GA_{51} \to GA_{51}$ catabolite. All of the gibberellins in the pathways are active, except GA_{29}, GA_{51}, and their catabolites. Hence, the interconversions lead ultimately to the inactivation of the endogenous gibberellins. At the early stages of seed development, however, the major gibberellins are the active ones, and the inactive ones are formed toward the end of seed maturation (Fig. 2.28). Clearly, if the endogenous gibberellins were detected solely on the basis of their biological activity, a broad peak would be found at days 18–22, owing to the combined effect of GA_9 and GA_{20}. This probably explains why in the early work on seed gibberellins, which relied on bioassay for detection of the hormones, single peaks of activity were generally found during seed development

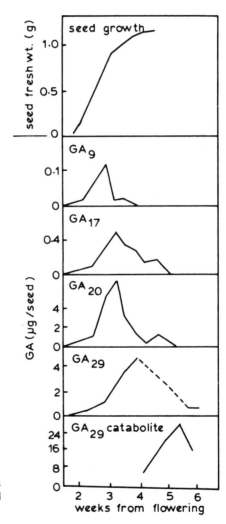

Figure 2.28. Gibberellins in developing pea seeds (cv. Progress no. 9). After Sponsel (1980).

in many species. Part of the drop in the free gibberellin content as seeds mature is due to the formation of various conjugates, the glucosyl esters and glucosides, and gibberellin catabolites, but the fate of the remainder is unknown.

2.5.1.3. Cytokinins

The first cytokinin to be identified in higher plants was zeatin, from developing maize kernels (Fig. 2.26). The chemically characterized cytokinins in immature seeds (see Table 2.7) are substituted adenines—zeatin, isopentenyl adenine, and their derivatives. The latter are various glycosylated forms containing ribose (e.g., ribosyl zeatin, zeatin ribotide, isopentenyl adenosine), glucose (e.g., zeatin glucoside), or both sugars (e.g., ribosyl zeatin glucoside). Cytokinins also occur in certain types of transfer RNA from which they can be liberated by hydrolysis.

It is not clear where the seed cytokinins are synthesized. Evidence points to two possible sources—the mother plant, for example, in the roots, and the seeds or fruits themselves; in some cases (e.g., tomato) the former seems the most likely source. The level of cytokinins increases markedly during seed development, particularly while the seed tissues are actively growing, and then declines with maturation (Fig. 2.29). This temporal distribution is consistent with the suggested role for cytokinins in the control of seed growth, which we will consider below.

2.5.1.4. ABA

ABA has been isolated from immature seeds of several species (see Table 2.7). The free form of the inhibitor can occur at relatively high concentrations,

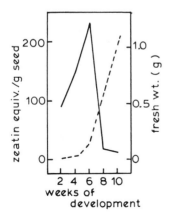

Figure 2.29. Cytokinin in developing seeds of *Lupinus albus*. ——, cytokinin; ----, growth of seed as increase in fresh weight. Cytokinin was determined by bioassay (soybean callus growth). Amounts are expressed as zeatin equivalents (i.e., amount of zeatin having comparable activity). Adapted from Davey and Van Staden (1979).

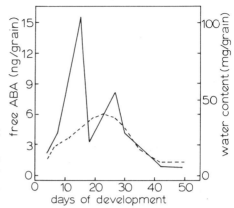

Figure 2.30 ABA in developing wheat grains (*Triticum aestivum* cv. Sappo). Total, extractable free ABA in the whole grain is shown (——), together with water content (----). From data of Mitchell (Ph.D. thesis, 1980).

especially in legumes, e.g., approximately 2 mg/kg fresh weight in soybeans, but more commonly the level is between 10^{-1} and 1 mg/kg fresh weight. Bound forms, the glucosyl ester and glucoside, are also common, and again at relatively high levels in legumes. Both free and bound ABA are located in the various parts of the seed—the embryo, the endosperm, and the enclosing tissues. Metabolities of ABA, phaseic and dihydrophaseic acid, have also been reported, at high concentrations in legumes.

   Like the other growth regulators in immature seeds, ABA rises in concentration during seed development, reaches one or two peaks, and generally then declines rapidly at about the time of seed drying (Fig. 2.30).

2.5.2. Possible Roles of Seed Hormones

   The endogenous growth regulators in developing seeds may be involved in several processes, as follows: (1) Growth and development of the seed, including the arrest of growth prior to seed maturation. (2) The accumulation of the storage reserves. (3) Growth and development of the extraseminal tissues. (4) Storage for later use during germination and early seedling growth. (5) Various physiological effects on tissues and organs close to the developing fruit.

2.5.2.1. Seed Growth and Development

   Most of the evidence linking the growth regulators with seed development comes from correlations between regulator content and embryo growth. The highest concentrations of the active gibberellins of dwarf pea seeds, for example, (GA_9, GA_{20}), occur during the maximum growth rate of the developing embryo (Fig. 2.28), and in the embryo of *Phaseolus coccineus,* the biologically active GA_{20} accompanies the early stages of growth. A further interesting feature of the latter species is that the suspensor may supply GA_1 to the embryo in the earliest phases of development. Very young, excised embryos do not develop further in culture if deprived of the suspensor, but development continues when gibberellin is added to the medium. Since the suspensor contains relatively high levels of GA_1 (approximately 1 $\mu g/100$ mg tissue), it seems likely that it is normally the source of gibberellin for the embryo.

The period of active cell division and enlargement in the seed (in both the embryo and endosperm) is also when the cytokinins are at their highest level. Cytokinins are known to promote cytokinesis (cell division) in certain plant tissues, and it is not unlikely that they have this role in the developing seed too. Since it contains cytokinins, the suspensor has been suggested as a supplier of these hormones to the young embryo, for example, in *Phaseolus coccineus* and *Lupinus albus.*

As might be expected, the inhibitor ABA is associated more with the arrest of embryo growth than with its promotion. There is some evidence, however (in barley, for example), that normal embryogenetic development can occur in the presence of ABA, but that germinative growth (i.e., the great longitudinal extension of the axis) cannot. ABA might therefore be the factor that prevents the embryo from passing directly from embryogenesis to germination, while still on the mother plant, without an intervening rest period; i.e., ABA imposes a period of quiescence on the embryo.

One of the first observations to connect ABA with the prevention of germinative growth of the premature embryo was the finding that developing embryos removed from cotton bolls slowly lose their capacity to germinate, a change that is accompanied by an increase in ABA content. The inhibitor is also thought to prevent the translation of certain mRNAs in the immature cotton embryo—those for the proteins synthesized just after germination—but although applied ABA is effective in this respect, it has not yet been proved that endogenous ABA is similarly active.

Correlations between ABA content and germinability of the developing embryo are indicated in several species. Immature seeds in the early stages of development generally cannot germinate, but the ability to do so is acquired at ages closer to full maturity. ABA levels are higher in young, nongerminable seeds than in older ones, e.g., in pea, soybean, wheat, and dwarf bean. In the latter, when young seeds first become able to germinate they do so with a long delay or lag period, but the lag period becomes progressively shorter as the seeds mature, until after they are 7–8 wk old, they germinate within a day or so after removal of the pod. This decline in the length of the lag period is accompanied by a steep decrease in the seeds' ABA content (Fig. 2.31), and

Figure 2.31. Germinability
and ABA content of devel-
oping seeds of dwarf bean
(Phaseolus vulgaris). Seeds
were removed from the pod
at intervals during development and tested for germinability. Note that for the
first 4½ wk seeds were not
germinable. The time taken
for radicle emergence to
begin in germinable seeds
was recorded (the lag phase)
(---△---). ABA is expressed
as ng/seed (—●—) and ng/
g seed (—○—). Adapted
from Van Onckelen *et al.*
(1980).

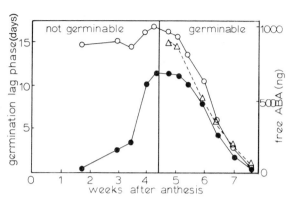

so it seems, therefore, that germinability is related to the ABA level within the young seed.

Further evidence suggesting that premature germination is suppressed by ABA comes from cases of vivipary, where there is virtually uninterrupted progression from embryogenesis to germination of the seed on the mother plant. The embryos of most viviparous mutants of maize have 25–50% of the ABA concentrations of that in the normal, nonviviparous wild type. One viviparous mutant, however, contains the same order of ABA concentration as in the wild type, but in this case the embryo of the mutant is much less sensitive to its own ABA. Relative insensitivity to ABA appears to be the case in mangroves (e.g., *Rhizophora mangle*) where vivipary is a normal occurrence. Much higher concentrations of ABA than are normally required are needed to inhibit the growth of excised embryos of this species.

Despite the evidence presented, however, it is still too soon to conclude that the control of germinability of the developing seed universally rests with ABA. There are several cases where no clear correlations can be found between ABA levels and changes in the ability of young seeds to germinate. And so while we have a tantalizing glimpse of the regulatory system that may operate to control germinability of the immature seed, we still need to know more about several aspects—particularly how sensitivity to ABA might change during seed development—before we can fully assess its plausibility.

A large component of growth of the developing seed is, of course, associated with the laying down of storage reserves, and hormones have been implicated in this process. It has been suggested, for example, that the cytokinins in the liquid endosperm might participate in the mobilization of assimilates to that tissue. Somewhat surprisingly, a role for the inhibitor ABA has also been suggested. Application of ABA to isolated cotyledons of *Phaseolus vulgaris* promotes the synthesis of storage protein, and in grapes the accumulation of sugars is enhanced. Grain filling in wheat may be promoted by the ABA in the grain, but the evidence is still inconclusive.

2.5.2.2. Fruit Growth and Development

It has long been known that the growth of fruit flesh is in many cases linked to the activity of the developing seeds. For example, fruit size in certain cucurbits (e.g., melon) is positively correlated with seed number. In the botanically false fruit the strawberry (where the pips or so-called seeds on the fleshy receptacle are the true fruits), the growth of the flesh is greatly reduced when the young, developing seeds are removed. The "seeds" are relatively rich in auxins, and application of auxins to a "deseeded" strawberry to a large extent restores flesh growth. It seems probable, therefore, that flesh growth is promoted by auxin from the developing seeds. The strawberry is a convenient fruit with which to do experiments of this kind because the "seeds" are borne externally. But although they are more difficult, similar experiments have been carried out on true fruits, such as the pea, where the seeds are, of course, internal. By inserting a fine needle through the pod wall (the ovary wall) the seeds inside 2- to 3-day-old pea fruits can be punctured and killed. The pod of fruits treated in this way fails to grow, but virtually normal growth is restored by application of gibberellin or the synthetic auxin, naphthyl acetic acid (Fig. 2.32). It has not yet been established that seeds younger than about 10 days contain gibberellin, but older pea seeds do have GA_9, GA_{20}, and GA_{29}, and these GAs are active in promoting the growth of seedless pods, GA_{20} being most effective. This evidence strongly suggests that gibberellins (and possibly auxin) from the developing seeds control growth of the ovary wall. Supporting this interpretation is the fact that the maximum concentrations of auxins and gibberellins in immature seeds coincide with the period of pod growth in peas and with flesh growth in other species.

2.5.2.3. Hormones for Germination and Growth

We have seen that seeds of several species, e.g., maize, contain auxin conjugates, many of which furnish IAA upon hydrolytic cleavage. There is no evidence that the IAA participates in the germination process or in radicle emergence, but later events—seedling growth—might be controlled by the hormone. Thus, IAA from the conjugates in the endosperm of germinated maize (and probably other cereals) is translocated to the tip of the coleoptile from which position it regulates growth of the cells of the elongating zone of that organ.

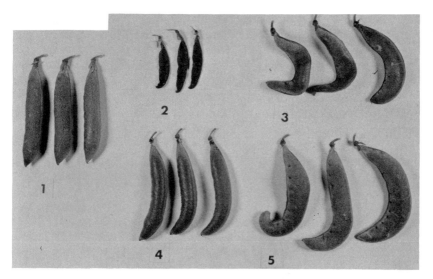

Figure 2.32. Effect of growth regulators on the growth of the ovary wall in peas (*Pisum sativum* cv. Progress no. 9). (1) Control (seeds alive). (2) Seeds killed. (3) Naphthylacetic acid treatment (100 μg/pod). (4) Gibberellic acid (GA₃) treatment (100 μg/pod). (5) Naphthylacetic acid plus GA₃ treatment (each at 100 μg/pod). Growth regulators were applied as solutions in 50% aqueous ethanol. The solvent alone was added to (1) and (2). From Sponsel (1982). Photograph kindly supplied by the author.

Radioactively labeled GA_1 fed to developing pea seeds is incorporated into various conjugates, probably glucosides. When the seeds are mature, dried, and then set to germinate, a proportion of the radioactive GA is liberated. Gibberellin A_{20} is conjugated by maturing maize grains and is liberated, together with GA, upon germination. Thus, it seems that the conjugates may be laid down in the developing seed as a store of GA for future use.

2.5.2.4. Other Effects of the Seed Hormones

When young seeds inside a pea pod are killed (Section 2.5.2.2) abscission of the fruit stalk is promoted. Treatment of the seedless pod with naphthylacetic acid or gibberellin (including the native GA_9, GA_{20}, and GA_{29}) stops the formation of the abscission layer. One of the effects of these seed hormones, therefore, might be to inhibit the abscission zone. On the other hand, abscission of fruits and neighboring leaves has been suggested to be stimulated by ABA from developing seeds, and although some evidence favors this view, the experimental findings are, on the whole, inconclusive.

An interesting effect which has been attributed to the seed gibberellins occurs in apple. Some cultivars are biennial bearers; i.e., they flower and consequently bear fruit only every 2 years. Now, flower buds are normally initiated on the spurs while the fruit is developing, and they remain dormant until the following year; so the flowers appearing in one year are those formed in the previous year. In the biennial bearers, flower buds fail to form while fruit is also developing, and hence a flowering season is omitted. Removal of developing fruits allows flower buds to be initiated, but the inhibitory effect of fruits can be mimicked by application of gibberellin. Since developing apple seeds are rich in GAs (see Table 2.7), these regulators are thought to be the cause of flower bud suppression.

USEFUL LITERATURE REFERENCES

SECTIONS 2.1 AND 2.2

Evans, L. T. (ed.), 1975, *Crop Physiology. Some Case Histories,* Cambridge University Press, London, New York, pp. 374 (overview of seed formation).

Feller, U. K., Soong, T.-S. T., and Hageman, R. H., 1977, *Plant Physiol.* **59**:290–294 (corn leaf senescence and proteolysis).

Flinn, A. M., and Pate, J. S., 1970, *J. Exp. Bot.* **21**:71–82 (carbon transfer from pea leaf and pod).

Flinn, A. M., Atkins, C. A., and Pate, J. S., 1977, *Plant Physiol.* **60**:412–418 (pea seed respiration and pod photosynthesis).

Hardham, A. R., 1976, *Aust. J. Bot.* **24**:711–721 (vascular transport in pea seed).

Jenner, C. F., and Rathjen, A. J., 1972, *Ann. Bot.* **36**:729–741 (sucrose modifications entering wheat grain).

Johansen, D. A., 1950, *Plant Embryology. Embryology of the Spermatophyta,* Chronica Botanica, Waltham, Massachusetts, pp. 306 (developmental patterns).

Maheshwari, P., 1950, *An Introduction to the Embryology of Angiosperms,* McGraw-Hill, New York, pp. 453 (developmental patterns).

Marinos, N. G., 1970, *Protoplasma* **70**:261–279 (embryogenesis of the pea).

Milthorpe, F. L., and Moorby, J., 1979, *An Introduction to Crop Physiology,* 2nd ed., Cambridge University Press, Cambridge, U.K., pp. 244 (overview of seed formation).

Pate, J. S., 1975, in: *Crop Physiology* (L. T. Evans, ed.), Cambridge University Press, Cambridge, U.K., pp. 191–224 (pea fruit formation).

Pate, J. S., Sharkey, P. J., and Atkins, C. A., 1977, *Plant Physiol.* **59**:506–510 (legume C and N assimilation).

Pate, J. S., Atkins, C. A., and Perry, M. W., 1980, *Aust. J. Plant Physiol.* **7**:283–297 (photosynthesis and reserve formation in *Lupinus*).

Sakri, F. A. K., and Shannon, J. C., 1975, *Plant Physiol.* **55**:881–889 (sugar translocation into wheat grains).

Shannon, J. C., 1972, *Plant Physiol.* **49**:198–202 (sugar translocation into maize kernels).

Simmonds, D. H., and O'Brien, T. P., 1981, *Adv. Cereal Sci. Technol.* **4**:5–70 (review of wheat endosperm development).

Sofield, I., Evans, L. T., Cook, M. G., and Wardlaw, I. F., 1977, *Aust. J. Plant Physiol.* **4**:785–797 (factors influencing wheat grain filling).

Thorne, J. H., 1979, *Agron. J.* **71**:812–816 (assimilate redistribution in pods).

Waters, S. P., Peoples, M. B., Simpson, R. J., and Dalling, M. J., 1980, *Planta* **148**:422–428 (nitrogen redistribution in wheat).

Yoshida, S., 1972, *Ann. Rev. Plant Physiol.* **23**:437–464 (physiology of grain yield: review).

SECTION 2.3

Adams, C. A., Novellie, L., and Liebenberg, N. vdW., 1976, *Cereal Chem.* **53**:1–12 (cereal protein bodies).

Adams, C. A., Rinne, R. W., and Fjerstad, M. C., 1980, *Ann. Bot.* **45**:577–582 (starch deposition in soybean).

Appleby, R. S., Gurr, M. I., and Nichols, B. W., 1974, *Eur. J. Biochem.* **48**:209–216 (lipid synthesis in *Crambe*).

Bailey, D. S., DeLuca, V., Durr, M., Verma, D. P. S., and Maclachlan, G. A., 1980, *Plant Physiol.* **66**:1113–1118 (glycoprotein synthesis and lipid intermediates).

Bain, J. M., and Mercer, F. V., 1966, *Aust. J. Biol. Sci.* **19**:49–67 (patterns of reserve deposition in pea).

Blobel, G., Walter, P., Chang, C. N., Goldman, B. H., Erickson, A. H., and Lingappa, V. R., 1979, in: *Secretory Mechanisms,* Volume 33 (C.R. Hopkins and C. J. Duncan, eds.), Symposium of the Society for Experimental Biology, Cambridge University Press, Cambridge, U.K., pp. 9–36 (signal hypothesis and protein passage through membranes).

Brandt, A., 1976, *Cereal Chem.* **53**:890–901 (protein synthesis in high-lysine barley).

Browder, S. K., and Beevers, L., 1980, *Plant Physiol.* **65**:924–930 (glycoprotein synthesis in pea).

Burr, B., Burr, F. A., Rubenstein, I., and Simon, M. N., 1978, *Proc. Natl. Acad. Sci. USA* **75**:696–700 (zein mRNA and protein bodies).

Buttrose, M. S., 1960, *J. Ultrastruct. Res.* **4**:231–257 (cereal starch granule formation).

Campbell, J. McA., and Reid, J. S. G., 1982, *Planta* **155**:105–111 (galactomannan biosynthesis).

Canvin, D. T., 1965, *Can. J. Bot.* **43**:63–69 (environmental effects and lipid deposition).

Chrispeels, M. J., Higgins, T. J. V., Craig, S., and Spencer, D., 1982, *J. Cell Biol.* **93**:5–14, 306–313 (legumin and vicilin deposition in pea).

Craig, S., Goodchild, D. J., and Hardham, A. R., 1979, *Aust. J. Plant Physiol.* **6**:81–98 (vacuole changes and protein deposition in pea).

Craig, S., Goodchild, D. J., and Miller, C., 1980, *Aust. J. Plant Physiol.* **7**:327–337 (vacuole changes and protein deposition in pea).

Croy, R. R. D., Lycett, G. W., Gatehouse, J. A., Yarwood, J. N., and Boulter, D., 1982, *Nature* **295**:76–79 (storage protein mRNAs in legumes).

Da Silva, W. J., and Arruda, P., 1979, *Phytochemistry* **18**:1803–1805 (lysine catabolism in maize lines).

Duffus, C. M., 1979, in: *Recent Advances in the Biochemistry of Cereals* (D. L. Laidman and R. G. Wyn Jones, eds.), Academic Press, London, pp. 209–238 (carbohydrate anabolism in cereals: review).

Dure, L. S., III, and Galau, G. A., 1981, *Plant Physiol.* **68**:187–194 (processing of cottonseed proteins).

Gayler, K. R., and Sykes, G. E., 1981, *Plant Physiol.* **67**:958–961 (conglycinin synthesis in soybean).

Gifford, D. J., Greenwood, J. S., and Bewley, J. D., 1982, *Plant Physiol.* **69:**1471–1478 (crystalloid protein synthesis in castor bean).

Goldberg, R. B., Hoschek, G., Ditta, G. S., and Breidenbach, R. W., 1981, *Dev. Biol.* **83:**218–231 (mRNA abundances in developing soybean).

Greene, F. C., 1983, *Plant Physiol.* **71:**40–46 (mRNA levels in developing wheat).

Greenwood, J. S., and Bewley, J. D., 1984, *Planta,* **160:**113–120 (site of phytin biosynthesis).

Gurr, M. I., 1980, in: *The Biochemistry of Plants,* Volume 4 (P. K. Stumpf, ed.), Academic Press, New York, pp. 205–248 (lipid biosynthesis: review).

Gurr, M. I., Blades, J., Appleby, R. S., Smith, C. G., Robinson, M. P., and Nichols, B. W., 1974, *Eur. J. Biochem.* **43:**281–290 (lipid synthesis in *Crambe*).

Hall, T. C., Ma, Y., Buchbinder, B. U., Pyne, J. W., Sun, S. M., and Bliss, F. A., 1978, *Proc. Natl. Acad. Sci. USA* **75:**3196–3200 (mRNA from *Phaseolus: in vitro* translation).

Higgins, T. J. V., and Spencer, D., 1981, *Plant Physiol.* **67:**205–211 (precursor forms of pea vicilin).

Ichihara, K. -I., and Noda, M., 1980, *Phytochemistry* **19:**49–54 (lipid synthesis in safflower).

Jennings, A. C., and Morton, R. K., 1963, *Aust. J. Biol. Sci.* **16:**318–331 (reserve synthesis in wheat).

Larkins, B. A., and Hurkman, W. J., 1978, *Plant Physiol.* **62:**256–263 (zein deposition in protein bodies).

Larkins, B. A., Pedersen, K., Handa, A. K., Hurkman, W. J., and Smith, L. D., 1979, *Proc. Natl. Acad. Sci. USA* **76:**6448–6452 (synthesis of zein in *Xenopus* oocytes).

Manteuffel, R., Muntz, K., Puchel, M., and Scholz, G., 1976, *Biochem. Physiol. Pflanzen.* **169:**595–605 (DNA, RNA, and protein accumulation in *Vicia faba*).

Meier, H., and Reid, J. S. G., 1977, *Planta* **133:**243–248 (galactomannan formation in fenugreek).

Meinke, D. W., Chen, J., and Beachy, R. N., 1981, *Planta* **153:**130–139 (protein subunit synthesis in soybean).

Miflin, B. J., and Shewry, P. R., 1981, in: *Nitrogen and Carbon Metabolism* (J. D. Bewley, ed.), Martinus Nijhoff/Dr. W. Junk, The Hague, pp. 195–248 (cereal storage proteins: review).

Miflin, B. J., Burgess, S. R., and Shewry, P. R., 1981, *J. Exp. Bot.* **32:**199–219 (protein bodies in cereals and legumes).

Millerd, A., and Spencer, D., 1974, *Aust. J. Plant Physiol.* **1:**331–341 (RNA and nuclei in pea cotyledons).

Pate, J. S., and Flinn, A. M., 1977, in: *The Physiology of the Garden Pea* (J. F. Sutcliffe and J. S. Pate, eds.), Academic Press, London, pp. 431–468 (pea seed development: review).

Roberts, L. M., and Lord, J. M., 1981, *Eur. J. Biochem.* **119:**31–41 (lectin synthesis in castor bean).

Sengupta, G., Deluca, V., Bailey, D. S., and Verma D. P. S., 1981, *Plant Mol. Biol.* **1:**19–34 (posttranscription processing in soybean).

Simcox, P.D., Reid, E. E., Canvin, D. T., and Dennis, D. T., 1977, *Plant Physiol.* **59:**1128–1132 (proplastids in developing castor bean endosperm).

Spencer, D., and Higgins, T. J. V., 1979, *Curr. Adv. Plant Sci.* **11:**34.1–34.15 (seed protein biosynthesis: review).

Stumpf, P. K., 1977, in: *Lipids and Lipid Polymers in Higher Plants* (M. Tevini and H. K. Lichtenthaler, eds.), Springer-Verlag, Berlin, pp. 75–84 (lipid biosynthesis in seeds: review).

Stymne, S., and Appelqvist, L. -A., 1978, *Eur. J. Biochem.* **90:**223–229 (linoleate synthesis in safflower).

Thompson, J. F., Madison, J. T., Waterman, M. A., and Meunster, A.-M. E., 1981, *Phytochemistry* **20:**941–945 (added methionine effects on legume proteins).

Tsai, C.-Y., and Nelson, O. E., 1966, *Science* **151:**341–343 (ADPGlc pyrophosphorylase maize mutant).

Turner, J. F., 1969, *Aust. J. Biol. Sci.* **22**:1321–1327 (pathways of starch synthesis).
Wanner, G., Formanek, H., and Theimer, R. R., 1981, *Planta* **151**:109–123 (lipid body formation).
Weber, E. J., 1980, in: *The Resource Potential in Phytochemistry. Recent Advances in Phytochemistry,* Volume 14, Plenum Press, N.Y., New York, pp. 97–137 (corn mutants).
Wu, X.-Y., Moreau, R. A., and Stumpf, P. K., 1981, *Lipids* **16**:897–902 (jojoba seed wax biosynthesis).
Yamagata, H., Tanaka, K., and Kasai, Z., 1982, *Agric. Biol. Chem.* **46**:321–322 (rice glutelin precursors).

SECTION 2.4

Carlier, A. R., Manickam, A., and Peumans, W. J., 1980, *Planta* **149**:227–233 (developmental specific mRNAs in mung bean).
Crouch, M. L., and Sussex, I. M., 1981, *Planta* **153**:64–74 (developmental synthesis of rape proteins).
Dasgupta, J., Bewley, J. D., and Yeung, E. C., 1982, *J. Exp. Bot.* **33**:1045–1057; *Plant Physiol.* **70**:1224–1227 (desiccation tolerance and intolerance and role in development).
Dure, L. S., III, and Galau, G. A., 1981, *Plant Physiol.* **68**:187–194 (ABA, mRNA, and cotton storage proteins).
Dure, L. S., III, Capdevila, A. M., and Greenway, S. C., 1980, in: *Genomic Organization and Expression in Plants* (C. J. Leaver, ed.), NATO Advanced Study Institutes Series. Series A: Life Sciences, Plenum Press, New York, pp. 127–146 (mRNA and cotton embryogenesis).
Evans, M., Black, M., and Chapman, J., 1975, *Nature* **258**:144–145 (induction of α-amylase following drying).
Gordon, M. E., and Payne, P. I., 1976, *Planta* **130**:269–273 (mRNA content of dry seeds).
Long, S. R., Dale, R. M. K., and Sussex, I. M., 1981, *Planta* **153**:405–415 (desiccation and culture of *Phaseolus* embryos).
Obendorf, R. L., Ashworth, E. N., and Rytko, G. T., 1980, *Crop Sci.* **20**:483–486 (drying and soybean germination).
Sussex, I., 1975, *Am. J. Bot.* **62**: 948–953 (vivipary in mangrove seeds).
Triplett, B. A., and Quatrano, R. S., 1982, *Develop. Biol.* **91**:491–496 (ABA and wheat embryo proteins).

SECTION 2.5

Brenner, M. L., Burr, B., and Burr, F., 1977, *Plant Physiol.* **59**:S-76 (ABA and vivipary in maize).
Davey, J. E., and Van Staden, J., 1979, *Plant Physiol.* **63**:873–877 (cytokinins in developing lupin seeds).
Eeuwens, C. J., and Schwabe, W. W., 1975, *J. Exp. Bot.* **26**:1–14 (growth regulators in developing pea seeds).
Khan, A. A., 1982, in: *The Physiology and Biochemistry of Seed Development, Dormancy and Germination* (A. A. Khan, ed.), Elsevier Biomedical Press, Amsterdam, pp. 111–136 (gibberellins in seed development).
King, R. W., 1982, in: *The Physiology and Biochemistry of Seed Development, Dormancy and*

Germination (A. A. Khan, ed.), Elsevier Biomedical Press, Amsterdam, pp. 157–181 (abscisic acid in seed development).

Sponsel, V. M., 1980, in: *Gibberellins—Chemistry, Physiology and Use.* British Plant Growth Regulator Group Monograph 5 (J. R. Lenton, ed.), Wessex Press, Wantage, U.K., pp. 49–62 (gibberellins in developing pea seeds).

Sponsel, V. M., 1982, *J. Plant Growth Regul.* **1**:147–152 (seed and hormonal control of pod growth in peas).

Ueda, M., Ehman, A., and Bandurski, R. S., 1970, *Plant Physiol.* **46**:715–718 (bound auxin in maize).

Van Onckelen, H., Caubergs, R., Horemans, S., and DeGreef, J. A., 1980, *J. Exp. Bot.* **31**:913–920 (ABA in developing bean seeds).

Storage, Imbibition, and Germination

3.1. THE LONGEVITY OF SEEDS

Considerable controversy has surrounded the claims that seeds remain viable for many hundreds or even thousands of years. A list of some of the more spectacular claims is presented in Table 3.1, along with reasons for accepting or rejecting them. Perhaps the most persistent myth concerning seed longevity is that viable grains of wheat and barley were uncovered during archaeological excavations of ancient Egyptian buildings. Reports that "mummy" grains could germinate and produce seedlings were given considerable publicity and credence during the nineteenth and early twentieth centuries. But recent, scientifically rigorous studies show unequivocally that most ancient grains (particularly the embryos) have undergone severe morphological and physiological degradation (including carbonization) with accompanying total loss of viability. Some stored grains retain their original shape and even much of their cell fine structure, although upon hydration considerable disintegration occurs. It is worth reemphasizing, then, that there is no scientific proof for the retention of viability by ancient cereal grains.

The most extreme claim for longevity is for the arctic lupin *(Lupinus arcticus)*, seeds of which were discovered frozen and buried in the Yukon Territory in Canada. These seeds were removed from ancient rodent burrows containing remnants of a nest, fecal matter, and the skull of a lemming species. Dating of nests and remains of arctic ground squirrels found buried under similar conditions in central Alaska showed them to be over 10,000 years old, and hence it was concluded that the seeds in the Yukon must be of this age too. The highly circumstantial nature of the "evidence" militates against its acceptance, and the putative longevity of the lupin seeds still requires confirmation using direct dating techniques. In the absence of reliable evidence from radiocarbon or alternative dating methods, other claims for extended longevity, e.g., *Nelumbium nucifera* (1000–3000 years) and *Chenopodium album* and *Spergula arvensis* (~1700 years), must be regarded with considerable skepticism (Table 3.1).

The longevity of a *Canna compacta* seed has been put at about 600 years. The seed was collected from a tomb in Argentina, enclosed in a *Juglans aus-*

Table 3.1. An Examination of the Claims for Extended Longevity of Seeds

Species	Location	Age	Status of seed	Comments
Stored in dry conditions				
Barley	Tomb of King Tutankhamen	ca. 1350 B.C.	Nonviable	Extensively carbonized
Wheat	Various ancient Egyptian tombs	1000 B.C. or earlier	Not known	Claims for their viability have been made, but age and source of grains never authenticated
Wheat	Thebes	3000–2000 B.C.	Nonviable	Some cell fine structure conserved, although degrades upon rehydration
	Feyum	4400 B.C.	Nonviable	
Canna compacta	Santa Rosa de Tastil, Argentina	ca. 600 years	Viable	Enclosed in a nutshell forming part of rattle. Shell dated at 600 years and seed probably of same age
Albizzia julibrissin	China to British Museum	1793	Viable in 1940	Germination started accidentally. Viable seedlings produced
Cassia multijuga	Museum of Natural History, Paris	158 years	Viable	Wholly authenticated history from collection to sowing
Buried in soil or water				
Arctic lupin (Lupinus arcticus)	Miller Creek, Yukon	>10,000 years	Viable	Doubtful. Age was derived indirectly from geological data which could be wrong. No direct dating evidence
Indian or sacred lotus (Nelumbium nucifera)	Pulantien basin, Southern Manchuria	150–several thousand years	Viable	Evidence of age from indirect geological and lake-draining data. Radiocarbon analysis of seeds shows them indistinguishable from modern plants
Indian lotus	Kemigawa, near Tokyo	3000 years	Viable	Seeds found on submerged boat radiodated at 3000 years. No direct measurements of age of seed, which could have settled in sediments after shedding from modern plants
Chenopodium album	Denmark and Sweden archaeological digs	>1700 years	Viable	No direct dating of seeds. Could be modern seeds dispersed into archaeological digs
Spergula arvenis				

tralis nutshell forming part of a rattle necklace. Radiocarbon analysis of the nutshell and surrounding charcoal remnants was used to date the seed. Insufficient seed material was available for analysis, but the only way in which the *C. compacta* seed could have arrived inside the *J. australis* nutshell is for it to have been inserted there through the still-developing nutshell, while it was soft. Then the nutshell hardened and dried with the seed inside, forming a rattle. Hence the seed must have been at least the same age as the shell, and possibly older.

Some seeds stored in museums and herbaria are known to have survived for more than 100 years. Specimens of *Albizzia julibrissin* collected in China in 1793 and deposited in the British Museum, London, germinated after attempts to quench a fire started by an incendiary bomb that hit the museum in 1940. Three of the resulting seedlings were planted nearby, but two eventually met their end during a bombing raid the following year. Tests on old collections of seeds from the Museum of Natural History in Paris in 1906 and 1934 showed that their viability ranged from 55–158 years (Table 3.2)—most of the long-lived species were members of the Leguminosae. It is noteworthy that the long life-span of these and other specimens was attained even though the storage conditions used were arbitrary (generally, warm temperatures and low relative humidity). Longevity might have been greater if different storage conditions had been used.

A number of projects have been carried out, or are in progress, to determine longevity of seeds buried in soil. A study of buried weed seed, due to last for 50 years, was initiated in 1972, and the results of germination and viability tests after 2.5 years of burial have been published: only 4 of the 20 species retained more than 50% viability. The longest controlled burial experiment to

Table 3.2. Viability Record of Some Old Seeds from the Museum of Natural History, Paris[a]

Species	Date collected	% Germinated in 1906	% Germinated in 1934	Longevity (years)
Mimosa glomerata	1853	50	50	81
Melilotus lutea	1851	30	0	55
Cytisus austriacus	1843	10	0	63
Dioclea pauciflora	1841	10	20	93
Trifolium arvense	1838	20	0	68
Stachys nepetifolia	1829	10	0	77
Cassia bicapsularis	1819	30	40	115
Cassia multijuga	1776	—	100	158

[a]After Becquerel (1934).

Table 3.3. Longevity of Some of the Seed Species from W. J. Beal's Buried-Seed Experiment Started in 1879[a]

Species	Longevity (years)
Agrostemma githago	< 5
Amaranthus retroflexus	40
Brassica nigra	50
Capsella bursa-pastoris	35
Euphorbia maculata	< 5
Lepidium virginicum	40
Polygonum hydropiper	50
Trifolium repens	5
Oenothera biennis	80 (10%)[b]
Verbascum blattaria	100 (42%)[b]

[a]After Kivilaan and Bandurski (1981).
[b]Figures in parentheses are percent germinated after the time indicated.

date is one initiated by W. J. Beal in 1879. He selected seeds of 21 different species of plants common in the vicinity of Michigan Agricultural College in East Lansing. Fifty seed lots of each species were mixed with moist sand in unstoppered bottles and buried in a sandy knoll. At regular intervals since, bottles have been unearthed and germinability of the seeds tested. The one-hundredth anniversary of the experiment has just passed, and of the 21 species originally buried only *Verbascum blattaria* seeds have retained their viability throughout, although many species were capable of germinating after 35–50 years (Table 3.3).

3.2. VIABILITY OF SEEDS IN STORAGE

The majority of seed species retain their viability when dried—in fact, drying is the normal final phase of maturation for most seeds. Hence, it is common for seeds to be stored in a "dry" state or, more correctly, with a low moisture content. There are, however, some seed species that must retain a relatively high moisture content during storage in order to maintain maximum viability. These are the so-called "recalcitrant" seeds, which will be discussed in Section 3.2.5.

3.2.1. The Relationship between Temperature and Moisture during Storage

Seeds that can be stored in a state of low moisture content are called "orthodox," and their viability under certain storage conditions conforms to some general rules, as follows:

1. For each 1% decrease in seed moisture content the storage life of the seed is doubled.
2. For each 10°F (5.6°C) decrease in seed storage temperature the storage life of a seed is doubled.
3. The arithmetic sum of the storage temperature in degrees F and the percent relative humidity (RH) should not exceed 100, with no more than half the sum contributed by the temperature.

These "rules of thumb" clearly indicate that temperature and moisture content of the seed are major factors in determining viability in storage. Let us now consider these factors, along with others that might play a role, as follows:

1. Moisture content. This is defined by International Seed Testing Association (ISTA) practices as:

$$\% \text{ Moisture content} = \frac{\text{Fresh weight of seed} - \text{dry weight of seed}}{\text{Fresh weight of seed}} \times 100$$

When the moisture content is high (> 30%), nondormant seeds may germinate, and from 18–30% moisture content rapid deterioration by microorganisms can occur (see Section 3.2.3). Seeds stored at moisture contents > 18–20% will respire, and in poor ventilation the generated heat will kill them. Below 8–9% moisture content there is little or no insect activity, and below 4–5% moisture content seeds are immune from attack by insects and storage fungi, but they may deteriorate faster than those maintained at a slightly higher moisture content. The activities of seed storage fungi are ultimately more influenced by the relative humidity (RH) of the interseed atmosphere than by the moisture content of the seeds themselves. This is because the moisture content of some seeds (e.g., oil seeds) may be different from that of others (e.g., starchy seeds) even though both are in equilibrium with the same atmospheric RH. For example, all cereals and many of the legumes that are high in starch and low in oil content have a moisture content of about 11% at 45% RH, whereas oil-containing seeds (e.g., rape) have a moisture content of only 4–6% at this RH (Fig. 3.1).

2. Temperature. Cold storage of seeds at 0–5°C is generally desirable, although this may be unadvisable unless they are sealed in moistureproof containers or stored in a dehumidified atmosphere. Otherwise the RH of storage could be high, causing the seeds to gain moisture; if they are then brought out to a higher temperature (e.g., for transport), they might deteriorate because of their high moisture content. At moisture contents below 14% no ice crystals form within cells upon freezing; so storage of dry seeds at subzero temperatures after freezing in a dry atmosphere should improve longevity. Freeze–drying of certain seeds improves their longevity in storage, but others may be killed by

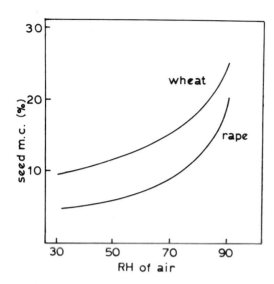

Figure 3.1. Changes in equilibrium moisture content of wheat and rape seeds with relative humidity of the ambient air. Based on work of Kreyger (1972) cited in Thomson (1979).

this treatment. A method that is increasing in popularity, particularly for small batches of seeds for gene banks (Chapter 9), is to keep them immersed in liquid nitrogen. Under this condition seeds should survive indefinitely.

3. Interrelationships. Several mathematical models have been proposed to relate the viability of seeds to their storage environment. Discussion of these models is beyond the intended scope of this text, but the reader may obtain details by reading the publications and reviews of E. H. Roberts (Roberts, 1972, 1973; Ellis and Roberts, 1980), some of which are referred to at the end of this chapter. His studies of the survival curves of seven species (wheat, rice, barley, pea, broad bean, onion, and tomato) have shown that under a given set of constant conditions a sample of seeds has a particular mean viability period, and that there is a random distribution of the viability periods in a population around this mean value. This is illustrated in Fig. 3.2, from which it can be seen that although the spread of distribution (of which σ, the standard deviation, is a measure) increases if the mean viability period ($\bar{p}$) of the seed lot is improved by improving the storage conditions, the coefficient of variation (which is the standard deviation expressed as a percentage of the mean value of distribution—$\sigma/\bar{p} \times 100$) remains the same. The resulting survival curves are sigmoid curves (negative cumulative normal distributions, or ogives). When percentage viability is plotted on a probability scale, survival curves are transformed to straight lines. This implies that under any constant storage conditions the frequency distribution of seed deaths in time is normal, even in adverse storage environments where death is rapid.

The relationship between mean storage temperature, moisture content, and mean viability periods is relatively simple, and the data for *Vicia faba* are plotted in Fig. 3.3.

3.2.2. Other Factors That Affect Seed Viability during Storage

Besides the interrelationships between temperature, moisture content, and time, other factors must be considered when attempting to determine the optimum storage conditions for a particular species.

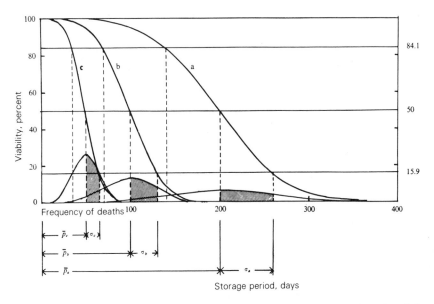

Figure 3.2. The relationship between the mean viability period and standard deviation of the distribution of the viability periods of individual seeds of a seed lot. The figure shows that an estimate of mean viability period ($\bar{p}$) may be made by noting the point on the time scale at which the survival curve intersects the 50% level of germination. The standard deviation (σ) may be estimated by noting the point at which the survival curve intersects the 15.9% (or 84.1%) level and measuring the distance from this point on the time scale to the point representing the mean viability period. This is based on the fact that the area under the normal curve between the mean and 1 S.D. contains 34.1% of the area under the whole curve (shaded areas). It is assumed that three seed samples from the same lot have been stored under different conditions such that the mean viability period of sample a ($\bar{p}_a$) is twice that of sample b ($\bar{p}_b$) and four times that of sample c ($\bar{p}_c$). It can be seen that the ratio $\bar{p}:\sigma$ is the same for all samples. The value $\bar{p}$ can be estimated quite accurately graphically when the data for percentage viability are transformed to probit values or when they are plotted on a probability scale, since survival curves then become straight lines. After Roberts (1973).

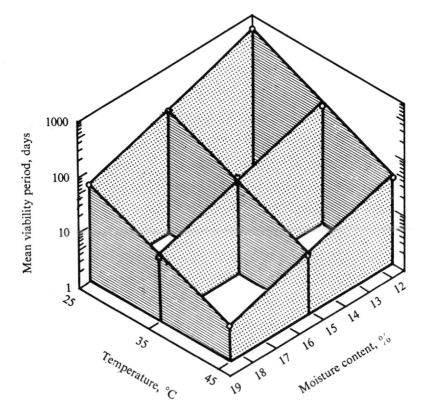

Figure 3.3. Isometric three-dimensional graph showing the relationship between moisture content, temperature in storage, and mean viability period for broad bean *(Vicia faba)*. Time (days) is plotted on a log scale. The open circles represent the experimental points for the observed times for germination to decline to 50%. After Roberts (1973).

3.2.2.1. Cultivar and Harvest Variability

Different cultivars and harvests of a particular species may show different viability characteristics under the same storage conditions. Differences between harvests are relatively small under good storage conditions, but under adverse conditions (elevated temperature and RH) they can be quite large.

3.2.2.2. Pre- and Postharvest Conditions

Environmental variation during seed development usually has little effect on the viability of seeds, unless the ripening process is interrupted by premature harvesting. Variation in the environment of seeds at about the time of completion of maturation and harvesting can result in different potential viability periods. Viability of cereal grains is generally lower in years when ripening and harvesting conditions are poor. Mechanical damage inflicted during harvesting can severely reduce the viability of some seeds, e.g., certain large-seeded legumes. Cereals are largely immune from mechanical injury, presumably because of the protective outer structures, the palea and lemma. Small seeds tend to escape injury during harvest, and seeds that are spherical tend to suffer less damage than elongated or irregularly shaped ones. During storage, injured or deeply bruised areas may serve as centers for infection and result in accelerated deterioration. Injuries close to vital parts of the embryonic axis or near the point of attachment of cotyledons to the axis usually bring about the most rapid losses of viability. High temperatures during drying, or drying too quickly or excessively, can dramatically reduce viability.

3.2.2.3. Oxygen Pressure during Storage

If seeds are not maintained in hermetic (airtight) storage at low moisture contents, then even under conditions of constant temperature and moisture the gaseous environment may change owing to respiratory activity of the seeds and associated microflora. For example, it takes pea seeds about 11 wk, when stored at 18.4% moisture content at 25°C (rather extreme conditions) in a closed atmosphere, to decrease the oxygen from 21% to 1.4% and increase the carbon dioxide from 0.03% to 12%. During this time, there is a 50% loss of viability. In open storage, seeds maintained in an atmosphere of nitrogen may retain their viability considerably longer than those placed in replenished air or oxygen, although there is probably no advantage over storing seeds hermetically sealed in air. In fact, for storage at relatively low temperatures and moisture there is probably little benefit in using controlled atmospheres, i.e., reduced oxygen pressures. There may be advantages for short-term storage in conditions of high temperature or moisture content, but it is probably simpler to reduce either the temperature or moisture content. Hermetic sealing provides a simple and convenient method of controlling seed moisture content after the seeds have been dried adequately.

3.2.2.4. Fluctuating Storage Conditions

Onion and dandelion seeds stored under alternating high and low RH conditions lose viability proportionately to the length of time that the seeds are subjected to the high RH. During lengthy (e.g., 8-wk) cycles of high and low RH, however, seeds deteriorate as rapidly as when kept only at the higher RH. In contrast, increases in temperature *per se* at intervals during cold storage do not necessarily have deleterious effects on viability. For example, batches of red clover *(Trifolium pratense)* seeds stored at -5 to $-15\,^{\circ}$C for 13 years and thawed for 24 h at annual, semiannual, monthly, or weekly (in all, 670 times) intervals do not show any appreciable loss of viability compared with controls that were not subjected to periodic thawing. Thus fluctuating storage conditions may be harmful to some species but not others; this must be determined empirically.

3.2.3. Microflora and Seed Deterioration

Bacteria probably do not play a significant role in seed deterioration, for germination is rarely reduced unless infection has progressed beyond the point of obvious decay. Since bacterial populations require free water to grow, they are unlikely to increase in stored seeds because the latter are usually too dry. If conditions were moist enough, this would encourage growth of fungi which would suppress bacterial growth.

Two types of fungi invade seeds: field fungi and storage fungi. The former invade seeds during their development on plants in the field or following harvesting while the plants are standing in the field. Field fungi need a high moisture content for growth (as high as 33% for cereals) and hence are infective only under conditions where seeds fail to follow their normal pattern of maturation drying. Hence, a period of high rainfall at harvest time can result in extensive grain deterioration. The main fungal species found associated with wheat or barley in the field are *Alternaria, Fusarium,* and *Helminthosporium* spp., although several others have been recorded. Seeds that are sheltered from airborne pathogens by pods, fleshy fruits, or other surrounding structures (e.g., pea, tomato, melon, maize) are generally less susceptible to field fungi than seeds that are more exposed (e.g., wheat, oat, barley).

Storage fungi, almost exclusively of the genera *Aspergillus* and *Penicillium,* infest seeds only under storage conditions and are never present beforehand, even in seeds of plants left standing in the field after harvesting. Each species of storage fungus has a sharply defined minimum of seed moisture content below which it will not grow, although other factors also determine virulence, e.g., ability to penetrate the seed, condition of seed, nutrient availability, and temperature. The major deleterious effects of storage fungi are to (1) decrease viability, (2) cause discoloration, (3) produce mycotoxins, (4) cause heat production, and (5) develop mustiness and caking. At seed moisture contents that are in equilibrium with an ambient RH below 68% fungi will not grow; hence, they are not responsible for deterioration that occurs at moisture contents below about 13% in starchy seeds and below 7–8% in oily seeds (Fig. 3.1).

Deterioration of seeds by insects and mites is a serious problem, particularly in warm and humid climates. Weevils, flour beetles, or borers are rarely active below 8% moisture content and 18–20°C, but they are increasingly destructive as the RH rises to 15% moisture content and the temperature to 30–35°C. Mites do not thrive below 60% RH, although they have temperature tolerance that extends close to freezing.

3.2.4. Seed Storage Facilities

It is obvious from the previous account that seed stores should incorporate various features into their design to minimize the chances for deterioration of their contents. Ideally, an establishment should contain rooms in which temperature and relative humidity can be closely controlled, and batches of seeds should be stored within after hermetic sealing in containers. But such establishments are impractical on logistic and economic grounds for storage of large quantities of one particular species of seed, especially if it has to be stored only from harvest until the next planting season. Storage of seeds under ambient conditions is possible if certain precautions are taken. The storage structure should be:

1. Protected from water. The roof and sidewalls must be free of holes and cracks that permit entry of rain and snow. Structures should have a waterproof floor. A wooden floor should be elevated and a concrete floor should have a moisture barrier beneath it.

2. Protected from cross contamination. Storage facilities should be constructed to provide maximum protection from chance contamination. For bulk storage, a separate bin should be provided for each cultivar. For bag storage, seeds of each cultivar should be stacked separately.

3. Ventilated and aerated. Fans or blowers are a useful addition to provide ventilation, although their covers should be tightly closed when not in use. Where it is impractical to provide electricity for forced-air ventilation, an arrangement of covered (insect-proof) ducts to allow flow-through of air is desirable. In the tropics, double roofs and heat-insulating materials are helpful. Seed or seed containers should not be piled against the walls since this reduces airflow.

4. Protected from rodents. In some countries, e.g., India, seed losses due to rodent attack are enormous. Metal and concrete buildings normally provide good protection from rodents. Small isolated stores should have a floor raised about a meter above the ground, a ratproof door, and a removable entrance ramp. Metal seed storage bins with tight covers also protect against rodents.

5. Protected against insects. A storage facility should be fumigated each time it is emptied, as should the bins and boxes in which seeds are stored. Areas

where bags and boxes are placed should be kept free of loose seeds and garbage at all times. An entrance constructed of a door leading into a small annex, from which there is a door entering into the main storage area, also cuts down direct access of insects to the seeds, as well as minimizing interior temperature and humidity fluctuations.

6. Protected against fungi. Since most fungi grow best under warm, humid conditions, storage structures should provide cool, dry conditions. Damage from storage fungi can be minimized by drying seeds to a safe moisture content and holding them under dry conditions. This is aided by ventilation to prevent the accumulation of translocated moisture. Treatment of seeds with fungicide may control some storage fungi (although many fungicides are most effective on soil fungi), and spraying the storage area when empty may be advantageous.

Centers for the storage of germ plasm have much more stringent requirements, and these will be discussed in Chapter 9.

3.2.5. Recalcitrant Seeds

Recalcitrant or "unorthodox" seeds must maintain a relatively high moisture content in order to remain viable. But even when these seeds are stored under moist conditions, their life-span is frequently brief and only occasionally exceeds a few months. Some species that produce recalcitrant seeds are listed in Table 3.4. Included are a number of large-seeded hardwoods (e.g., spp. of *Corylus, Castanea, Quercus, Aesculus;* in addition, *Salix* and *Juglans* are recalcitrant) and important plantation crops like coffee, kola nut, cacao, and rubber. Seeds of most aquatic species (e.g., wild rice) also rapidly lose viability in dry conditions.

The inability to store seeds of recalcitrant species is a serious problem, for while vegetative propagation is possible for some species and is the usual practice, the retention of a viable seed stock is desirable in order to preserve maximum genetic diversity. Unfortunately, methods of storage other than drying may also be detrimental to recalcitrant seeds; e.g., low-temperature storage is inappropriate for some species, particularly for seeds of tropical plants. Grains of wild rice present a different problem, for although they may be successfully stored under moist conditions at low temperatures, they eventually lose dormancy and sprout.

At present, the optimal storage conditions for recalcitrant seeds can be determined only by trial and error, which is a tedious, time-consuming, and expensive approach. As our knowledge advances, it is apparent that some seeds (e.g., those of *Citrus limon*) that were once thought to be recalcitrant can now be classified as orthodox—either the original method of seed drying or germination testing was at fault. However, much work still lies ahead to establish

Table 3.4. Some Species That Produce Recalcitrant Seeds and Examples of Appropriate Storage Conditions[a]

Species	Longevity and (% germinated)	Storage conditions	Damaging conditions
Corylus avellana (hazel)	6 mo +	1°C in polyethylene bag	Drying
Castanea crenata (Japanese chestnut)	6 mo	0–3°C in ventilated can or polyethylene bag	<0°C; excessive water or drying
Quercus borealis (red oak)	20 mo + (50)	5°C in sealed tin	<20–40% moisture content
Juglans nigra (black walnut)	4 years	3°C in outdoor pit	Drying
Hevea brasiliensis (rubber)	4 mo (30)	7–10°C in damp sawdust in perforated polyethylene bag	<20% moisture content
Cocos nucifera (coconut)	16 mo	Ambient temperatures and high RH%	Drying
Coffea arabica (coffee)	10 mo (59)	25°C in moist charcoal at 92–98% RH	<8–35% moisture content and < 10°C
Cola nitida (kola nut)	5 mo (80)	Ambient, heaped in open and kept moist	Drying
Theobroma cacao (cacao)	8–10 wk	21–27°C in pod, coated with fungicide	<13°C, drying
Zizania aquatica (wild rice)	14 mo (86)	1°C in water	Drying

[a]Taken from information published by King and Roberts (1979).

clearly which species are truly recalcitrant and then to define quantitative relationships between storage conditions and viability. Moreover, we are still ignorant of the physiological and biochemical bases for desiccation intolerance in recalcitrant seeds.

3.3. THE BIOCHEMICAL BASIS OF DETERIORATION OF ORTHODOX SEEDS

Storage of seeds under adverse conditions results in the production of "aged" seeds exhibiting a variety of symptoms ranging from reduced viability or germinability (sometimes to zero germination) to more or less full viability (i.e., no obvious decline in germinability) but with abnormal development of the seedling (i.e., poor vigor). It can be appreciated, therefore, that it is difficult to make comparisons between deteriorated seeds that show such diversity in their final response, particularly since the metabolic lesions affecting viability might be manifest in the seedling stage in one case but prior to the completion of germination in another. In an attempt to simplify the situation, we will distinguish between the metabolic changes wrought in completely nonviable seeds and those occurring in populations of deteriorated seed showing all manner of viabilities and degrees of vigor.

As will be discussed in greater detail in Chapter 4, early and important events during normal germination include the establishment of respiration and an ATP energy-producing system and the commencement of RNA and protein synthesis. Let us, then, first consider the status of these aspects of metabolism in deteriorated and nonviable seeds.

3.3.1. Respiration and ATP Synthesis

In viable embryos and axes the activity and integrity of mitochondria increase with time after the start of imbibition and, with time, ATP production becomes more efficiently coupled to oxygen consumption. Production of ATP begins very early after the start of imbibition, and sufficient ATP is produced within a short time to allow other metabolic processes to commence (Section 4.2.4).

In nonviable embryos of rye, rice, and maize the mitochondria appear to be swollen, and their internal membrane structure is distorted. Whereas in viable embryos mitochondrial organization becomes more ordered as germination progresses, in nonviable embryos these organelles become increasingly disorganized after imbibition, eventually leading to their complete lysis. Rye embryos that fail to germinate after storage exhibit little respiratory activity (Fig. 3.4). In some species there appears to be some correlation between the decline in gaseous exchange and reduced activity of respiratory enzymes. In nonviable rice grains, for example, levels of cytochrome oxidase, succinic, glutamic, malic, and alcohol dehydrogenases, catalase, and peroxidase are all lower than in viable grains. It is not known, however, to what extent this decline in enzyme activity contributes to respiratory failure or if it is simply a manifestation of deterioration of mitochondria and loss of cell compartmentalization. Not surprisingly, the ATP content of nonviable seeds is considerably lower than that of viable controls (as little as 1% in *Trifolium incarnatum*), and presumably it is insufficient to support metabolic processes essential for germination.

Respiratory patterns of deteriorated but still viable (at least partially)

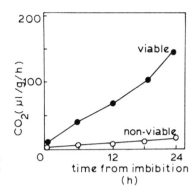

Figure 3.4. Respiratory activity, as measured by CO_2 output, of viable and nonviable embryos of rye *(Secale cereale)*. After Hallam *et al.* (1973).

populations of seeds are both complex and variable, and often they are difficult to interpret. One of the more clear-cut responses of seeds to deterioration is found in isolated mitochondria of aged soybean seeds. Mitochondria from axes of aged seeds take up 10–40% more oxygen than those from seedling axes from fresh seeds, but the amount of inorganic phosphate esterified into ATP per volume of oxygen consumed (P/O ratio) by mitochondria from fresh seeds is over twice that of mitochondria from old seeds (Table 3.5). Perhaps, then, mitochondria extracted from deteriorated seeds are at least partially uncoupled. Some attempts have been made to correlate ATP content of imbibed seed with seed vigor (i.e., seed weight, seedling weight, and hypocotyl length). In nine lettuce cultivars, including Great Lakes (Fig. 3.5), highly significant correlations have been found, as they have also for rape, ryegrass, and crimson clover seeds. Unfortunately, however, ATP measurements cannot be used universally as a test for seed viability and seedling vigor, for in several other species aged under a variety of conditions the correlation between seedling growth and ATP content of the imbibed seed is poor. Whether this is because of differences in seed species or differences in the aging conditions used remains to be determined.

Table 3.5. Activities of Mitochondria Isolated from 4-Day Dark-Grown Seedling Axes of 3-Year-Old (Old) and Freshly Harvested (Fresh) Seeds of Soybean *(Glycine max)*[a]

Seed age	Oxygen uptake (μ atoms)	Pi esterified (μ moles)	P/O ratio
Fresh	7.7 ± 0.5	23.1 ± 0.9	3 ± 0.2
Old	8.5 ± 0.5	11.1 ± 3.7	1.4 ± 0.4

[a]The standard deviation of four replications is given. After Abu-Shakra and Ching (1967).

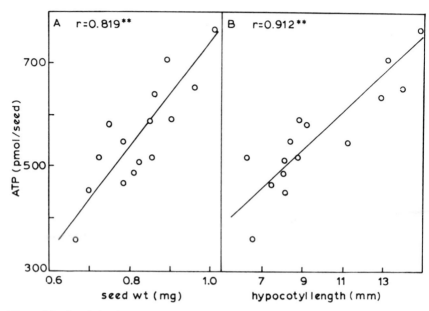

Figure 3.5. Correlation between the ATP content of 4-h imbibed lettuce (*Lactuca sativa* cv. Great Lakes) and (A) seed weight and (B) 5-day emerged hypocotyl length. r, correlation coefficient; **, significant at 1% level. After Ching and Danielson (1972).

3.3.2. Protein and Ribonucleic Acid Synthesis

An account of protein and RNA synthesis in viable seeds and embryos during germination and subsequent seedling growth is to be found in Sections 4.4.2 and 4.4.3. The reader should consult these for explanations of appropriate terminology and for methods of measurement of *in vivo* and *in vitro* protein synthesis.

Nonviable embryos of cereals and nonviable embryonic axes of legumes (Fig. 3.6, A) and several other dicot families fail to conduct protein synthesis upon imbibition. Moreover, even aged but viable embryos from certain cereals exhibit signs of reduced protein synthesis. For example, rye embryos from a population possessing 86% viability (which show reduced root growth after germination, i.e., loss of vigor) have only about 20% of protein-synthetic capacity (as measured by incorporation of radioactive amino acids) of those with 95% viability (and which show no signs of vigor loss). Loss of protein-synthetic capacity in some species of seeds can be correlated with a decline in activity of various cytoplasmic components, including initiation and elongation factors. The ability of a supernatant fraction from wheat embryos to catalyze *in vitro* protein synthesis (Section 4.4.2) is reduced when endogenous mRNAs and tobacco mosaic virus (TMV)-RNA are used as messages, but not when the synthetic mRNA polyuridylic acid is used (Fig. 3.7). This suggests that the initiation factor components of the supernatant, which are essential for initiating the translation process at the 5′ end of natural (endogenous and TMV) mRNAs, lose activity as the seeds age. But the ability to read mRNAs that do not require the initiation reaction for translation to start (polyuridylic acid) is

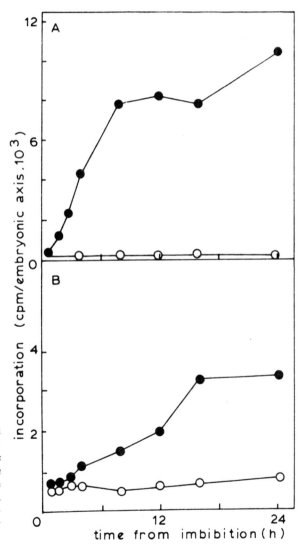

Figure 3.6. Protein and RNA synthesis in viable (●) and nonviable (○) embryonic axes of field pea *(Pisum arvense)*. (A) ^{14}C-leucine incorporation into protein. (B) ^{3}H-uridine incorporation into RNA. After Bray and Chow (1976).

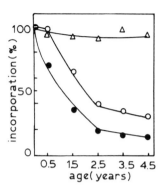

Figure 3.7. Relationship between age of wheat *(Triticum aestivum)* embryos and the activity of their supernatants in catalyzing protein synthesis *in vitro* using natural mRNAs [endogenous mRNA (●), TMV-RNA (○)] or synthetic mRNA [polyuridylic acid (△)]. After Peumans and Carlier (1981).

not reduced; i.e., in wheat, aminoacylation reactions and elongation of the nascent protein chain are not seriously impaired by aging.

In nonviable cereal embryos and dicot embryonic axes the inability to synthesize proteins is accompanied by a marked loss of capacity for RNA synthesis (Fig. 3.6, B). Moreover, in several seeds, including rye embryos, a gradual loss of viability is associated with decline in synthesis of all major classes of RNA and in the processing of ribosomal RNA (rRNA) into the subunits of the ribosome (Section 4.4.3). In aged but still viable seeds of tobacco, there appears to be some correlation between the integrity of rRNA and their rate of germination (Table 3.6).

The fate of mRNA in the dry seed during complete loss of viability is not fully understood. A fraction with mRNA-like activity has been extracted from nonviable rye embryos, and its activity is retained for 24 h after imbibition. These mRNAs might be imperfect, however, and although capable of catalyzing protein synthesis *in vitro,* the products may be defective or incomplete proteins. Such messages, therefore, might be ineffective as templates for translation *in vivo.* Aged, but viable, dry wheat embryos with "median" vigor have relatively lower amounts of poly(A)-containing mRNA (Section 4.4.2) than do high-vigor embryos (Table 3.7). In the high-vigor embryos the amount of

Table 3.6. A Correlation between the Integrity of Ribosomal RNA in Dry Seeds of *Nicotiana tabacum* (Tobacco) and Germinability and Vigor[a]

Year of harvest	% Germinated	Time to complete germination (h)	% rRNA integrity
1978	82	5.1	62
1973	72	17.1	23

[a]After Brocklehurst and Fraser (1980).

Table 3.7. Poly(A)-Containing RNA Levels and Synthesis in Wheat Embryos during Germination[a]

State of embryos	Vigor[b]	Poly(A) RNA as % of total cell RNA	% of level in dry high-vigor embryos
Dry	High	0.86	100
	Median	0.7	81
Imbibed 6 h, 20°C	High	1.08	125
	Median	0.53	62
Imbibed 24 h, 10°C	High	1.11	129
	Median	0.47	55

[a]Taken from data in Blowers *et al.* (1980).
[b]The vigor rating is a relative measure of seedling performance under suboptimal, osmotic-stress germination conditions. High-vigor seeds rated 80–100%; median-vigor seeds, 60–79% (C. M. Bray, personal communication).

mRNA increases during 6 h of germination but declines in median-vigor embryos. This decline is emphasized if the embryos are subjected to stress conditions, e.g., imbibition at a low temperature. Hence differences in messenger RNA levels could contribute to differences in vigor of seed lots.

Although there is substantial evidence that loss of seed viability is accompanied by a reduced capacity for protein synthesis, it would be unwise to suggest that one defective component (e.g., ribosomes, mRNAs, cytoplasmic factors) is more important than any other. Also, we should not forget that a reduction in ATP and GTP synthesis in non- or low-viable seeds (Section 3.3.1) will directly affect protein synthesis, and an important link exists between nucleoside triphosphate levels in seeds and their capacity to carry out other essential metabolic functions.

3.3.3. Damage to Chromosomes and DNA

Almost any combination of time, temperature, and moisture content that leads to loss of viability of seeds in storage will lead to some genetic damage in the survivors. In several species, including garden pea, broad bean, and barley, this damage is manifest as chromosomal breakages that appear during the anaphase stage of the first mitotic divisions of the root tips. The relationship between viability and the mean frequency of aberrant cells for broad bean seeds is shown in Fig. 3.8, A. There is an increase in chromosome damage with increase in the period of storage, as perceived by the increase in the number of aberrant cells (Fig. 3.8, B) and concomitant loss of viability (Fig. 3.8, C). Under the most severe storage conditions (45°C at 18% moisture content), where the mean viability period is less than 1 wk, the frequency of aberrant

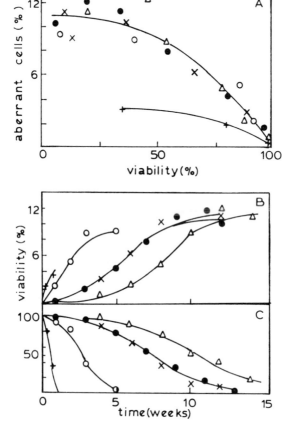

Figure 3.8. (A) The relationship between percentage viability and the mean frequency of aberrant cells, and (B) the increase in mean frequency of aberrant cells in surviving populations of seeds of *Vicia faba* stored under various conditions of temperature and moisture content (m.c.). (C) Seed survival curves for the same treatments. Conditions: 45°C, 18% m.c. (+); 45°C, 11% m.c. (●); 35°C, 18% m.c. (o); 35°C, 15% m.c. (X); 25°C, 18% m.c. (△). After Roberts (1975).

cells is considerably less than for all other treatments (Fig. 3.8, A). Perhaps under these unusually harsh conditions of aging some nonnuclear (cytoplasmic) lesions occur which are in themselves lethal and which lead to loss of viability at a faster rate than the breakage of chromosomes. With this exception, though, there is generally an excellent correlation between loss of viability and accumulation of chromosome damage in survivors.

Most aberrant cells in surviving seeds do not persist beyond the first cell division, and they are lost from the apical region before the roots grow extensively. Minor genetic damage such as recessive gene mutations may persist, but being masked by their dominant allele, they may have little obvious effect. Nevertheless, some recessive genes may be lethal in haploid cells, and their frequency can be related to the frequency of pollen abortion in mature plants grown from aged seeds. Accumulation of genetic damage in stored seed lots that are to be planted for food or feed production (i.e., whose progeny will not be replanted) may be of little consequence, unless the seeds germinate slowly or produce stunted seedlings. In these cases, the progeny may be more susceptible to pathogens, to adverse environmental changes, or to competition from other plants, e.g., weeds. Genetic changes in seeds of crops grown for germplasm stock may have more serious long-term implications. For example, a stored seed lot with little reduction in viability still could harbor a considerable number of mutations, which would not be expressed immediately in the crop grown from that seed but which would begin to segregate in the subsequent generation, and continue to do so in all generations thereafter. Thus poor storage of a "pure line" would soon result in a loss in desired purity.

Why aging seeds accumulate chromosome aberrations is not understood. One suggestion is that there is a buildup of chemical mutagens within seeds during aging, but the published evidence in favor of this is not convincing. Another possibility is that nuclear DNases are activated during aging, resulting in partially degraded DNA molecules. DNase activity in dry rye embryos is higher in nonviable than in viable material. How this increase in DNase occurs in the dry seed and how it is active therein are matters for conjecture. There could be activation of a preexisting latent enzyme or the removal of a DNase inhibitor during aging. If the now active DNase is present in the nucleus and closely associated with nuclear chromatin, then even at low moisture contents slow enzymic fragmentation of DNA might occur. The total amount of DNA in dry nonviable and viable embryos of rye is the same, although in the former the total amount of intact high-molecular-weight DNA is less (Table 3.8). This and other evidence is indicative of at least partial DNA fragmentation into lower-molecular-weight components during aging in dry storage. Minor repair to DNA might be possible when viable seeds are imbibed; loss of viability could be accounted for if the fragmentation of DNA in the dry seed is too extensive for repair to be effected upon subsequent hydration, or if the DNA repair system itself loses its integrity during aging.

When lettuce *(Lactuca sativa)* and ash *(Fraxinus americana)* seeds are

Table 3.8. Nuclear DNA Content and Extractable High-Molecular-Weight DNA in Rye Embryos of Different Viability[a]

Viability	Nuclear DNA (relative OD_{254})	High-molecular-weight DNA content ($mg.g^{-1}$ dry wt)
95	8.4	10.2
64	—	9.7
15	—	7.8
0	8.1	3.1

[a]After Osborne *et al.* (1980/81).

Table 3.9. Effects of 6-Mo. Storage at Different Water Contents on the
Germinability and Degree of Chromosome Damage in Radicle Tips of
Germinated Arctic King Lettuce[a]

Moisture content of stored seed (%)	Germination	Abberrant nuclear divisions
9.7	17	45
7	80	17
5.1	100	10
Fully imbibed	100	2
Not stored	100	2

[a]Based on Villiers (1974).

stored in a fully imbibed but dormant state, they maintain their full capacity for germination for long periods (at least 12 mo) and sustain very little chromosome damage. This is illustrated in Table 3.9 for Arctic King cultivar lettuce seeds. Loss of viability from seeds stored at a low moisture content (9.7%) might result from the inactivity of enzyme systems capable of repairing storage-induced damage to DNA and also to other essential macromolecules (and organelles) in the cytoplasm. At even lower water contents (5–7%) there is not enough damage to affect germination, for degradative enzymes cannot operate at such low water contents. Assuming that repair can only occur successfully in the fully imbibed state, then any damage suffered by the wet-stored seeds will be continuously repaired and will not accumulate. Although in the dry-stored seeds the repair mechanisms will become activated upon imbibition after storage, by this time the damage might be so extensive and beyond restitution that viability would be lost. This, for the moment, must remain only a hypothesis, for biochemical evidence in its favor has still to be obtained. Nevertheless, the concept of repair mechanisms is worthy of consideration, for many seeds may remain in soil in the imbibed state for long periods before conditions for germination become favorable.

3.3.4. Aging and the Deterioration of Membranes

Of the many factors involved in maintaining the control of metabolism within cells, spatial separation of metabolic components and the correct alignment of synthetic complexes are very important. Not only are cooperative enzymes of a metabolic pathway linked together within organelles (e.g., respiratory enzymes within mitochondria) but often they are intimately associated with or integrated into membrane structures. Thus, disruption of membranes due to aging could lead to diverse metabolic changes, all of which contribute to different extents to seed deterioration and loss of viability and vigor. As will be outlined in Section 3.5.3, imbibition by viable seeds is accompanied by a rapid but transient efflux of inorganic and organic compounds through the plasmalemma and tonoplast membranes and into the surrounding solution. Moreover, membrane integrity is incomplete for at least several minutes after water uptake. But the situation is reversed with time, the membranes either physically reverting to their most stable configuration or else being repaired by some still unidentified enzymic mechanism. In low- or nonviable seeds such repair mechanisms might be absent or inefficient, or the membranes might be so badly damaged that repair is impossible.

Loss of membrane integrity in deteriorated seeds is suggested by the observation that upon imibibition more substances leak into the medium from such seeds than from viable ones. Excess leakage of sugars may represent loss of respirable substrate from some seed species, whereas others leak more amino acids than sugars. To what extent this leakage affects germinability is not known, and it may be only one manifestation of the severe perturbations that occur to membrane systems within the cells. Increased leakage of organic metabolites from deteriorated seeds might indirectly enhance their demise by encouraging the growth of contaminating microorganisms.

How membranes and macromolecules become destabilized during storage under adverse conditions remains to be elucidated. It is possible that the aging of seeds at relatively high moisture contents brings about the hydrolysis of membrane components which cannot be replaced because of insufficient activity of the appropriate repair processes. In addition, free radical formation might occur at both high and low moisture contents. A variety of enzymatic and spontaneous oxidations generates the free superoxide radical (O_2^-), which is cytotoxic and which can react with H_2O_2 to produce singlet oxygen and the hydroxyl radical ($OH \cdot$). Both of these highly potent oxidants can bring about considerable destruction of large polymers, including membrane lipids. In well-hydrated tissues, free radical absorbents (e.g., tocopherols) or scavenging reactions limit the extent of undesirable oxidations. An important free radical scavenger is the enzyme superoxide dismutase; it converts the superoxide radical to H_2O_2, which in turn is removed by catalases. It is possible that in aging seeds the equilibrium between free-radical-producing and scavenging reactions is disturbed in favor of the former. This would lead to progressive inactivation of enzymes, denaturation of other proteins, and disruption of the integrity of DNA and RNA, due to uncontrolled free-radical activity. In addition, membranes would be rendered more permeable, *de novo* synthesis of enzymes impaired, and cell elongation and division slowed down or prevented. Thus symptoms of low vigor (slow growth or abnormal growth) and reduced viability (no growth) could result from fundamental changes in membranes and macromolecules.

From the standpoint of the above hypothesis it is perhaps unfortunate that evidence in favor of the free radical hypothesis is somewhat weak, and there is no direct evidence for the buildup of free radicals within aged seeds. One novel but highly circumstantial way of testing the free-radical hypothesis has been to examine the effect of storing maize kernels in a negatively charged conductor. This instrument generates electrons that react with free radicals and retard the aging process (Fig. 3.9). Accelerated aged maize kernels given this "cathodic protection" retain their viability and vigor longer than do unirradiated controls, leakage on imbibition is less, and the extent of cellular disruption and chromosome aberrations is reduced.

Biochemical evidence in favor of the free-radical hypothesis is ambivalent. On the basis of what is known about lipid oxidation it is likely that fatty acids with two or more unsaturated bonds are more labile than highly saturated fatty acids and more prone to form free radicals. Hence if fatty acid oxidation and free-radical formation are occurring during aging, the unsaturated fatty acid content of seeds should decline as deterioration progresses, but the saturated fatty acid levels should not. There is some evidence that in deteriorated axes of pea and of rapidly aged soybeans there is a loss of unsaturated fatty acids, particularly linolenic acid. On the other hand, in soybean seeds subjected to relatively mild aging conditions the loss of viability occurs without concomitant loss of unsaturated fatty acids, nor does the level of scavenging agents such as tocopherol decline during aging. Perhaps the safest conclusion at the present time is that fatty acid peroxidation can occur during aging under certain conditions, but it probably is not an essential component in or cause of viability loss. Instead, perhaps, we have to invoke changes to the proteins that are structurally incorporated into the membrane as the reason for its loss of integrity.

A summary of the variety of metabolic lesions that can occur during storage and which result in a loss of viability is to be found in Fig. 3.10. The reader can appreciate that although any single lesion can result in the loss of viability or reduction or vigor, it is likely that aging elicits a number of changes. Which one occurs first or is the most important is impossible to tell, but it makes an interesting topic for debate!

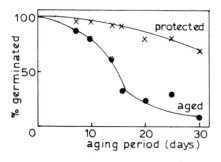

Figure 3.9. Viability of stored kernels of *Zea mays* in the presence (X) or absence (●) of irradiation with negative charges (cathodic protection). After Pammenter *et al.* (1974).

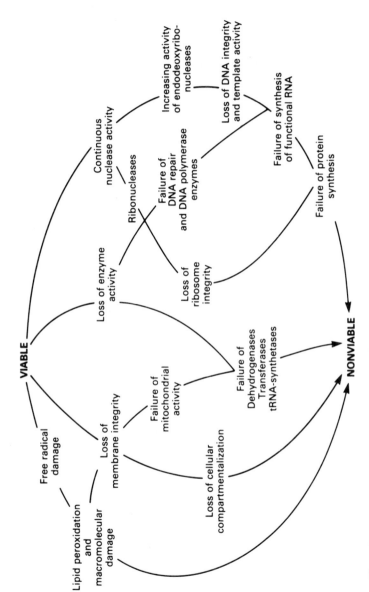

Figure 3.10. A scheme to illustrate the variety of causes for the loss of viability in stored seeds. Based on Osborne (1980).

3.4. METABOLISM OF THE DRY SEED

The status of dry seeds is described in several places in this book—in the preceding sections of this chapter the changes occurring during dry seed storage are described, and in later sections the cellular changes associated with the transition from the dry to the hydrated state will be outlined. The next chapter is devoted to the metabolic responses resulting from hydration of the dry seed.

Several ultrastructural studies have been conducted on dry seeds, but interpretation of the results of some must be treated with caution. In these, aqueous fixatives have been used prior to embedding and sectioning: cells imbibe water rapidly from such fixatives, resulting in partial to complete hydration before adequate fixation occurs. This can lead to the formation of artifacts owing to distortion of cellular contents, and the resultant picture obviously is not that of a dry cell. Far better fixation, giving a truer picture of the inclusions within dry cells, can be obtained by using nonaqueous fixatives or osmium vapor, or by freeze–etching. Cells of the coleoptile of dry rice grains (fixed in osmium vapor for 11 mo at room temperature) are shrunken and have highly folded walls. Nuclei, plastids, mitochondria, and lipid bodies have irregular outlines owing to shrinkage and compression. All cellular membranes, including the bounding plasmalemma, appear to be intact, however. Early during imbibition the mitochondria have a poorly defined inner structure (Section 4.2). But it is not possible to tell if this is also the case in the dry seed, or if the perturbation of this organelle is due to the rapid uptake of water (and, presumably, concomitant osmotic changes) immediately upon imbibition. From the little work that has been done on the structure of other dry seeds it appears that they also contain intact organelles and no discernible damage to membraneous structures within the cells.

Desiccated seeds, at oven-dry weight, are probably too dry to conduct any metabolic reactions, but the extent to which air-dry seeds metabolize is a function of their moisture content. Some unique problems must be overcome when studying dry seeds, because radioactive precursors that are commonly used to follow metabolism are usually in aqueous solution. Their application to dry seeds would result in hydration and initiate the intense metabolic activity associated with germination. Radioactive gas ($^{14}CO_2$) has been used, and experiments with charlock *(Sinapis arvensis)* seeds and embryos suggest that a little is incorporated into organic compounds in seeds of 6–15% water content, with an exponential increase in incorporation at higher levels of hydration. Dry, dormant wild oat grains fed with radioactive ethanol in quantities that do not change their water content can incorporate the radioactivity into sugars, amino acids, and proteins. Decarboxylation and transamination reactions occur in wheat grains with 18% water content. Probably, then, many seeds can metabolize at a very low rate when maintained above about 6% water content, and above about 15% water content the metabolism of "dry" seeds might be substantial.

3.5. IMBIBITION

In this and subsequent sections of this chapter we will concentrate on the physical and structural changes occurring in normal, viable seeds during the initial phases of imbibition and through germination to seedling establishment. Chapter 4 will deal with the metabolism that is invoked by hydration.

3.5.1. Uptake of Water from the Soil

The uptake of water by seeds is an essential, initial step toward germination. The total amount of water taken up during imbibition is generally quite small and may not exceed 2–3 times the dry weight of the seed. For subsequent seedling growth, which involves the establishment of the root and shoot systems, a larger and more sustained supply of water is required.

A number of factors governs the movement of water from the soil into the seed, but particularly important are the water relations of the seed and of the soil. Water potential (ψ) is an expression of the energy status of water, and net diffusion of water occurs down an energy gradient from high to low potential (i.e., from pure water to water containing solutes). Pure water has the highest potential, and by convention, it is assigned a zero value. Other potentials, therefore, have positive (i.e., > 0) or negative (i.e., < 0) values. The water potential of the cells in a seed can be expressed as follows:

$$\psi_{cell} = \psi_{\pi} + \psi_{c} + \psi_{p}$$

This means that the cell's water potential is affected by three components: (1) ψ_{π}—The osmotic potential. The concentration of dissolved solutes in the cell will influence water uptake, and the greater their concentration, the lower is the water potential and hence the greater the energy gradient along which water will flow. (2) ψ_{c}—The matric component. This is contributed by the ability of matrices (e.g., cell walls, starch, protein bodies) to be hydrated and bind water. (3) ψ_{p}—the pressure potential, which occurs because as water enters a cell the contents swell and exert a force on the external cell wall. Values for ψ_{π} and ψ_{c} are negative since they have a lower potential than pure water, and ψ_{p} is an opposite and hence positive force. The sum of the three terms, the water potential, is a negative number, except in fully turgid cells where it approaches zero. Water potential can be expressed in terms of pressure or energy, and the unit *bar* is often used (1 bar = 10^3 dynes.cm^{-2}, 10^2 J.kg^{-1}, or 0.987 atm). The term megaPascal (MPa) is now commonly in use, and -1 MPa = -10 bar.

The soil also has its own water potential (ψ soil), which is the sum of its ψ_π, ψ_c, and ψ_p, although of these only ψ_c plays a significant role (except in saline soils where ψ_π may be appreciable). The difference in water potential between seed and soil is one of the factors that determines availability and rate of flow of water to the seed. Initially, the difference in ψ between the dry seed and moist soil is very large owing to the high ψ_c of the dry coats, cell walls, and storage components of the former. But as the seed moisture content increases during imbibition and the matrices become hydrated, the water potential of the seed increases (i.e., becomes less negative) and that of the surrounding soil decreases as water is withdrawn. Hence, the rate of water transfer from soil to seed declines with time and declines more quickly in soils of low water-holding capacity (e.g., sandy soils) (Fig. 3.11). Continued availability of water to the seed depends on the water potential of the zones of soil immediately surrounding the seed and on the rate at which water moves through the soil, i.e., the hydraulic conductivity of the soil. Capillary and vapor movement of water near the seed is influenced considerably by soil compaction (bulk density), which may result in mechanical restraint of the swelling seed and decreased imbibition. Other factors may play a role in determining the rate and extent of water uptake regardless of the difference in ψ between the seed and soil, e.g., the

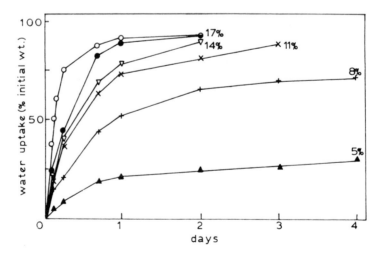

Figure 3.11. Water uptake by chickpea *(Cicer arietinum)* seeds in distilled water (o) or in soils at various moisture contents (on a percent dry weight basis). After Hadas (1970).

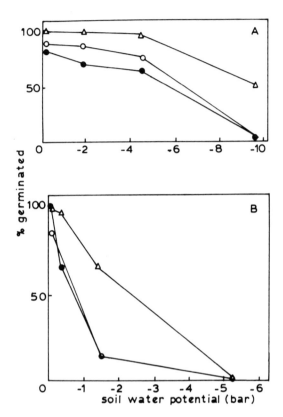

Figure 3.12. The germination of *Themeda australis* (●), *Danthonia* spp. (○), and ryegrass *(Lolium perenne)* (△) in (A) clay soil and (B) sandy soil at different water potentials. After Hagon and Chan (1977).

impedance of the soil matrix (due chiefly to surface and colloidal factors) and the degree of contact of the seed with soil moisture (i.e., seed–soil contact). The latter varies with seed size and shape and with the texture and compactness of the soil itself. Small seeds, seeds that produce mucilage, and seeds with relatively smooth coats tend to be the most efficient in absorbing water owing to their greater contact with soil and their larger surface area/volume ratio.

The influence of hydraulic conductivity and seed–water contact area on germination at a particular soil water potential varies between soil types, and so the germination response to soil water potentials in sandy soil is markedly different from that in clay soils (Fig. 3.12).

Unsaturated sandy soils have a lower hydraulic conductivity than clay soils, and because of the larger particle size of the former, the seed–water contact area is also reduced. Hence, at any given stress, germination is better in clay soils (Fig. 3.12, A) than in sandy soils (Fig. 3.12, B). This figure also illustrates that different seeds can utilize different soil moisture conditions for successful establishment; e.g., ryegrass can germinate better in lighter soil types than can other grass species. This may be related to variations in the efficiency with which seeds take up water or may be a reflection of the fact that some seeds can complete germination at lower water contents than others.

3.5.2. Kinetics of Water Uptake by Seeds

Under optimal conditions of supply the uptake of water by seeds is triphasic (Fig. 3.13).

Phase I. As noted in Section 3.5.1., the water potential of a mature dry seed is much lower than that of the surrounding moist substrate and can exceed -1000 bars owing to its high ψ_c. Phase I, or imbibition, is largely a consequence of these matric forces, and water uptake occurs regardless of whether the seed is dormant or nondormant, viable or nonviable. A wetting front is formed as water permeates the seed, and there is an abrupt boundary of water content between wetted cells and those about to be wetted (Fig. 3.14). Moreover, the average water content of the wetted area increases as a function of time. This initial pattern of water uptake is thus marked by three characteristics: (1) a sharp front separating wet and dry portions of the seed, (2) continued swelling as water reaches new regions, and (3) an increase in water content

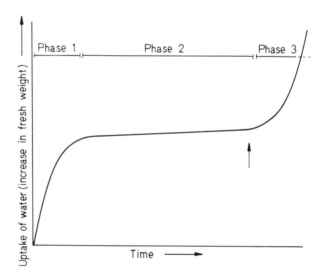

Figure 3.13. Triphasic pattern of water uptake by germinating seeds. Arrow marks the time when the first signs of radicle protrusion (i.e., completion of germination) occur.

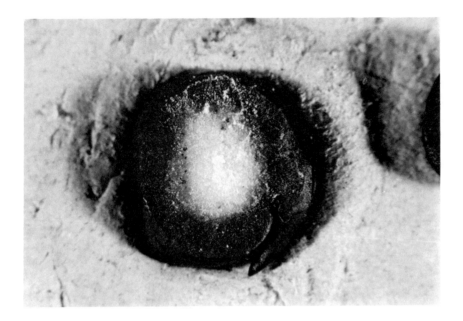

Figure 3.14. The wetting front in pea seeds, shown by cobalt chloride staining. Dark area shows extent of water penetration. By Waggoner and Parlange (1976).

of the wetted area. In some seeds, however, water uptake may be uneven owing to certain inherent structural features. In certain legumes (*Vicia* and *Phaseolus* spp., for example), water uptake through the micropyle may be greater than through the rest of the testa, and in some hard-coated seeds of the Papilionaceae (e.g., *Melilotus alba* and *Trigonella arabica*) a plug covering a special opening—the strophiolar cleft—must be loosened or removed before water can enter, and then only through that region. Imbibition is probably very rapid into the peripheral cells of the seed and into a tissue as small as the radicle. Hence metabolism can commence during this first phase, within minutes of introduction of the seed to water (see Chapter 4 for details).

Phase II. This is the lag phase of water uptake when the ψ_c no longer plays a significant role, and the ψ of the seed is largely a balance between ψ_π and ψ_p. In this phase, the value for ψ for many seeds probably does not exceed -10 to -15 bars. During this phase major metabolic events take place in preparation for radicle emergence from nondormant seeds; dormant seeds are also metabolically active at this time.

Phase III. Although dormant seeds may achieve phase II, only germinating seeds enter this third phase, which is concurrent with radicle elongation.

The increase in water uptake is initially related to the changes that cells of the radicle undergo as they extend (Section 3.5.3), marking the completion of germination. Then water uptake is influenced by decreases in ψ_π due to production of low-molecular-weight osmotically active substances resulting from the postgerminative hydrolysis of stored reserves. Endosperms and nonpersistent (hypogeal) cotyledons do not expand and hence do not achieve phase III of water uptake: eventually their water content declines as degeneration occurs.

The duration of each of these phases depends on certain inherent properties of the seed (e.g., hydratable substrate levels, seed coat permeability, seed size, oxygen uptake) and on the prevailing conditions during hydration (e.g., temperature, moisture levels, composition of substrate). Different parts of a seed may pass through these phases at different rates; e.g., an embryo or axis located near the surface of a large seed may commence elongation (i.e., enter phase III of water uptake) even before its associated bulky storage tissue has become fully imbibed (i.e., completed phase I). As an example, when the water content of whole dent corn *(Zea mays)* grains reaches 75%, the water content of the embryo on a dry weight basis is 261%, but that of the remainder of the grain is only 50%.

3.5.3. Soaking Injury, Solute Leakage, and Imbibitional Chilling Damage

As seeds start to take up water, there is a release from colloids of adsorbed gases (Section 4.1.2) and a rapid leakage into the surrounding medium of solutes such as sugars, organic acids, ions, amino acids, and proteins. In the field these solutes might stimulate the growth of fungi and bacteria in the soil, which will invade the seed and lead to its deterioration. Some seeds, however, (e.g., soybean) also leak certain proteinase inhibitors and lectins from their cell walls early upon imbibition, and these may function as protective agents against microbial or insect invasion.

When imbibed, pea embryos, i.e., seeds from which the testa has been removed, show an immediate and rapid leakage of potassium and of electrolytes, the latter being detected as an increase in conductivity of the surrounding water (Fig. 3.15, A). A similar rapid leakage of sugar and protein occurs (Fig. 3.15, B). This initial leakage lasts for up to about 30 min (Fig. 3.15, C) and occurs only from the outermost cell layers of the embryo at this time. Embryos that have been imbibed for 60 min, dried, and then rehydrated show the same rapid initial leakage (Fig. 3.15, C). Intact seeds do not leak so copiously into the surrounding medium. In some species this is because the uptake of water is restricted by the enclosing structures (e.g., testa, pericarp), or else they act as barriers to the efflux of solutes, or both. Seeds of certain species e.g., *Phaseolus vulgaris* and soybean (Fig. 3.17) germinate very poorly and show reduced

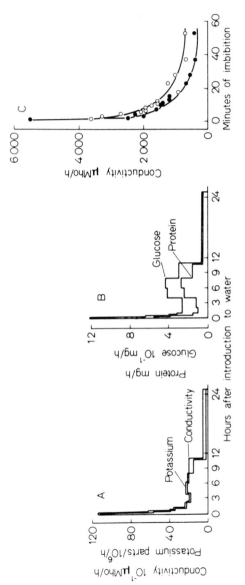

Figure 3.15. Time course of leakage of (A) electrolytes and potassium and (B) sugar and protein from pea embryos (i.e., seed minus testa) immersed in water. In (C) electrolyte leakage is followed from pea embryos immersed in water (●) and from embryos removed after 60 min in water, dried, and then returned to water (○). After Simon and Raja Harun (1972).

growth or vigor when hydrated without their testas. In these seeds, then, the coat plays an important protective role during imbibition. Damage to the coat during harvesting or planting could seriously reduce yield. Seeds that are hydrated wholly or partially in water-saturated atmospheres before introduction to liquid water, and seeds that do not undergo a drying process during their final stages of maturation, do not leak solutes when placed in water. Such seeds do not imbibe, however, since they are already hydrated.

Although some components are leaked from the cell wall, many are lost from within the cells themselves. Hence it is likely that the selectively permeable membranes of the tonoplast and plasmalemma that normally retain solutes within cells lose their integrity during drying and do not act as retentive barriers during the initial stages of imbibition. Membranes are composed chiefly of proteins and phospholipids arranged in a fluid bilayer. Individual phospholipid molecules are lined up side by side with their polar groups facing the aqueous phase on either side of the membrane and with their hydrocarbon chains forming the hydrophobic central region (Fig. 3.16, A). This molecular organization is stabilized by the relationship between membrane components and water. About 25% hydration above fully dried weight is essential for the maintenance of this membrane configuration. It has been claimed by some that at lower water contents the membrane bilayer becomes rearranged into a hexagonal phase which is largely hydrophobic but is pierced by long water-filled channels lined by the polar heads of the phospholipids (Fig. 3.16, A). The membrane proteins themselves appear to be unaffected by desiccation, but they may be displaced as the phospholipid components become rearranged (Fig. 3.16, B). Rapid imbibition of dry tissues results in the phospholipid and protein constituents spontaneously reconstituting the bilayer configuration once a critical degree of hydration is reached. Membrane components may be displaced, or further displaced if already so, by the rapid influx of water. Until such displacements are reversed (either by spontaneous reformation or by molecular repair mechanisms), leakage will occur, and the longer it takes for reconstitution to be completed, the greater the amount of leakage. Slow intake of water, from either vapor or liquid, may allow the membranes to return to their normal state in a controlled and orderly manner, without there being leakage. For the sake of completeness, it should be noted that not all workers are convinced that the phospholipid components when arranged in a membrane can undergo the aforementioned configurational changes upon desiccation. No alternative model for membrane changes has been proposed, however.

On the basis of electron microscope studies some workers have claimed that the plasmalemma of dry seeds has discontinuities in its structure, and that it takes some time for these to be eliminated upon imbibition. But these claims should be treated with considerable caution, for nonaqueous fixatives were used to prepare the tissues, and this could have led to distortions of the cells and

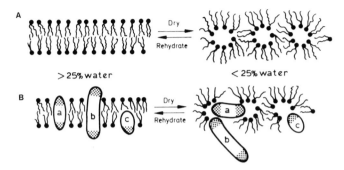

Figure 3.16. (A) Orientation of membrane phospholipids in relation to their degree of hydration. (Left) The lamellar configuration with the polar head groups facing the aqueous phase on either side of the bilayer. (Right) The hexagonal configuration adopted at water levels below approximately 25%. The polar head groups line a series of water-filled channels running through the relatively dry matrix. (B) Proposed orientation of membrane components in relation to their degree of hydration. (Left) The fluid mosaic model of a membrane under aqueous conditions showing three integral proteins (a)–(c), their polar and ionic moieties shaded. (Right) The configuration adopted with less than approximately 25% water. Note the displacement of proteins due to the reorientation of the phospholipids. Based on Simon (1978).

their contents due to limited imbibition (Section 3.4). Studies using nonaqueous fixatives have shown that the plasmalemma of dry seeds is intact.

Seeds of several species are injured if imbibed for as short a time as an hour at low temperatures (ca. 5°C) before being placed in normal incubation regimes (20–25°C), e.g., soybean *(Glycine max)*, cotton *(Gossypium hirsutum)*, some vegetable species, and seeds of many tropical crops and ornamental plants. Slow imbibition of water, such as in the presence of seed coats, in the cold is less injurious than rapid imbibition (by naked embryos), as illustrated for both soybean and the relatively chilling-insensitive pea seed (Fig. 3.17). At one time it was thought that chilling sensitivity might be related to the phospholipid composition of the membranes. It was suggested that the fatty acids of chilling-tolerant seeds are generally more unsaturated than those of intolerant ones. Consequently, the suggestion goes, their phospholipids are in a liquid–crystalline state at chilling temperatures and, being more fluid, are able to reorientate to more favorable configurations under these conditions. Phospholipids of intolerant seeds, having a more saturated fatty acid composition, would be in the crystalline state and have reduced mobility at low temperatures, which could lead to irreversible membrane disruption upon hydration. Recently, however, it has been shown that the membrane phospholipid composition of the chilling-tolerant pea seed is not different from that of the intolerant soybean seed. Moreover, the membrane fatty acid (saturated and unsaturated) composition of seeds of an intolerant soybean cultivar is almost

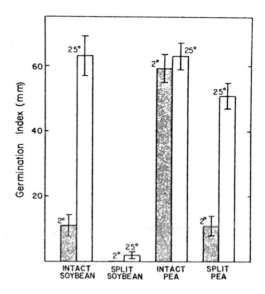

Figure 3.17. The vigor of soybean *(Glycine max)* and pea *(Pisum sativum)* seeds with intact or split seed coats after an initial 3-h imbibition at 2°C (shaded bars) or 25°C (clear bars) before incubation at 25°C for 4 days. Germination index (mean seedling length in millimeters of radicle and hypocotyl) was then measured. After Tully *et al.* (1981).

identical to that of a tolerant cultivar of the same species. Again, perhaps, we may have to invoke differences in the protein composition or differences in more minor membrane components (e.g., sterols) as factors in chilling tolerance, although proof is lacking. In some species there is evidence that chilling damage is most severe in seeds that already exhibit signs of reduced vigor (perhaps as a consequence of aging). When imbibition conditions are favorable, these seeds will germinate, but their competence to do so is reduced when they are subjected to a second stress, viz., imbibitional chilling.

3.6. THE COMPLETION OF GERMINATION: RADICLE ELONGATION AND ITS CONTROL

The cell walls of the embryo in mature dry seeds are shrunken, giving them a wavy or folded appearance under the electron microscope. Initially upon imbibition, the walls expand a little and straighten as they take up water, but this occurs in nonliving and in dormant seeds, as well as in viable nondormant ones. For germination to be completed, the radicle must expand and penetrate the surrounding structures; over the years, there has been considerable debate whether this occurs solely by cell expansion, or if cell division is required too. It is now generally conceded that cell division is neither correlated with nor is necessary for radicle expansion. Cereal grains bombarded with large doses of gamma rays to induce chromosome breakages still germinate and develop into seedlings ("gamma plantlets"), in the absence of any mitoses or increases in tissue cell numbers. Germination and early seedling establishment can therefore occur without cell divisions. In a number of other, nonirradiated seeds, cell extension clearly precedes cell division, e.g., in *Vicia faba* (Fig. 3.18) where the length of the radicle increases about 18 h prior to the increase in mitotic index. In other seeds, the difference in timing between cell division and cell expansion is much closer, but there is no strong evidence that the former ever precedes the latter during germination, nor do they occur simultaneously. From the pattern of cell elongation shown by *Vicia faba* (Fig. 3.18) it is evident that extension of the radicle occurs in two phases. There is initially a slow phase up until about 48 h (I), by which time the root has pierced the testa in about 30% of the seeds. Then a rapid phase ensues (II), which is accompanied later by an increase in the number of cells, and by mitoses in the apical region of the radicle. A similar biphasic mode of radicle elongation has been reported for other seeds.

That initial cell elongation is different from subsequent radicle growth can be demonstrated also by examining the sensitivity of these two events to water stress (Fig. 3.19). Seeds that have not commenced radicle elongation, i.e., not completed germination, can be prevented from doing so by incubation in solutions with a maximum water potential of −12 bar. On the other hand, seeds that have already entered the second phase of radicle elongation, i.e., com-

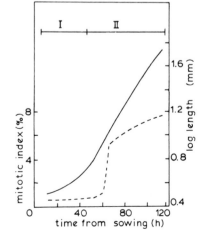

Figure 3.18. Length (———) and mitotic index (------) of roots of germinating broad bean *(Vicia faba)* seeds. I and II refer to phases of radicle elongation (see text). Based on Rogan and Simon (1975).

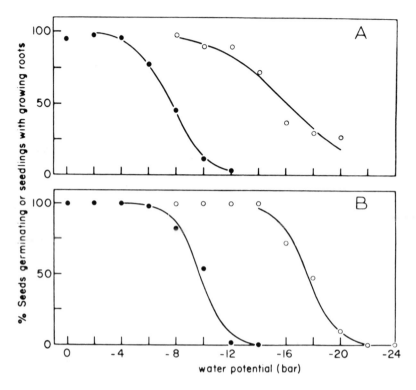

Figure 3.19. Germination (●) and seedling radicle growth after germination (○) of (A) calabrese (*Brassica oleracea* var. *italica*) and (B) cress *(Lepidium sativum)* incubated in solutions of polyethylene glycol at different water potentials. After Hegarty and Ross (1978).

menced growth, need lower water potentials (at least −20 bar) for complete arrest. This implies that the initiation of cell elongation is controlled in a different manner from later events. But what is the nature of this control? What specific events are essential for radicle elongation to commence, and what internal signals are responsible for setting this process in motion? Unfortunately, definitive answers cannot be given to these fundamental questions, nor to the one we consider in later chapters, i.e., what is the difference between dormant and nondormant seeds such that the latter can commence radicle extension, but the former cannot?

Germination, by definition, culminates in radicle emergence without an increase in cell numbers; maybe, then, the key to the control of germination lies in the control of cell extension. Maintenance of turgor (ψ_p) is an obvious requirement for cell expansion, to apply pressure on the walls to cause their extension. Water uptake by cells depends on their internal water potential, and it can be argued that germination can be controlled and promoted by decreasing the osmotic potential (ψ_{π}) of the radicle cells. But, at best, evidence for such a control mechanism is fragmentary (Section 5.5.1.10). Equally plausible is the possibility that "loosening" of the cell wall is a prerequisite (or corequisite) for the cell elongation associated with the completion of germination. The mechanism of cell wall loosening to allow for subsequent expansion has been studied not in seeds, but rather in growing tissues, such as the coleoptiles of cereals and the young stems of certain dicots. A current hypothesis is that expanding cells secrete protons (hydrogen ions) into their walls, resulting in their acidification; this leads to a breaking of the hydrogen bonds between adjacent chains of wall carbohydrates, or the low pH optimizes the conditions for limited enzymatic hydrolysis of the wall. The resultant change in wall rigidity, i.e., increased stretchability, thus allows for cell extension. Auxins and certain other chemicals (e.g., the fungal product fusicoccin) can promote various plant tissues to increase their growth rate. This is accompanied by a stimulation of hydrogen ion secretion to such an extent that not only are the cell walls acidified but the surrounding medium also shows a significant pH drop. But current evidence strongly suggests that auxin does not play any role in seed germination. Fusicoccin will stimulate the germination of certain dormant seeds, and the addition of this compound to the medium containing radish embryos leads to a decline in pH, indicating the release of protons (Fig. 3.20). But much of the decline in pH can be attributed to expanded, i.e., already germinated, radicles, and there is no strong evidence that fusicoccin acts to initiate radicle elongation by *prior* lowering of the pH. It is attractive to postulate that other hormones promote radicle elongation by increasing cell membrane permeability and stimulating the proton pump, but, again, evidence is lacking.

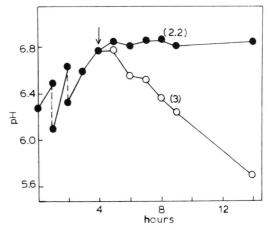

Figure 3.20. Changes in the pH of the incubation medium induced by fusicoccin treatment of radish embryos (o) and water-imbibed controls (●). Time of introduction of fusicoccin to the incubation medium is indicated by an arrow. Numbers in brackets indicate the extent of radicle elongation (in millimeters) after 9 h. After Lado *et al.* (1975).

We must conclude, then, that the events specific to the completion of germination—the extension of the cell walls—remain elusive. It is axiomatic that without an understanding of these events we cannot hope to gain any insight into their control.

3.7. SEEDLING DEVELOPMENT

The first sign that germination has been completed is usually that there is an increase in length and fresh weight of the radicle. In many seeds the radicle penetrates the surrounding structures as soon as growth commences, but in others (e.g., *Vicia faba* and other beans) there is considerable radicle growth before the testa is ruptured. There are some seeds, however, from which the hypocotyl is the first structure to emerge; this occurs in some members of the Bromeliaceae, Palmae, Chenopodiaceae, Onagraceae, Saxifragaceae, and Typhaceae. In some grains (e.g., rice and oryzicola) germinated under anoxic or hypoxic conditions coleoptile growth precedes that of the radicle (Fig. 4.14); such an unusual germination pattern is also sometimes found in wheat when grains begin to sprout on the mother plant (Section 9.3).

Seedlings can be conveniently divided into two types on the basis of the fate of their cotyledons following germination: (1) epigeal, in which the cotyledons are raised out of the soil by expansion of the hypocotyl and often become foliate and photosynthetic, and (2) hypogeal, in which the hypocotyl remains short and compact, and the cotyledons remain beneath the soil. The epicotyl expands to raise the first true leaves out of the soil (Fig. 3.21, Table 3.10).

In seedlings showing the epigeal mode of seedling growth, e.g., in castor bean *(Ricinus communis)* the endosperm may be carried above ground by the

Table 3.10. Some Species Exhibiting Hypogeal and Epigeal Seedling Growth

	Hypogeal	Epigeal
Endospermic	*Triticum aestivum*	*Ricinus communis*
	Zea mays	*Fagopyrum esculentum*
	Hordeum vulgare	*Rumex* spp.
	Phoenix dactylifera	*Allium cepa*
	Tradescantia spp.	*Trigonella foenum-graecum*
	Hevea spp.	
Nonendospermic	*Pisum sativum*	*Phaseolus vulgaris*
	Vicia faba	*Cucumis sativus*
	Phaseolus multiflorus	*Cucurbita pepo*
	Aponogeton spp.	*Sinapis alba*
	Tropaeolum spp.	*Crambe abyssinica*
		Arachis hypogaea

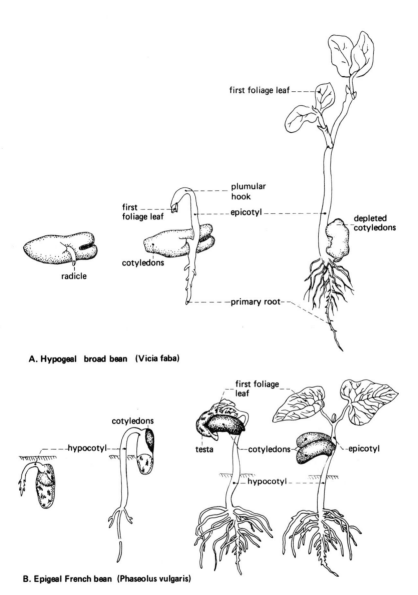

A. Hypogeal broad bean (Vicia faba)

B. Epigeal French bean (Phaseolus vulgaris)

Figure 3.21. The hypogeal and epigeal type of seedling development shown by two species in which the cotyledons are the storage organs.

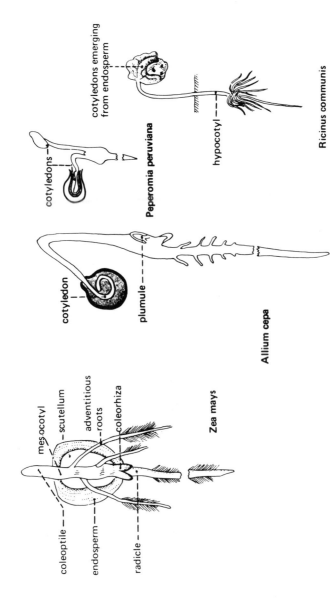

Figure 3.22. Patterns of seedling development and the different roles played by the cotyledons (see text for details). Not drawn to scale.

cotyledons as they utilize its food stores. In onion *(Allium cepa)*, another epigeal type, the absorptive tip of the single cotyledon may remain embedded in the degrading endosperm, while the rest of the cotyledon turns green (Fig. 3.22). The cotyledon in monocots may become highly specialized for absorption; in the Gramineae, for example, it is modified to form the scutellum, which may become extended as an absorptive haustorium (Fig. 7.2). Highly developed haustorial cotyledons are found in the Palmae. When the small embryo of the date palm *(Phoenix dactylifera)* commences growth, the cotyledon tip enlarges to form an umbrella-shaped body buried within the endosperm, from which it absorbs the hydrolyzed reserves. Likewise, the absorptive cotyledon of the coconut *(Cocos nucifera)* enlarges to invade the endosperm.

An interesting pattern of growth is shown by certain species of *Peperomia* (a dicot) in which one cotyledon emerges from the seed, and the other remains as an absorptive organ (Fig. 3.22). It has been suggested that the monocotyledonous condition evolved from this pattern of emergence, although this remains a matter for conjecture.

USEFUL LITERATURE REFERENCES

SECTION 3.1

Becquerel, M. P., 1934, *Compt. Rend. Acad. Sci. (Paris)* **199**:1662–1664 (viability records).

Egley, G. H., and Chandler, J. M., 1978, *Weed Sci.* **26**:230–239 (50-year-buried-seed study).

Godwin, H., and Willis, E. H., 1964, *New Phytol.* **63**:410–412 (dating Indian lotus seeds).

Kivilaan, A., and Bandurski, R. S., 1981, *Am. J. Bot.* **68**:1290–1292 (Beal's 100-year burial experiment).

Lerman, J. C., and Cigliano, E. M., 1971, *Nature* **232**:568–570 (600-year-old *Canna* seeds).

Odum, S., 1965, *Dansk. Bot. Arkiv.* **24**:2–70 (seeds in archaeological sites).

Porsild, A. E., Harington, C. R., and Mulligan, G. A., 1967, *Science* **158**:113–114 (arctic lupin longevity).

SECTION 3.2

Barton, L. V., 1961, *Seed Preservation and Longevity,* Leonard Hill, London, pp. 215 (comprehensive review of earlier literature).

Chin, H. F., and Roberts, E. H. (eds.), 1980, *Recalcitrant Crop Seeds,* Tropical Press, Malaysia (several review articles).

Christensen, C. M., 1972, in: *Viability of Seeds* (E. H. Roberts, ed.), Chapman and Hall, London, pp. 59–93 (microflora and seed decay).

Ellis, R. H., and Roberts, E. H., 1980, *Ann. Bot.* **45**:13–30 (seed longevity equations).

Harrington, J. F., 1973, in: *Seed Ecology* (W. Heydecker, ed.), Butterworths, London, pp. 251–263 (viability: rules of thumb).

James, E., Bass, L. N., and Clark, D. C., 1967, *Proc. Am. Soc. Hort. Sci.* **91**:521–528 (varietal differences and longevity).

Justice, O. L., and Bass, L. N., 1979, *Principles and Practices of Seed Storage,* Castle House, Beccles and London, U.K., pp. 289 (review book).

King, M. W., and Roberts, E. H., 1979, *The Storage of Recalcitrant Seeds,* Report for the International Board for Genetic Resources Secretariat, Rome, pp. 96 (recalcitrant seed storage).

Moore, R. P., 1972, in: *Viability of Seeds* (E. H. Roberts, ed.), Chapman and Hall, London, pp. 94–113 (mechanical injury and viability).

Roberts, E. H. (ed.), 1972, *Viability of Seeds,* Chapman and Hall, London (review book).

Roberts, E. H., 1973, *Seed Sci. Technol.* **1**:499–514 (predicting longevity in storage).

Roberts, E. H., and Abdalla, F. H., 1968, *Ann. Bot.* **32**:97–117 (storage in different atmospheres).

Stanwood, P. C., and Bass, L. N., 1981, *Seed Sci. Technol.* **9**:423–437 (liquid nitrogen storage).

Thomson, J. R., 1979, *An Introduction to Seed Technology,* Wiley, New York, Toronto, pp. 252 (seed storage methods).

Woodstock, L. W., Simkin, J., and Schroeder, E., 1976, *Seed Sci. Technol.* **4**:301–311 (freeze-drying for storage).

SECTION 3.3

Abdalla, F. H., and Roberts, E. H., 1968, *Ann. Bot.* **32**:119–136 (storage conditions and chromosome damage).

Abu-Shakra, S. S., and Ching, T. M., 1967, *Crop Sci.* **7**:115–118 (mitochondrial activity).

Aspinall, D., and Paleg, L. G., 1971, *J. Exp. Bot.* **22**:925–935 (differential aging of seed parts).

Blowers, L. E., Stormonth, D. A., and Bray, C. M., 1980, *Planta* **150**:19–25 (mRNA integrity and vigor).

Bray, C. M., and Chow, T. -Y., 1976, *Biochim. Biophys. Acta* **442**:1–13 (protein and RNA in nonviable peas).

Brocklehurst, P. A., and Fraser, R. S. S., 1980, *Planta* **148**:417–421 (rRNA integrity and vigor).

Ching, T. M., and Danielson, R., 1972, *Proc. Assoc. Official Seed Analysts* **62**:116–124 (ATP content and vigor).

Conger, A. D., and Randolph, M. L., 1968, *Radiat. Bot.* **8**:193–196 (free radicals in seeds).

Ellis, R. H., and Roberts, E. H., 1981, *Seed Sci. Technol.* **9**:373–409 (quantification of aging and survival: review).

Hallam, N. D., Roberts, B. E., and Osborne, D. J., 1973, *Planta* **110**:279–290 (respiration in nonviable rye).

Harman, G. E., and Mattick, L. R., 1976, *Nature* **260**:323–324 (lipid oxidation and seed aging).

Krasnook, N. P., Morgunova, E. A., Vishnyakova, I. A., and Povarova, R. I., 1976, *Soviet Plant Physiol.* **23**:130–134 (rice enzymes and viability).

Linko, P., and Milner, M., 1959, *Plant Physiol.* **34**:392–396 (transaminases in aged wheat).

Osborne, D. J., 1980, in: *Senescence in Plants* (K. V. Thimann, ed.), CRC Press, Boca Raton, Florida, pp. 13–37 (review of seed aging).

Osborne, D. J., Sharon, R., and Ben-Ishai, R., 1980/81, *Isr. J. Bot.* **29**:259–272 (DNA integrity and repair).

Pammenter, N. W., Adamson, J. H., and Berjak, P., 1974, *Science* **186**:1123–1124 (cathodic protection).

Peumans, W. J., and Carlier, A. R., 1981, *Biochem. Physiol. Pflanzen* **176**:384–395 (cytoplasmic factors and aging).

Priestley, D. A., McBride, M. B., and Leopold, C., 1980, *Plant Physiol.* **66**:715–719 (lipid oxidation and aging in soybean).

Roberts, E. H., 1975, in: *Crop Genetic Resources for Today and Tomorrow*, Volume 2, I. B. P., Cambridge University Press, Cambridge, U.K., pp. 269–296 (storage and genetic changes).

Sen, S., and Osborne, D. J., 1977, *Biochem. J.* **166**:33–38 (RNA and protein synthesis in nonviable rye).

Styer, R. C., Cantliffe, D. J., and Hall, C. B., 1980, *J. Am. Soc. Hort. Sci.* **105**:298–303 (noncorrelation between ATP and vigor).

Throneberry, G. O., and Smith, F. G., 1955, *Plant Physiol.* **30**:337–343 (respiration and vigor loss).

Villiers, T. A., 1974, *Plant Physiol.* **53**:875–878 (storage in hydrated state).

SECTIONS 3.4 AND 3.5

Blacklow, W. M., 1972, *Crop Sci.* **12**:643–646 (differential imbibition by corn).

Chen, S. S. C., 1978, in: *Dry Biological Systems* (J. H. Crowe and J. S. Clegg, eds.), Academic Press, New York, pp. 175–184 (dry oat grain metabolism).

Collis-George, N., and Melville, M. D., 1978, *Aust. J. Soil Res.* **16**:291–310 (seed surface–soil interactions).

Currie, J. A., 1973, in: *Seed Ecology* (W. Heydecker, ed.), Butterworths, London, pp. 463–480 (seed–soil system, review).

Dasberg, S., 1971, *J. Exp. Bot.* **22**:999–1008 (soil–water movement to seeds).

Edwards, M., 1976, *Plant Physiol.* **58**:237–239 (dry charlock seed metabolism).

Flentje, N. T., and Saksena, H. K., 1964, *Aust. J. Biol. Sci.* **17**:665–675 (pathogen growth and leakage).

Hadas, A., 1970, *Isr. J. Agr. Res.* **20**:3–14 (soil moisture stress and germination).

Hadas, A., 1977, *J. Exp. Bot.* **28**:977–985 (soil moisture stress and germination).

Hagon, M. W., and Chan, C. W., 1977, *Aust. J. Exp. Agric. Anim. Husb.* **17**:86–89 (soil moisture stress and germination).

Harper, J. L., and Benton, R. A., 1966, *J. Ecol.* **54**:151–166 (seed–surface contact).

Hwang, D. L., Yang, W.-K., Foard, D. E., and Lin, K.-T.-D., 1978, *Plant Physiol.* **61**:30–34 (protein leakage upon imbibition).

McKersie, B. D., and Stinson, R. H., 1980, *Plant Physiol.* **66**:316–320 (membrane configuration during dehydration).

Opik, H., 1980, *New Phytol.* **85**:521–529 (dry seed structure).

Rogers, R. B., and Dubetz, S., 1980, *Can Agric. Eng.* **22**:89–92 (seed–soil contact).

Simon, E. W., 1978, in: *Dry Biological Systems* (J. H. Crowe and J. S. Clegg, eds.) Academic Press, New York, pp. 205–224 (membranes in dry and imbibing seeds).

Simon, E. W., and Raja Harun, R. M., 1972, *J. Exp. Bot.* **23**:1076–1085 (leakage during imbibition).

Stewart, R. R. C., and Bewley, J. D., 1981, *Plant Physiol.* **68**:516–518 (soybean imbibitional chilling).

Tully, R. E., Musgrave, M. E., and Leopold, A. C., 1981, *Crop Sci.* **21**:312–317 (seed coats and imbibitional chilling).

Waggoner, P. E., and Parlange, J. -Y., 1976, *Plant Physiol.* **57**:153–156 (water diffusivity and imbibition).

SECTIONS 3.6 AND 3.7

Haber, A. H., and Luippold, H. J., 1960, *Plant Physiol.* **35**:168–173 (lettuce cell division and elongation).

Haber, A. H., Carrier, W. L., and Foard, D. E., 1961, *Am. J. Bot.* **48**:431–438 (gamma plantlets of wheat).

Hegarty, T. W., and Ross, H. A., 1978, *Ann. Botany (London)* **42**:1003–1005 (water stress; germination and growth).

Lado, P., Rasi-Caldogno, F., and Colombo, R., 1975, *Physiol. Plant.* **34**:359–364 (fusicoccin and proton extrusion).

Rogan, P. G., and Simon, E. W., 1975, *New Phytol.* **74**:273–275 (root elongation and mitosis).

Chapter 4

Cellular Events during Germination and Seedling Growth

Respiration, enzyme and organelle activity, and RNA and protein synthesis are fundamental cellular activities intimately involved in the completion of germination and the preparation for subsequent growth. It is not surprising, therefore, that most research into the biochemistry of germination and growth have concentrated on these events, to which we now turn our attention.

4.1. RESPIRATION—OXYGEN CONSUMPTION

4.1.1. Pathways, Intermediates, and Products

Three respiratory pathways are assumed to be active in the imbibed seed, namely, glycolysis, the pentose phosphate pathway, and the citric acid (Krebs or tricarboxylic acid) cycle. Glycolysis, catalyzed by cytoplasmic enzymes, operates under aerobic and anaerobic conditions to produce pyruvate, but in the absence of O_2 this is reduced further to ethanol, plus CO_2, or to lactic acid if no decarboxylation occurs. Anaerobic respiration, also called fermentation, produces only two ATP molecules per molecule of glucose respired, in contrast to six ATPs produced during pyruvate formation under aerobic conditions. In the presence of O_2, further utilization of pyruvate occurs within the mitochondrion: oxidative decarboxylation of pyruvate produces acetyl CoA, which is completely oxidized to CO_2 and water via the citric acid cycle to yield a further 30 ATP molecules per glucose molecule respired. The generation of ATP molecules occurs during oxidative phosphorylation when electrons are transferred to molecular O_2 along an electron transport (redox) chain via a series of electron carriers (cytochromes) located on the inner membrane of the mitochondrion. An alternative pathway for electron transport, which does not involve cytochromes, may also operate in mitochondria (see Section 4.2.2). The pentose phosphate pathway is an important source of NADPH, which serves as a hydrogen and electron donor in reductive biosynthesis, especially of fatty acids.

135

Intermediates in this pathway are starting compounds for various biosynthetic processes, e.g., synthesis of various aromatics and perhaps nucleotides and nucleic acids.

4.1.2. Respiration during Imbibition and Germination

Respiration by mature "dry" seeds (usual moisture content: 10–15%) is extremely low when compared with developing or germinating seeds, and often measurements are confounded by the presence of a contaminating microflora. When dry seeds are introduced to water, there is an immediate release of gas. This so-called "wetting burst," which may last for several minutes, is not related to respiration, but the gas that is released from colloidal adsorption as water is imbibed. This gas is released also when dead seeds or their contents, e.g., starch, are imbibed.

Keto acids (e.g., α-ketoglutarate, pyruvate), which are important intermediates in respiratory pathways, are chemically unstable and may be absent from the dry seed. A very early metabolic event during imbibition, occurring within the first few minutes after water enters the cells, is their re-formation from amino acids by deamination and transamination reactions (e.g., of glutamic acid and alanine).

The consumption of O_2 by many seeds follows the basic pattern outlined in Fig. 4.1 although, as indicated, the pattern of consumption by the embryo differs ultimately from that by storage tissues. Respiration is considered to involve three or four phases.

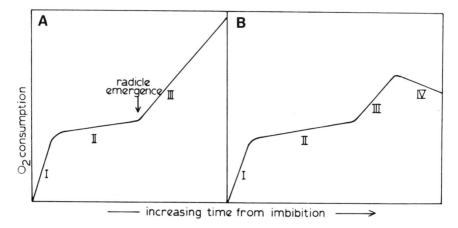

Figure 4.1. Pattern of oxygen consumption by the embryo (A) and storage tissues (B) of seeds during and after germination.

4.1.2.1. Phase I

Initially there is a sharp increase in O_2 consumption, which can be attributed in part to the activation and hydration of mitochondrial enzymes involved in the citric acid cycle and electron transport chain. Respiration during this phase increases linearly with the degree of hydration of the tissue.

4.1.2.2. Phase II

This is characterized by a lag in respiration as O_2 uptake is stabilized or increases only slowly. Hydration of the seed parts is now completed and all preexisting enzymes are activated. Presumably there is little further increase in levels of respiratory enzymes or in the number of mitochondria during this phase. The lag phase in some seeds may be in part because the coats or other surrounding structures limit O_2 uptake to the imbibed embryo or storage tissues, leading temporarily to partially anaerobic conditions. Removal of the testa from imbibed pea seeds, for example, diminishes the lag phase appreciably. Another possible reason for this lag is that the activation of the glycolytic pathway during germination is more rapid than the development of mitochondria. This could lead to an accumulation of pyruvate due to deficiencies in the citric acid cycle or the oxidative phosphorylation (electron transport) chain; hence, some pyruvate would be diverted temporarily to the fermentation pathway, which is not O_2 requiring.

Between phases II and III in the embryo the radicle penetrates the surrounding structures: germination is completed.

4.1.2.3. Phase III

There is now a second respiratory burst. In the embryo, this can be attributed to an increase in activity of newly synthesized mitochondria and respiratory enzymes in the proliferating cells of the growing axis. The number of mitochondria in storage tissues also increases, often in association with the mobilization of reserves. Another contributory factor to the rise in respiration in both seed parts could be an increased O_2 supply through the now punctured testa (or other surrounding structures).

4.1.2.4. Phase IV

This occurs only in storage tissues and coincides with their senescence following depletion of the stored reserves.

The lengths of phases I–IV vary from species to species owing to such factors as differences in rates of imbibition, seed coat permeability to oxygen, and metabolic rates. Moreover, the lengths of the phases will vary considerably with the ambient conditions, especially the temperature. In a few seeds, e.g., *Avena fatua,* there is no obvious lag phase (II), in oxygen uptake (Fig. 4.2). The reasons for its absence are not known, but it could be because efficient respiratory systems become established early following imbibition, including the development of newly active mitochondria, thus ensuring a continued increase in O_2 utilization. Also, coat impermeability might not restrict O_2 uptake prior to the completion of germination.

During germination a readily available supply of substrate for respiration must be available. This is not provided by hydrolysis of the major reserves, since this event usually commences only after germination is completed. Most dry seeds contain sucrose, and many contain one or more of the raffinose-series oligosaccharides: raffinose (galactosyl sucrose), stachyose (digalactosyl sucrose), and verbascose (trigalactosyl sucrose), although the latter is usually present only as a minor component. The distribution and levels of these sugars within seeds are very variable, even between different varieties of the same species. To introduce a brief digression, the raffinose series oligosaccharides have attracted some interest in relation to human nutrition in that they are reported to cause flatulence in man and domestic animals. These sugars escape digestion and absorption in the upper gastrointestinal tract but are degraded and fermented by microbes in the colon to yield H_2 and CO_2. The odious and malodorous consequences to human beings of the relatively high levels of stachyose and raffinose in digested beans are legendary.

During germination, sucrose and the raffinose-series oligosaccharides are hydrolyzed, and in several species the activity of α-galactosidase, which cleaves the galactose units from the sucrose, increases as raffinose and stachyose levels decline. Although there is little direct evidence that the released monosaccha-

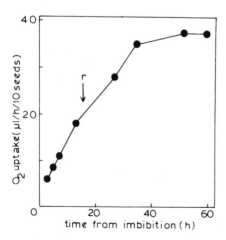

Figure 4.2. Time course of respiration of nondormant grains of wild oat *(Avena fatua).* r, time of radicle emergence. After Chen and Varner (1970).

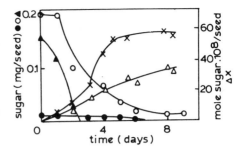

Figure 4.3. Decline in stachyose (▲), sucrose (○), and raffinose (●) in seeds of *Sinapis alba* accompanied by an increase in glucose (X) and fructose (△). A separate analysis showed that most of the stored stachyose, sucrose, and raffinose is located in the cotyledons, but that most of the glucose and fructose is in the root and stem. After Gould and Rees (1964).

rides are utilized as respiratory substrates, there is strong circumstantial evidence. Free fructose and glucose may accumulate in seeds during the hydrolysis of sucrose and the oligosaccharides, but there is no buildup of galactose (e.g., in mustard, *Sinapis alba,* Fig. 4.3). Hence it is probably rapidly utilized, perhaps through incorporation into cell walls or into galactolipids of the newly forming membranes in the cells of the developing seedling.

4.2. MITOCHONDRIAL DEVELOPMENT AND OXIDATIVE PHOSPHORYLATION

4.2.1. Site of ATP Production during Early Imbibition

Mitochondria in dry and freshly imbibed seeds are functionally and structurally deficient, and it has been debated whether they can effectively conduct oxidative phosphorylation during the first few hours after hydration. Other pathways for the production of ATP during this time have been proposed. One was suggested to involve the transfer of phosphate from phytin (*myo*-inositol hexaphosphate, see Section 7.6) to ADP to produce ATP. But this seems unlikely on three accounts: (1) ATP production is an early event during germination, but phytin breakdown is a postgermination phenomenon; (2) phytic acid does not possess a phosphate ester bond of sufficient "high energy" to act as donor of a phosphoryl group for ATP formation; and (3) the total amount of phosphate that can be released from phytin would produce only enough ATP to sustain metabolism for a few minutes.

Enzymatic production of ATP in the cytosol during the early stages of imbibition also has been suggested as an alternative to oxidative phosphorylation. The proposed synthetic process, which does not require O_2, involves at least the following steps:

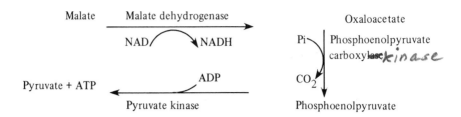

This scheme has been devised from experiments carried out *in vitro* and utilizing rehydrated powders of dry seeds: whether it operates *in vivo* remains to be demonstrated. It may not be necessary, however, to invoke a mechanism for ATP synthesis other than oxidative phosphorylation utilizing the normal cytochrome pathway within the mitochondrion. There is increasing evidence that this pathway operates soon after the start of imbibition to produce ATP. But before considering ATP synthesis itself, let us examine how mitochondria function and their status in dry and imbibed seeds.

4.2.2. The Route of Electrons between Substrate and Molecular Oxygen

Many studies on the development of mitochondrial activities in seeds have been conducted on the cotyledons of legumes, and for the moment we will have to assume that similar changes occur within the axis. Mitochondria in dry seeds are characteristically poorly differentiated and lack cristae. To understand the changes that occur in the mitochondria during germination we must first review the major routes by which oxidizable substrates enter this organelle and then how they become involved in oxidative phosphorylation. Glycolysis and various other metabolic pathways produce NADH, from which NAD has to be regenerated. The outer membrane of the mitochondrion is permeable to NADH (called exogenous NADH because it is produced outside of the mitochondrion), and it passes to the intermembrane space (Fig. 4.4). Pyruvate also enters the mitochondrion, but it is transported through the membranes to the inner matrix, which contains the enzymes of the citric acid cycle. This cycle produces NADH also (called endogenous NADH because it is formed within the mitochondrion). Both endogenous and exogenous NADH are cleaved by membrane-bound NADH dehydrogenase complexes, although the released reducing equivalents (as electrons) enter the electron transport chain at different sites. Electrons originating from endogenous NADH are transferred to ubiquinone via a coupled proton efflux site (site I) which permits energy conservation by driving phosphorylation; i.e., ATP is produced. Electrons released from exogenous NADH are not transferred via site I but are passed directly

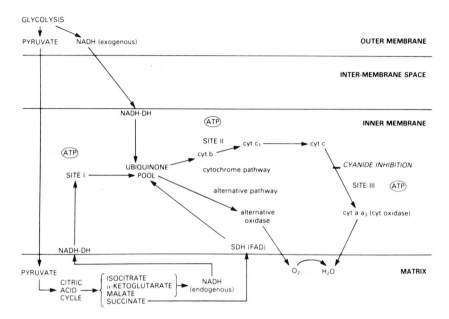

Figure 4.4. The pathways for utilization of reducing power and electrons in plant mitochondria. NADH–DH, NADH dehydrogenase; SDH (FAD), flavoprotein-linked succinate dehydrogenase; sites I–III, sites of proton extrusion where ATP (circled) is produced; cyt, cytochrome.

to the ubiquinone pool, thus eliminating the first phosphorylation step. The oxidation of succinate to malate in the citric acid cycle is not NAD linked; succinate dehydrogenase (SDH) is inserted into the inner membrane on the matrix side, and the released electrons are transferred to ubiquinone. There is no site of energy conservation along this pathway (Fig. 4.4).

The electrons are transferred from the ubiquinone pool via cytochromes b, c_1, and c and finally to O_2 via cytochrome oxidase (aa_3). This part of the chain is very important because it incorporates two sites of proton extrusion and, consequently, of energy conservation (Fig. 4.4). Mitochondria of many higher plant tissues possess an alternative electron transfer pathway which, unlike the cytochrome chain, is insensitive to cyanide (and hence it is often called the cyanide-insensitive pathway). The alternative pathway branches between ubiquinone and cytochrome b, and electrons are passed via an "alternative oxidase" system, as yet unidentified, to O_2. There are no sites of energy conservation linked to this pathway, and hence when exogeneous NADH and succinate are oxidized by this route, there is no ATP production. When an NAD-linked substrate from the mitochondrial matrix, such as malate and α-ketoglutarate, is oxidized through the alternative pathway, only one ATP is produced (at site I) compared to three (at sites I, II, and III) when electrons pass through the cytochrome pathway (Fig. 4.4).

Isolated mitochondria must be used to study their phosphorylating activity. Their ability to oxidize various substrates and to phosphorylate ADP is monitored using an oxygen electrode, and a typical trace of oxygen consumption by a mitochondrial preparation consuming O_2 and conducting oxidative phosphorylation is shown in Fig. 4.5. Initially the isolated mitochondria are incubated in the absence of both ADP and added substrate (α-ketoglutarate, malate, succinate, or NADH) to test activity driven by the purely endogenous components. A low amount of oxygen consumption may occur: this is state 1 respiration. Then substrate (in this case malate) is added and O_2 consumption occurs, which is the state 2, or substrate rate, of O_2 uptake. Next ADP is added to the mitochondrial preparation, and the rate of O_2 uptake increases appreciably as oxidative phosphorylation occurs (state 3 respiration). The rate of uptake then begins to level off as ADP becomes limiting (state 4). State 3 respiration can be reinduced by adding more ADP, until such time as the available O_2 in the incubation mixture becomes limiting and anaerobiosis (state 5 respiration) is achieved. Two important calculations can be derived from this type of experiment: (1) the ADP/O ratio, which tells how many μmoles of ADP are used (to make ATP) per μmole of O_2 consumed and is a measure of efficiency of phosphorylation, and (2) the respiratory control ratio (RCR), which is the ratio of state 3 to state 4 respiration and is an indication of how tightly respiration is controlled by ADP availability. If phosphorylation is efficient (i.e., tightly coupled to O_2 consumption) and utilizes the cytochrome pathway, then, in reference to Fig. 4.4, α-ketoglutarate and malate should elicit an ADP/O ratio close to 3, and succinate and added NADH a ratio close to 2. Substantial lowering of these ratios indicates that mitochondria are deficient

Figure 4.5. Pattern of oxygen consumption by a preparation of mitochondria measured with an oxygen electrode. Arrows indicate the time of addition of the mitochondrial extract (mito.), substrate (malate), and ADP. A well-aerated mitochondrial reaction mixture is 250 μM with respect to O_2: at the end of the experiment all O_2 is depleted.

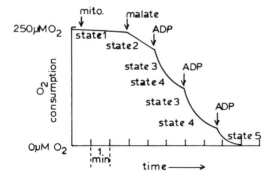

$$RCR = \frac{\text{State 3 rate of } O_2 \text{ consumption}}{\text{State 4 rate of } O_2 \text{ consumption}}$$

$$\text{ADP/O ratio} = \frac{\mu\text{moles ADP added}}{\mu\text{atoms } O_2 \text{ consumed}}$$

in their cytochrome pathway and/or phosphorylating ability. A clear transition from state 3 to state 4 is expected in efficiently phosphorylating mitochondria; failure to observe this transition can result from damage to the inner membrane (allowing leakage of protons from the sites of proton extrusion), or if electron transport is diverted to the nonphosphorylating alternative pathway.

Now let us consider mitochondrial changes in the imbibed seed in relation to this information.

4.2.3. Mitochondrial Development in Imbibed Seeds

As noted previously, mitochondria in unimbibed embryos and storage tissues are poorly differentiated internally. As might be expected, they exhibit deficiencies in their capacity to utilize reducing power from NADH and succinate dehydrogenase (SDH) and to conduct oxidative phosphorylation. Generally, mitochondria extracted from dry seeds can oxidize supplied succinate and NADH (i.e., can achieve state 3 respiration). Hence the electron pathways between succinate and ubiquinone, and between exogenous NADH and ubiquinone, are present and are immediately active upon hydration. Although these pathways may initially be low in activity, they become increasingly prominent with time after imbibition (Fig. 4.6). In contrast, mitochondria from early-imbibed seeds oxidize malate or α-ketoglutarate only poorly, and the development of oxidizing capacity using endogenous NADH occurs more slowly during germination (Fig. 4.7). During the early stages of germination, mitochondria may not respond clearly to ADP addition when malate and α-ketoglutarate are used as substrate, and a transition from state 3 to state 4 respiration does not occur. Hence both the RCR and ADP/O ratio are extremely low, indicating a block to the transfer of electrons through the pathway to ubiquinone via site I. Note that the results presented in Figs. 4.6 and 4.7 were obtained using pea cotyledons, and hence the changes that occur are not directly related to germination, which is an axial phenomenon. However, similar changes appear to occur in the axis, although in some species the time required for mitochondria to develop a response to added malate and α-ketoglutarate may be considerably shorter.

Mitochondria from dry and early-imbibed peanut embryonic axes are deficient in cytochrome c, and this deficiency remains until about the end of the lag period. But mitochondrial respiration during this time probably still takes place via a cytochrome pathway, since oxidation of succinate is inhibited by cyanide, a specific inhibitor of the terminal oxidation (cytochrome oxidase) reaction (Fig. 4.4). Hence, although there is only a low level of cytochrome c present in mitochondria until germination is completed, it might still be important in preventing completely uncontrolled loss of respiratory substrate, and

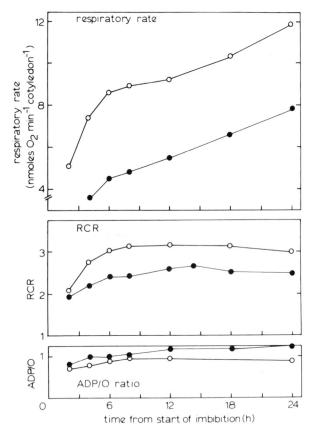

Figure 4.6. Changes in activities of mitochondrial fractions isolated from pea cotyledons (cv. Alaska) over 24 h from the start of imbibition. Substrate: succinate (●); NADH (o). After Morohashi and Bewley (1980).

some oxidative phosphorylation can occur. Cyanide sensitivity of the electron transport chain has been observed in the mitochondria from axes and storage tissues of other seeds also. This tends to argue against the suggestion made by some that the alternative (cyanide-insensitive) pathway is the predominant electron transport pathway during the early stages of germination. It is evident, however, that the alternative pathway can become more active with time after the start of imbibition, and in pea cotyledons (Fig. 4.8) it reaches a maximum several days after SDH and cytochrome oxidase do. In storage tissues, the alternative pathway may play an important role in the reoxidation of NADH produced by such catabolic pathways as β-oxidation and the glyoxylate cycle (see Section 7.4.4). Reducing equivalents are passed to the mitochondria from the glyoxysome and need to be oxidized to maintain an adequate supply of NAD for operation of important enzymes in this latter organelle. But transfer of electrons through the normal cytochrome pathway could lead to excess ATP production and result in a negative feedback and a shutdown of metabolism. Hence to maintain a high rate of metabolism and an optimum supply of ATP, a balance of electron flow between the cytochrome and alternative pathways is maintained in times of high NADH production.

A general feature of mitochondrial oxidation in seeds is that it increases

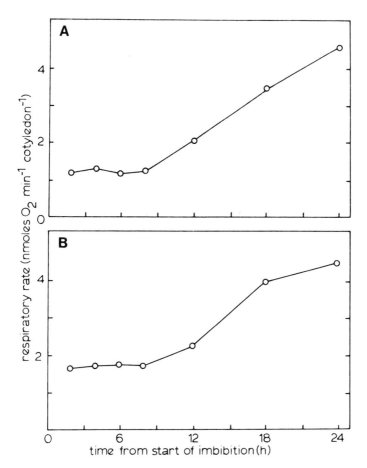

Figure 4.7. Changes in respiration rate of mitochondrial fractions from pea cotyledons (cv. Alaska) over 24 h from the start of imbibition. Substrate: malate (A) or α-ketoglutarate (B). After Morohashi and Bewley (1980).

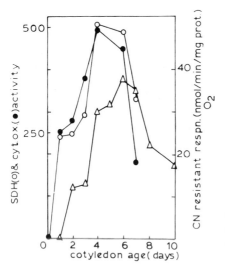

Figure 4.8. Changes in activities of succinate dehydrogenase (o), cytochrome oxidase (●) (both expressed in nmol substrate reduced/min/mg protein) and cyanide-resistant respiration, with succinate as substrate (△) in mitochondria from pea (cv. Homesteader) cotyledons. After James and Spencer (1979).

with time from imbibition. This may be due to increased efficiency of existing mitochondria and/or an increase in the number of mitochondria. It is now evident that there are two distinct patterns of mitochondrial development in imbibed seeds: (1) repair and activation of organelles already existing within the mature dry seed (as typified by the pea cotyledon) and (2) biogenesis of new mitochondria (as typified by the peanut cotyledon). An insufficient number of species has been studied to determine how widespread is each of these patterns, nor is it known in any species what pattern of development is followed in the axis prior to germination. After germination, there must be considerable biogenesis of mitochondria as new cells are formed in the growing axes by mitotic divisions.

In pea cotyledons, some development of mitochondria occurs within the first day of imbibition, and by the end of the second day cristae appear to be more or less fully developed. The structurally deficient mitochondria from pea cotyledons are also enzymically deficient, having only low levels of malate dehydrogenase and cytochrome oxidase. SDH is present but, unlike in normal mitochondria, it is not tightly bound to the inner membrane. Incorporation of applied radioactive leucine into mitochondrial proteins can occur within 6 h of the start of imbibition and may continue for at least another 18 h. Inhibition of protein synthesis on cytoplasmic ribosomes prevents this synthesis of mitochondrial proteins, whereas inhibition of protein synthesis on mitochondrial ribosomes does not. Therefore, the new mitochondrial proteins are cytoplasmic in origin. But even in the absence of synthesis of new mitochondrial proteins, cytochrome oxidase and malate dehydrogenase levels rise, and mitochondria increase in their respiratory efficiency. Moreover, SDH activity becomes tightly associated with the inner membrane when both cytoplasmic and mitochondrial protein synthesis is inhibited. Thus, protein synthesis is not a prerequisite for the development of mitochondrial activity. Instead, it appears that preformed proteins, both structural and enzymic, are transferred into preformed, immature mitochondria after imbibition to produce active and efficient mitochondria with complete membrane and oxidative systems.

Mitochondria in peanut cotyledons are difficult to detect early after imbibition, but they increase in quantity and quality (i.e., in internal structure) over the first few days after imbibition (Fig. 4.9). Mitochondrial development is accompanied by an increase in state 3 respiration, ADP/O ratio, RCR, and cytochrome oxidase and other enzyme activity, and by an increased protein–nitrogen and phospholipid content of the mitochondrial membrane fraction (Table 4.1). Moreover, there is an increase in incorporation of radioactive leucine into mitochondrial proteins, which ceases when cytoplasmic protein synthesis is inhibited. These observations are all supportive of the contention that the increase in mitochondrial activity in peanut cotyledons is due to *de novo* synthesis of these organelles. This does not preclude the possibility that there is also improvement in the low number of mitochondria preexisting in the dry seed.

As the stored reserves in the cotyledons become depleted, the mitochondria become disorganized and gradually lose their respiratory efficiency, enzyme complement, and activity. For example, in cotyledons of dark-grown Alaska peas there is a marked decline in RC and ADP/O ratios, although cytochrome oxidase and malate dehydrogenase activities remain fairly constant. Degenerative changes to the inner membrane may occur, for SDH is no longer tightly bound to it in aging cotyledons, and some activity is lost (Table 4.2). The fall in respiratory activity that accompanies the overall decline in stored reserves might not occur uniformly in all parts of dicot cotyledons. In

Table 4.1. Content and Activities of Mitochondria Isolated from Peanut Cotyledons during 5 Days after Imbibition[a]

| Cotyledon age | P. cotyledon^{-1} (μg) | N. cotyledon^{-1} (μg) | ADP/O | RCR | State 3 respiration (nmol O_2.min^{-1} cotyledon^{-1}) | | Relative activites | | |
					+ Malate	+ Succinate	SDH	MDH	cyt ox
6 h	0.4	7.2	nd	nd	1	4	100	100	100
3 days	1.4	22.5	1.9	2.8	6.8	38	495	550	340
5 days	2.3	27	2.1	3	11.7	61	610	—	675

[a]SDH, succinate dehydrogenase; MDH, malate dehydrogenase; cyt ox, cytochrome oxidase; nd, not detectable—no clear transition from state 3 to state 4;—not determined. From data in Morohashi *et al.* (1981a, 1981b).

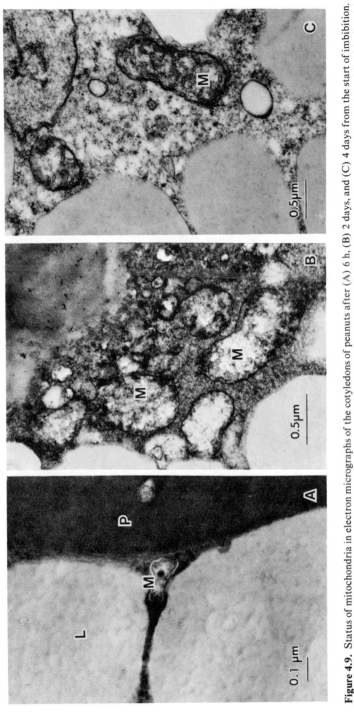

Figure 4.9. Status of mitochondria in electron micrographs of the cotyledons of peanuts after (A) 6 h, (B) 2 days, and (C) 4 days from the start of imbibition. (A) Cells are packed with protein (P) and lipid (L) storage products, and it is difficult to distinguish other organelles among the protein and lipid bodies. M is possibly a mitochondrion. (B) Developing mitochondria (M) are visible although the internal membrane structure is not well developed. (C) Well-developed mitochondria (M) with obvious internal cristae. Photographs courtesy of Dr. E. C. Yeung.

Table 4.2. Decline in Repiratory Activities of Mitochondria in Cotyledons of Alaska Pea Owing to Senescence[a]

Cotyledon age	State 3 respiration + succinate (nmoles $O_2 \cdot min^{-1} \cdot cotyledon^{-1}$)	RCR	ADP/O	Cyt. ox.	SDH
1 day	118	2.7	1.75	100	100
5 days	32	1	nr	102	38

Relative activity (spanning RCR, ADP/O, Cyt. ox., SDH)

[a]Abbreviations as Table 4.1. nr, no response to added ADP. The cyt. ox. and SDH activities are those associated with the inner membrane. Note that pea cotyledons after 5 days show signs of senescence, but peanut cotyledons of the same age do not (Table 4.1). From data in Nakayama et al. (1978).

squash *(Cucurbita maxima)* cotyledons, for example, which persist and become photosynthetic, the mitochondria are considerably reduced in numbers in storage cells as the reserves are expended and chloroplasts are formed, but not in these cells close to the veins. Thus there is a redistribution of respiratory activity in the cotyledons with age, it being fairly uniform during reserve mobilization but becoming more restricted to vascular tissue as the organs assume a photosynthetic role. The vein cell mitochondria might be important for the vein loading of catabolites and the subsequent products of photosynthesis. In dark-grown seedlings, which do not become photosynthetic, vein localization of mitochondria is less distinct.

4.2.4. ATP Synthesis and Adenylate Energy Charge during Germination

Synthesis of ATP during germination has been studied in only a few seed species, but the same general pattern emerges. The level of ATP in dry seeds is extremely low compared to other adenine nucleotides (ADP and AMP), but upon hydration it increases very rapidly (e.g., in lettuce, Fig. 4.10, A and B). During the lag phase of O_2 consumption ATP levels remain fairly constant; a further rise in ATP occurs after germination. The increase in ATP in the seed is accompanied by an increase in the total pool of adenine nucleotides, which suggests an early requirement for their *de novo* synthesis. The ATP pool during the lag phase is not static, for if ATP synthesis is inhibited, e.g., by placing seeds in an atmosphere of N_2 to prevent terminal oxidation in the mitochondria, then the ATP pool is quickly used up, and levels diminish (Fig. 4.10, B). The ATP pool returns rapidly to control levels upon reintroduction of the seeds to O_2 at atmospheric concentrations. Thus, during the lag phase, although there is no net increase in ATP, the rate of its synthesis is balanced exactly by its rate of utilization.

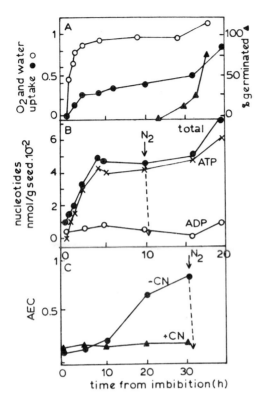

Figure 4.10. (A) and (B) Time courses of increases in (A) water uptake, O_2 consumption, and germination and in (B) total adenine nucleotides, ADP, and ATP in lettuce seeds. (A): o, water uptake in g/g seed; •, O_2 uptake in μ mol/min; ▲, percent germinated. (B): o, ADP; X, ATP; •, total adenine nucleotides. The arrow indicates time of transfer of the seeds to an atmosphere of nitrogen, and the dotted line the immediate and rapid decline in ATP levels (to zero within 6 min). (C) Increase in adenylate energy charge over the first 30 min from the start of imbibition of lettuce seeds on water (•) or 5×10^{-4}M KCN (▲). Arrow indicates the time of transfer of seeds to an atmosphere of nitrogen, and the dotted line the immediate and rapid decline in AEC. Based on data in Pradet *et al.* (1968), and Hourmant and Pradet (1981).

In most living cells the adenine nucleotides are maintained under near equilibrium conditions by a very active adenylate kinase

$$2ADP \rightleftharpoons AMP + ATP$$

and so it follows that the adenine nucleotide ratios ATP/ADP and ATP/AMP are connected. The adenylate energy charge (AEC) integrates these ratios and is expressed as follows:

$$AEC = \frac{[ATP] + 0.5\,[ADP]}{[ATP] + [ADP] + [AMP]}$$

The resultant value is an indication of the amount of available energy within the adenylate pool and varies from 0, when the pool contains only AMP, to 1, when only ATP is present. Such extremes are not found in living cells, but it is generally recognized that the AEC of normally metabolizing cells is usually in excess of 0.8. Adenylate energy charges of 0.5 or less are indicative of quiescent or senescent cells.

The adenylate energy charge increases rapidly in imbibed seeds as water is taken up and ATP synthesis is activated (e.g., in lettuce, Fig. 4.10, C). This increase is prevented by cyanide (an inhibitor of terminal oxidation, Fig. 4.4), an observation that supports the contention that ATP synthesis at the earliest times after imbibition is routed through cytochrome oxidase; i.e., it is supplied by mitochondrial oxidative phosphorylation. Predictably, the increase in ATP content and AEC is accompanied by an increase in metabolic activity, e.g., RNA and protein production within the germinating seed, as the energy status of the cells changes to favor synthetic reactions.

4.2.5. The Synthesis and Utilization of Reducing Power: Pyridine Nucleotides

Reduced pyridine nucleotides (NADH and NADPH) are essential coenzymes for several key metabolic pathways and hence are required by the germinating seed, the developing seedling, and the storage organs. They participate, for example, in glucose respiration, amination reactions, deoxynucleotide synthesis, and fatty acid catabolism. Metabolic pathways might be controlled by the availability of these coenzymes or by their rates of reoxidation within the cell.

The oxidized pyridine nucleotide NADP is a coenzyme of glucose-6-phosphate dehydrogenase (Glc-6-Pdh), which is a key link between the glycolysis pathway and the pentose phosphate pathway (PPP). Glycolysis requires a supply of NAD for the intermediate step involving glyceraldehyde-3-phosphate dehydrogenase (Fig. 4.11). When NADP is in limited supply, glycolysis proceeds at the expense of the PPP, whereas in conditions of limiting NAD, the reverse is the case. The relative activities of the two pathways have been studied in only a few seeds, and more usually in the cotyledons than in the axis; nevertheless, the following pattern of events emerges:

During the first few hours to several days (depending on the species) after the start of imbibition, it appears that the glycolytic pathway predominates, favoring pyruvate formation and some synthesis of ATP (Fig. 4.11). Pyruvate is converted to ethanol under the partially anaerobic conditions that tend to prevail in intact seeds prior to emergence of the radicle. By channeling respirable substrate through the glycolytic pathway under these conditions the cells can to some extent offset the reduced ATP production by mitochondria resulting from limited O_2 availability due to natural anaerobiosis (Section 4.3), or from their inherent structural or enzymic deficiencies (Section 4.2.3). Later, the PPP predominates, and as mitochondria become more active and O_2 more available, the seeds enter a period of energy production for increased metabolic activity associated with growth and reserve mobilization, with NADPH production increasing to provide the necessary reducing power. Some indication of the type of changes in status of the glycolytic and PPP pathways is illustrated, using cotton *(Gossypium hirsutum)* as the example, in Fig. 4.12. Aldolase and phosphofructokinase, both glycolytic pathway enzymes, decline in activity in the radicles and to a lesser extent in the cotyledons, during the first 1–3 days after sowing. Glc-6-Pdh, a PPP enzyme, increases in activity after imbibition, particularly in the growing regions. Although activities of extracted enzymes assayed *in vitro* are not always an accurate measure of their *in vivo* activity, the illustrated changes are at least consistent with the generally accepted view that carbohydrate catabolism via the glycolytic pathway precedes that via the PPP. Incidentally, the rise in aldolase activity in the cotyledons after 3 days (Fig. 4.12) could be related to the onset of gluconeogenesis associated with fat catabolism, which requires the operation of most of the glycolytic pathway in reverse (Fig. 7.11), but not that part involving phosphofructokinase.

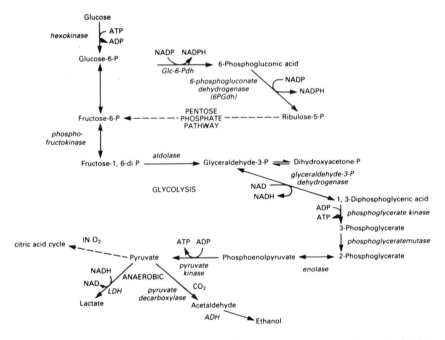

Figure 4.11. Some important steps in the glycolysis and pentose phosphate pathways in relation to NADH, NADPH, and ATP production. Enzymes are in italics.

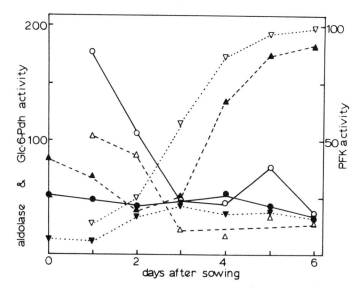

Figure 4.12. Changes in activities (expressed in μmol product/mg protein/min) of the PPP enzyme glucose-6-phosphate dehydrogenase (▼, ▽) and the glycolysis enzymes phosphofructo-kinase (●, ○) and aldolase (▲, △) in radicles (open symbols) and cotyledons (closed symbols) of cotton seeds. Taken from data in Purvis and Fites (1979).

4.3. RESPIRATION UNDER ANAEROBIC CONDITIONS

As outlined in Section 4.1.2., many seeds experience conditions of temporary anaerobiosis during lag phase II of respiration (Fig. 4.1). Consequently, both ethanol and lactic acid, products of anaerobic respiration (Fig. 4.11), accumulate within the seed; they are found in different proportions in different species. Upon penetration of the enclosing structures by the radicle, the levels of these anaerobic products decline as they are metabolized under conditions of increasing aerobiosis. Lactate dehydrogenase (LDH) and alcohol dehydrogenase (ADH) are responsible for both the synthesis and the removal of lactate and ethanol, respectively. Under aerobic conditions, the former enzyme converts lactate back to pyruvate, which can then be utilized by the citric acid cycle, and the latter converts ethanol to acetaldehyde, which is oxidized to acetate by acetaldehyde dehydrogenase. The acetate is activated to acetyl CoA, which can be used in many metabolic processes. Not surprisingly, then, ADH and LDH tend to be present in seeds during germination, and often they are even present in the dry seed, e.g., ADH in pea cotyledons (Fig. 4.13). In some species of seeds the activity of LDH and ADH increases appreciably (ten-fold or more) during germination, perhaps owing to *de novo* synthesis. After ger-

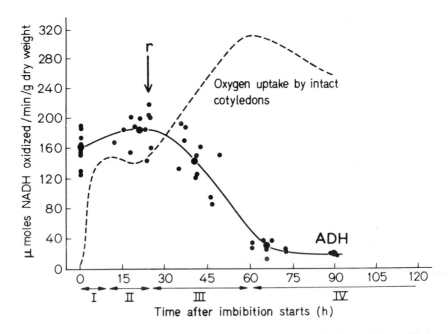

Figure 4.13. Alcohol dehydrogenase activity (●) in, and oxygen uptake by, cotyledons of dark-germinated intact pea seeds cv. Rondo. Arrow (r) marks the time of penetration of the radicle through the testa. After Kollöffel (1968).

mination, and when conditions are more aerobic, ADH (Fig. 4.13) and LDH decline to negligible levels: loss of both ethanol and lactate from the seed parallels this decline.

The period of natural anaerobiosis in seeds during germination can last from a few hours to several days. Anaerobic conditions can be prolonged by sowing seeds under water. Many species will actually complete germination when submerged, although subsequent growth of the radicle is stunted, and if the germinated seeds are maintained in water they will die. Seeds of certain aquatic species germinate better under conditions of reduced O_2 tension, e.g., *Typha latifolia,* and some, e.g., *Juncus effusus,* will germinate readily even after submergence for up to 7 years. Grains of aquatic species such as rice *(Oryza sativa)* and barnyard grass (*Echinochloa crus-galli* var. *oryzicola*), a noxious weed in rice fields, germinate and grow under water and show some interesting adaptations to these conditions of reduced oxygen availability. Both rice and barnyard grass will germinate in a totally O_2-free environment, although only the coleoptile elongates; root growth is inhibited, as is emergence of the leaves from the coleoptile (Fig. 4.14). The seedlings of both species produce considerable quantities of ethanol (shown for barnyard grass in Fig. 4.15, A), much of which (up to 95% in rice) diffuses into the surrounding water. Even so, under anoxic conditions the seedlings may contain close to 100 times more ethanol than aerated controls. ADH activity rises along with ethanol production (Fig. 4.15, B). Increases in lactate levels occur also, but they are very small in comparison to ethanol. Mitochondrial development in both rice and barnyard grass seedlings under anoxic conditions is not very different from that in fully aerated seedlings. Although oxidative phosphorylation obviously does not occur in grains imbibed in nitrogen, the mitochondria develop and maintain their capacity for ATP production in the absence of O_2. Upon return of seedlings to atmospheres containing O_2, oxidative phosphorylation commences almost immediately. Synthesis of ATP takes place in the absence of O_2 (Fig.

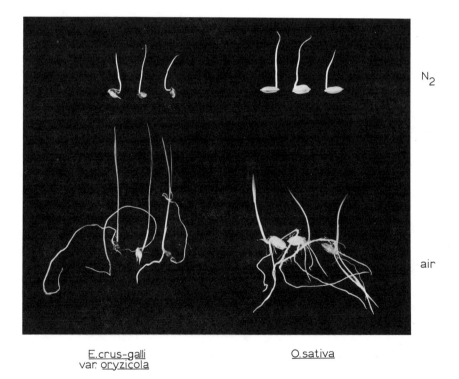

E.crus-galli
var. oryzicola   O.sativa

Figure 4.14. Seven-day-old seedlings of (left) barnyard grass (*Echinochloa crus-galli* var. *oryzicola*) and (right) rice *(Oryza sativa)* imbibed in N_2 (top row) or air (bottom row) since sowing. Note differences in root morphology and extent of leaf emergence from the coleoptile. Courtesy of Dr. R. A. Kennedy.

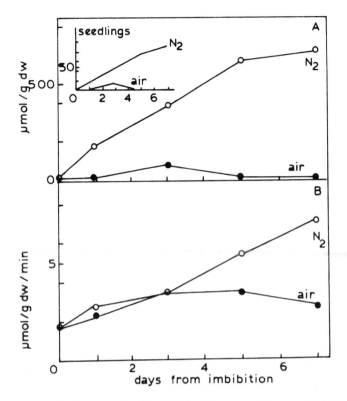

Figure 4.15. (A) Ethanol content of grains of *Echinochloa crus-galli* var. *oryzicola* imbibed in atmospheres of N_2 or air. The main graph is the sum of the ethanol content of the grains and the surrounding liquid, and the inset is the content of the seedlings alone. (B) Alcohol dehydrogenase activity. Germination was completed (i.e., coleoptile elongation commenced) by the third day from the start of imbibition. After Rumpho and Kennedy (1981).

4.16) from the earliest stages of imbibition, probably by operation of the glycolytic pathway, terminating in the conversion of pyruvate to ethanol (Fig. 4.11). Even the adenylate energy charge will adjust under anaerobic conditions to maintain a high value (0.75–0.8).

4.4. PROTEIN AND RIBONUCLEIC ACID SYNTHESIS

Protein synthesis is essential for germination to be completed and for the radicle to emerge. Its commencement after imbibition is largely independent of prior RNA synthesis; before the completion of germination, however, newly formed RNA is probably needed for protein synthesis. On the other hand,

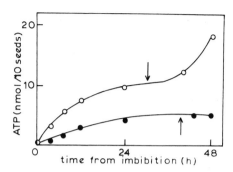

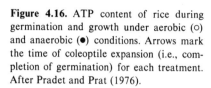

Figure 4.16. ATP content of rice during germination and growth under aerobic (o) and anaerobic (•) conditions. Arrows mark the time of coleoptile expansion (i.e., completion of germination) for each treatment. After Pradet and Prat (1976).

DNA synthesis in most seeds occurs only after germination and is an integral part of growth of the axis.

4.4.1. The Mechanism and Measurement of Protein Synthesis

It is not our intention here to give a detailed account of the mechanism of protein synthesis in plants; only a brief overview will be presented. Protein synthesis commences in the cytoplasm of a hydrated cell when ribosomes (80 S) become associated with messenger RNA (mRNA) (Fig. 2.16). Initiation of protein synthesis requires attachment of the small ribosomal subunit (40 S) and the initiating transfer RNA (tRNA) molecule (methionyl tRNA) to the starting point—the initiation site—on the mRNA. After formation of this initiation complex, the large ribosomal subunit (60 S) becomes attached and protein synthesis commences. Formation of the initiation complex requires soluble protein factors (initiation factors), GTP, and perhaps ATP. Other soluble protein factors (elongation factors) are required for transfer of aminoacylated tRNAs to their appropriate codon on the ribosome–mRNA complex and to move the message through the ribosome; termination factors are involved in the disassembly of the complex when protein synthesis is completed.

Experimentally, the ability of a seed to conduct protein synthesis can be determined using three techniques. The status of the protein-synthesizing system at any time can be deduced by analysis of the ratio of polysomes (those ribosomes associated with mRNA) to ribosomes (those not attached to mRNA and hence not involved in protein synthesis). Most commonly, this is achieved by extraction of the total ribosomal pellet, followed by separation into ribosomes and polysomes by centrifugation on a sucrose density gradient. Alternatively, the activities of polysomes and ribosomes can be determined using an appropriate *in vitro* assay for protein synthesis. That for polysomes allows the extracted native mRNAs to be translated by the attached ribosomes; for estimation of extractable ribosome activity a synthetic mRNA (usually poly U) is added to the assay mixture. The results obtained from polysome analysis and *in vitro* assays generally give a good indication of a seed's capacity for protein synthesis at the time of extraction, rather than an indication of the amount of protein synthesis actually in progress. The best indication of *in vivo* protein synthesis can be obtained by following the incorporation of radioactive amino acids into protein over a defined and usually brief time period. There are problems to be contended with when using this technique, however; e.g., not all seeds take up amino acids with equal facility owing to variations in thickness of seed coats and other factors. Also, in large seeds diffusion of radioactive precursors to the inner tissues takes considerably longer than to the outer tissues.

4.4.2. Protein Synthesis during Germination and Its Dependence on Messenger RNA Synthesis

The transition from development to germination necessitates fundamental changes in the control of gene expression within a seed. At some stage the expression of genes coding for reserve proteins, and for enzymes involved in the synthesis of stored material and other activities, has to be switched off, whereas genes coding for enzymes involved in germination and subsequent reserve mobilization must be switched on. Concomitant with germination, there must be met the increasing demand for proteins essential to this event and later the synthesis of proteins for seedling growth. How, then, is the change in direction of protein synthesis brought about? In this section we will concentrate on studies using germinating tissues, viz., isolated embryos of cereals or axes of legumes and other dicots. In Section 4.4.5. we will consider changes occurring within the seed storage tissues.

Polysomes are absent from dry seeds, but they rapidly increase as soon as the cells become hydrated. Accompanying this increase in polysomes is a decline in the number of single (unattached or free) ribosomes as these become recruited into the protein-synthesizing complex. An example of early polysome formation is found in the wheat embryo (Fig. 4.17). Here the extracted ribosomal pellet from the embryo of the mature dry grain is incapable of catalyzing *in vitro* protein synthesis in the absence of added mRNA, because there are no polysomes present. But after a lag period of only 10–15 min, during which the embryo rapidly takes up water, the protein-synthesizing capacity of the extracted pellet is evident. Thereafter its activity increases as more and more polysomes are found.

As discussed in Chapter 2, considerable protein synthesis occurs during seed development, but this ceases as desiccation proceeds during the final

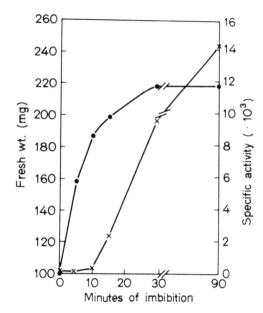

Figure 4.17. Comparison of water uptake (●) and development of polysomal activity assayed *in vitro* (X) during the early stages of imbibition by isolated wheat embryos. After Marcus *et al.* (1966).

stages of maturation. It is pertinent to ask, therefore, whether some or all of the components involved in protein synthesis during development are destroyed as a consequence of desiccation, or whether they are conserved in the dry seed and then reutilized upon subsequent rehydration. A necessary ingredient of many *in vitro* protein-synthesizing systems is a supernatant prepared from mature dry wheat embryos or wheat germ. From this supernatant have been isolated a number of the components including ribosomes, transfer RNA (tRNA), initiation and elongation factors, amino acids, and aminoacyl tRNA synthetases. Many of these components have been extracted from dry embryos, axes, and storage tissues of other species too, and hence there can be no doubt that they are present in a potentially active form in the cells of dry seeds. Moreover, they appear to be present in sufficient quantities for protein synthesis to commence immediately when the seeds are hydrated.

More heatedly debated has been the question as to whether protein synthesis can commence in the imbibing seed without the necessity for any prior RNA synthesis. Although it has been accepted for nearly two decades that active ribosomes (containing rRNA) and tRNA are present in the dry seed, the status of translatable mRNA has taken longer to resolve. Isolated wheat embryos synthesize mRNA very soon after they start to imbibe water, and within 30–90 min newly synthesized mRNA becomes incorporated into polysomes. However, this new mRNA synthesis might not be required for the

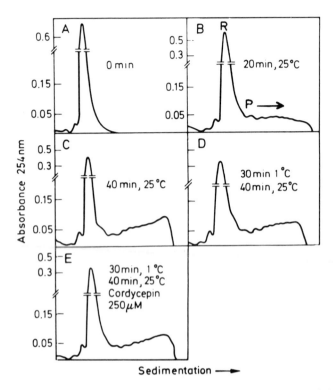

Figure 4.18. Polysome formation during early germination of wheat embryos and the effect of RNA synthesis inhibitors. Isolated embryos were imbibed in water at 25°C for (A) 0 min; (B) 20 min; (C) 40 min; (D) 1°C 30 min, then 40 min at 25°C, both in water; or (E) 1°C 30 min, then 40 min at 25°C, both in cordycepin. Cordycepin inhibited uridine incorporation by 82%. R, area of single ribosomes; P, area of polysomes. After Spiegel and Marcus (1975).

resumption of protein synthesis upon imbibition, for during the first hour polysomes are still formed when over 90% of the initial RNA synthesis, including mRNA, is prevented by chemical inhibitors (Fig. 4.18). The addition of a sequence of poly(A) to the 3′ end of the mRNA may be a necessary step in the processing of most mRNAs prior to their involvement in cellular protein synthesis. Both RNA synthesis and this polyadenylation reaction are inhibited by cordycepin. Since imbibition of wheat embryos in effective concentrations of this inhibitor does not prevent rapid polysome formation, this suggests that mRNA is conserved in the dry embryo and does not require further polyadenylation before being recruited for early protein synthesis. Similar conclusions have been drawn from studies on embryos and axes from other species of seeds. In wheat embryos, other mRNA-processing events (e.g., capping and internal methylation) do not appear to be required for conserved (or newly synthesized) messages to be recruited into polysomes.

Although protein synthesis may proceed in the absence of RNA synthesis during early imbibition, it eventually comes to depend on it—before germination is completed. For example, protein synthesis in radish embryonic axes is insensitive to cordycepin for the first 2–3 h from the start of imbibition (Fig. 4.19, A), indicating that new RNA synthesis (presumably including mRNA synthesis) is not required during this period. Poly(A)-containing RNA is lost from the embryos germinating on cordycepin during the first few hours, although the total poly(A) content of embryos increases after an initial decline (Fig. 4.19, B). Thus, it appears that there is a loss of conserved mRNA during the first few hours of germination and an increase in new messages, with protein synthesis becoming increasingly dependent on the latter with time of germination. In wheat embryos, the change in mRNA status with time is reflected by a qualitative change in the types of protein being synthesized (Fig. 4.20). Here, the *in vitro* translation products of mRNAs extracted from dry wheat embryos (Fig. 4.20, A) and from early imbibed embryos (Fig. 4.20, B) are quite similar, although there are some quantitative differences. At later times, in this case after 17–18 h (Fig. 4.20, C), the types of proteins determined by extracted mRNAs are considerably different from those at earlier times. In particular, mRNAs for ribosomal proteins are present early during imbibition and not later: since new ribosomes are not required for the commencement of protein synthesis, this observation might be part of the explanation why early mRNA synthesis is not essential. This is not to imply, of course, that new mRNAs for ribosomal proteins are not required later during germination or

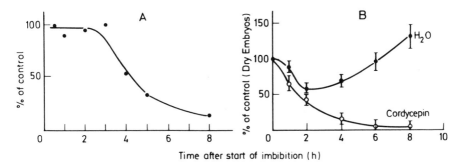

Figure 4.19. (A) Decline in the amount of *in vivo* protein synthesis coded for by mRNA present in the dry embryonic axes of radish *(Raphanus sativus)*. New RNA synthesis during the 8-h experimental time period was prevented by cordycepin. (B) Poly(A)-containing RNA (mRNA) content of radish embryos during germination in the presence (o) or absence (●) or cordycepin. After Delseny *et al.* (1977).

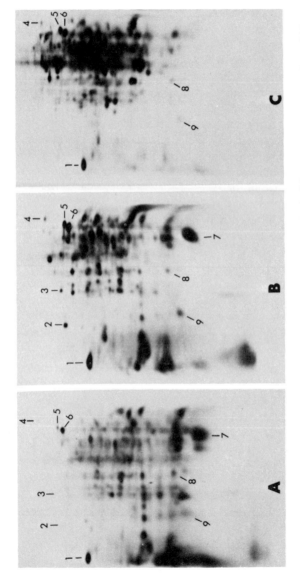

Figure 4.20. Separation by two-dimensional electrophoresis of proteins synthesized in an *in vitro* system using mRNAs extracted from (A) dry wheat embryos, (B) 0-to-40-min-imbibed embryos, and (C) 17-to-18-h-germinated embryos. Some proteins have been given numbers to aid identification of their presence or absence. From Cuming and Lane (1979).

seedling growth. Other studies have now shown that in wheat embryos qualitative changes in their mRNA complement occur 5–6 h from the start of imbibition, or even earlier.

It is evident from the pattern of *in vitro* translation products in Fig. 4.20, A that dry wheat embryos contain a variety of stored mRNAs, and this has been confirmed for other species too, e.g., rye and cotton embryos, pea axes, radish, and rape seeds. There are probably two classes of mRNA present within the dry embryo or axis (see also Section 2.4): (1) Residual mRNAs— these are mRNAs produced during seed development and which are not destroyed during late maturation and desiccation. They are not essential for germination and may be degraded early after imbibition: The early ribosomal protein messages in wheat might fall into this category. (2) Stored or conserved mRNAs—these are laid down during development so that they are available immediately upon hydration and hence can quickly be translated into proteins that are an integral part of germination. These may be subcategorized into mRNAs for (a) enzymes essential for intermediary metabolism, i.e., for essential cellular processes not necessarily unique to germination, and (b) proteins essential for the successful completion of the germination process *per se,* culminating in the elongation of the radicle.

But how, then, do mRNAs come to reside within the dry seed, and how do the cells of the imbibed seed discriminate between mRNAs to be utilized in germination and those to be destroyed? The answers to these questions are by no means clear. One possible explanation is that mRNAs to be conserved in the dry seed are protected by being stored in the nucleus during maturation, either in their activated (polyadenylated) state or ready to be activated, although the former appears to be the more likely. These stored mRNAs are then released from the nucleus to the protein-synthesizing sites within the cytoplasm when required. There is some evidence that in cotton embryos and rape seeds the subcellular location of polyadenylated messages is within the nucleus. On the other hand, the accumulation of mRNAs in cytoplasmic informosomes [called messenger ribonucleoproteins (mRNPs), which describes more accurately the status of these mRNAs as being associated with proteins and hence being surrounded and protected by them] during development of wheat and rye embryos has been advocated. For example, in dry rye embryos most of the mRNA is in the form of cytoplasmic mRNPs ranging in size from 25–104 S. When the protein moiety is removed, the mRNAs are revealed to range in size from about 7–35 S, with the majority being between 15 S and 25 S. In dry wheat embryos, the mRNPs and mRNAs may be considerably smaller.

To summarize this section, then, there can be little doubt that most of the essential components of the protein-synthesizing complex are conserved in the dry seed including the vital templates for synthesis, the mRNAs. These mRNAs are in a protected form in the dry seed, perhaps in the nucleus or in the cytoplasm associated with proteins as mRNP particles. Unprotected messages are presumably hydrolyzed during late maturation or early germination. Protein synthesis begins within minutes of hydration of the seed, and some of the conserved mRNA is utilized to direct this. There is an excess of conserved mRNA in relation to the early requirements of the cell, but the mRNAs that are not translated might not be qualitatively different from those that are. Within a few hours there is synthesis of new mRNAs, and as the conserved ones are degraded, protein synthesis, leading to the completion of germination, probably becomes increasingly dependent on them. Some of the newly synthesized mRNAs may code for the same proteins as the conserved messages, whereas others are for distinctly different products, perhaps proteins that are essential for the commencement of radicle elongation. Such proteins remain to be identified, however. Before moving on to the next section it is pertinent to caution the reader that the published literature on protein synthesis during germination is far from conclusive, and that there are many apparent contradictions to the generalized scheme of events that has been presented here. What we have presented is a sequence of events that is most compatible with current ideas: a perusal of more detailed reviews than this will reveal some of the contentious issues.

4.4.3. Ribosomal and Transfer RNA Synthesis

Synthesis of rRNA in imbibing wheat embryos has been detected as early as that of mRNA and within the first 1–2 h of imbibition. The new rRNA becomes incorporated into ribosomes soon after its synthesis. Maturation of rRNA in rye embryos also occurs quickly after imbibition, with the following sequence of events taking place:

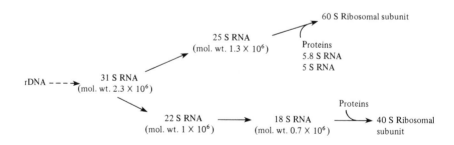

After 40 min from the start of imbibition the 31 S rRNA precursor is evident, and by 60 min it has started to mature into the 25 S and 18 S rRNA of the larger and smaller subunits, respectively. Between the third and sixth hours from the start of imbibition, processing of rRNA is virtually complete and large amounts of the mature rRNAs are present in the embryo. The large 31 S rRNA precursor may accumulate in the nucleolus (the site of the rRNA genes) prior to its migration into the cytoplasm: such appears to be the case in pea embryos. In rye, it is not known when the newly synthesized rRNAs become associated with ribosomes actively participating in protein synthesis. On the basis of studies on other embryos and axes, however, there is little reason to expect that its incorporation into ribosomes is delayed. As yet, it is not known in any seed when and how the low-molecular-weight RNA components (5.8 S and 5 S) of the large ribosomal subunit are synthesized.

Amplification of rRNA genes occurs during certain developmental processes in animals, but current evidence is against such an event occurring during seed germination and seedling establishment.

Transfer RNAs and their aminoacylating enzymes are present in dry seeds, and it has been surmised that there is sufficient of each present to support protein synthesis during germination. Even so, tRNA synthesis commences soon after imbibition, e.g., within 20 min in rye embryos and within 1 h in dissected axes of *Phaseolus vulgaris*. It is also possible that a proportion of the tRNAs present in dry seeds are defective at the −CCA end (to which the amino acid becomes attached), and these molecules must be repaired before they can be used in protein synthesis. The enzyme responsible for the repair of tRNA, nucleotidyltransferase, is present in the dry axes of lupin *(Lupinus luteus)* and increases in activity after imbibition. In wheat, however, the low activity present in the dry embryo does not increase until germination is completed.

4.4.4. Enzymes and Precursors of RNA Synthesis

Synthesis of RNA requires the involvement of three RNA polymerases: (1) RNA polymerase I, a nuclear polymerase that transcribes genes coding for 18 S and 25 S rRNA; (2) RNA polymerase II, localized within the nucleoplasm and responsible for transcription of mRNA; and (3) RNA polymerase III, found within the nucleoplasm (and possibly the cytoplasm), which transcribes genes coding for tRNA and 5 S RNA. Very little is known about these polymerases in seeds. Dry wheat germ, rye embryos, and soybean axes have been shown to contain DNA-dependent RNA polymerases, and presumably these enzymes are present in sufficient quantities to catalyze RNA synthesis as soon as is necessary upon imbibition. Polymerase activity may increase during germination, but the largest increase probably occurs after germination is completed. This, at least, appears to be the case with RNA polymerases in soybean axes (Fig. 4.21).

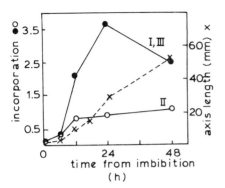

Figure 4.21. RNA polymerase I–III activities (expressed in nmol ³H-UMP incorporated/mg DNA) isolated from nuclei of soybean axes during germination and growth (---X---). After Guilfoyle and Malcolm (1980).

The initiation and continuation of RNA synthesis are dependent on a ready supply of ribonucleoside triphosphate precursors. ATP, CTP, GTP, and UTP are present in very low quantities in the dry wheat embryo (Table 4.3), and all four increase rapidly during the early stages of germination. The rise in the pyrimidine nucleoside triphosphates (UTP and CTP) is slower, however, although they increase substantially by 3 h and 5.5 h after the start of imbibition. Hence the rate of RNA synthesis at these later times might be controlled by pyrimidine nucleoside triphosphate levels alone. But, in truth, the nature of the quantitative and qualitative control mechanisms governing RNA synthesis in germinating seeds is unknown. Many levels of control are possible, e.g., chromatin template availability, polymerase activity, availability of substrates, processing of RNA (particularly mRNA and rRNA), but at the moment there is no substantive evidence for the operation of any of them during germination.

4.4.5. Protein and RNA Synthesis in Storage Tissues

A major event in any storage tissue is the mobilization of stored reserves, which requires the participation of many enzymes, a substantial number of

Table 4.3. Levels of Ribonucleoside Triphosphates Present in Wheat Embryos during Germination[a]

Time	ATP	GTP	UTP	CTP
0	8	46	1	1
40 min	1288	320	132	84
3 h	2000	304	420	200
5.5 h	2072	425	540	320

[a]Values in pmol.mg embryo⁻¹. After Cheung and Suhadolnik (1978).

which are synthesized *de novo*. The synthesis of specific hydrolytic enzymes will not be dealt with here, but in later chapters (7 and 8). Instead, we will concern ourselves briefly with the establishment of the protein-synthesizing complex in storage tissues during and after imbibition. Since the cells of the starchy endosperm of cereals are dead at maturity, they are incapable of synthetic processes and hence will receive no attention here; changes in the aleurone layer are considered in Chapters 7 and 8.

Total RNA levels in imbibed storage tissues may increase over several days before eventually declining, e.g., in peanut and cucumber cotyledons and in castor bean endosperms. In other tissues, e.g., pea and broad bean *(Vicia faba)* cotyledons, however, the levels remain constant until there is a decline associated with organ senescence. The RNA content of the megagametophyte of red pine *(Pinus resinosa)* changes little over 2 wk after initial imbibition, even though radicle emergence occurs on day 5, the cotyledons expand on the tenth day, and by the fourteenth day the megagametophyte and seed coat are attached to the end of the expanding cotyledon and are about to be shed.

Not all regions of a storage organ necessarily exhibit the same pattern of RNA metabolism. Nuclear RNA appears to increase in the outer storage tissues of field pea *(Pisum arvense)* cotyledons during the first 5 days after imbibition and then declines to a low level before any changes occur in the nuclei of cells in the inner storage region. Although RNA synthesis has been detected in cotyledons of rape seed as early as 2 h from the start of imbibition, it is confined to localized regions. More regions commence RNA synthesis with time. Hence measurements of RNA synthesis (and also protein synthesis) made on whole storage organs must be regarded as an aggregate of changes occurring at different locations. An increase in RNA synthesis in whole organs with time from imbibition might, therefore, reflect an increase in the number of cells commencing synthesis (at least initially) rather than an increased synthetic capacity of individual cells.

Mature dry storage tissues contain ribosomes that can catalyze protein synthesis in an *in vitro* system, and it is safe to assume that other components of the protein-synthesizing complex are present also. Certainly dry cotyledons of several species have been found to contain all of the tRNA species and aminoacyl tRNA synthetases necessary for protein synthesis to commence. As in the axis, protein synthesis commences quickly upon imbibition—indicated by rapid polysome formation and *in vivo* incorporation of radioactive amino acids.

The presence of mRNA in dry cotyledons has been a topic of interest and controversy for some time. This was considered in some detail in Chapter 2. Suffice it to say here that mRNAs are present in the dry storage tissues of seeds, and that they may become involved in, and may be essential for, early protein synthesis in much the same way as in axes (Section 4.4.2). More doubtful are the claims that mRNAs for specific hydrolytic enzymes are conserved in the dry state (Chapter 2). There is ample evidence, on the other hand, for the production of new messages for enzymes involved in the mobilization of stored reserves and their conversion to other useful catabolites (Chapter 7).

4.5. DNA SYNTHESIS AND CELL DIVISION

Expansion of the radicle within the seed occurs initially by cell elongation, and its subsequent emergence through the seed coat may or may not be accompanied by cell division. Hence, DNA synthesis and cell division are largely postgermination phenomena, concerned with axis growth and establishment of the seedling.

The relationship of the status of DNA to the cell cycle is illustrated in Fig. 4.22. After mitosis (M), when chromosomes are in their two-stranded (2C) configuration, there is a period of normal cell growth (Gap 1 or G_1) during which synthetic events, including those for subsequent DNA synthesis, take place. DNA synthesis (S) results in a doubling of the chromosomes to a four-stranded (4C) configuration without cell division, and a second preparative growth period (G_2) occurs before mitosis. The reader will appreciate that if the G_2 phase is long, then so will be the lag between DNA synthesis and cell division; some cells, in fact, synthesize DNA but never divide, whereas in others there may be a pause of many hours between synthesis and division.

If all the nuclei of a dry embryo contained chromosomes in the 4C state, then DNA synthesis would not be a prerequisite for the first cell division. To date, however, there has been no recorded instance of all nuclei being exclusively 4C, although some species contain both 2C and 4C and others only 2C (Table 4.4). Within a particular species the proportions of 2C and 4C nuclei

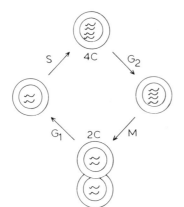

Figure 4.22. Simplified diagram to illustrate the states of DNA during the cell cycle. For an explanation of terms, see text.

Table 4.4. Some Species of Seeds Which in the Mature
Dry State Have Embryo or Radicle Nuclei in the 2C, or 2C
and 4C, Stage of the Cell Cycle.

2C only	2C and 4C
Allium cepa	Crepis capillaris
Dactylis glomerata	Hordeum vulgare
Festuca arundinacea	Pisum sativum
Lactuca sativa	Triticum durum
Pinus pinea	Vicia faba
Tradescantia paludosa	Zea mays

may vary between parts of the seed (Table 4.5). The significance of nuclei in the dry seed being at different stages of the cell cycle is unclear. But in those species whose nuclei are exclusively in the G_1 phase there must presumably be some stringent control of nuclear events to ensure that all nuclei enter the final developmental and drying phases of maturation at the same stages of their cell cycle.

During mitosis in the radicles of germinated onion *(Allium cepa)* seeds the nuclei first enlarge from a diameter of about 7 μ to more than double this size. DNA synthesis commences prior to mitotic cell division, for the former occurs when the root length is 1.4 mm (and the nuclei are 15 μ in diameter), but the latter is not evident until the roots are at least 2.8 mm long (Fig. 4.23). Between DNA synthesis and mitosis (S to G_2 phase) there is an interlude of approximately 9 h. Similar observations have been made on other seeds, although the onset of DNA synthesis in different tissues of a germinated seed can be quite variable; e.g., in *Zea mays* embryos DNA synthesis can be detected first in the root cap, coleorhiza, and scutellar node, then in the root, and finally in the mesocotyl, shoot apex, and leaves.

Little is known about the regulation of DNA synthesis and why it is

Table 4.5. Percentage of Cells in the Roots and Shoots
(Leaf Region) of Mature Dry Embryos of *Zea mays*
Containing Different Amounts of Nuclear DNA in the 2C,
2C–4C, and 4C Stages of the Cell Cycle[a]

Cell stage	Root	Shoot
2C (G_1)	43	70
2C–4C (S)	5	9
4C (G_2)	46	14

[a]Based on data of Conger and Carabia (1976).

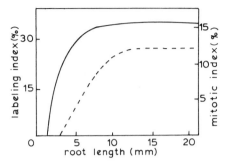

Figure 4.23. Labeling index of radioactive thymidine incorporation into DNA (——) and the mitotic index (----) as a function of root length following emergence from quiescent *Allium cepa* seeds. After Bryant (1969).

delayed until the terminal stages of germination and/or early stages of growth. The possibility has been considered that the availability of the precursor deoxyribonucleoside triphosphates plays a role (for they are not present in dry seeds), or the availability of enzymes that process them into DNA, or both. Thymidine must first be converted to its triphosphate (TTP) before incorporation as the deoxyribonucleotide (dTTP) into the DNA molecule. Two enzymes must be present for thymidine to be utilized: (1) thymidine kinase for phosphorylation of thymidine, and (2) DNA polymerase for insertion of deoxyribonucleotides onto the DNA template. Thymidine kinase activity is very low in dry wheat embryos but is markedly active by 36 h after the start of imbibition—the time course of its increasing activity is unknown. The enzyme consists of two components (P and T), and both are required for its activity. In corn *(Zea mays)* embryos, the T component (but not the P) is absent during the first 36 h from imbibition and then increases in activity. It is possible, then, that thymidine kinase (and perhaps also the kinases for the other deoxyribonucleotides) limit DNA synthesis during the early stages after imbibition by restricting the utilization of the endogenous pool of precursors. But DNA polymerase levels could also be a controlling factor, for activity of this enzyme is generally low in dry seeds and increases up to the time when germination is completed. In other seeds (e.g., pea), however, polymerase activity does not appear to be limiting at any time prior to the actual onset of DNA synthesis. A third enzyme, ribonucleoside reductase (which converts ribonucleotides to deoxyribonucleotides), increases in activity in *Vicia faba* seeds coincidentally with increased DNA synthesis; in wheat embryos, however, ribonucleotide reduction lags well behind. At the present time, then, it is not possible to invoke any one enzyme or any one event as the limiting or controlling step in DNA synthesis—it may vary from species to species or, simply, there may not be just one point of control. For successful synthesis of DNA a large number of other factors and enzymes are required, e.g., unwinding proteins, gyrases, RNA polymerases, DNases, and primases, and studies on some of these have yet to be initiated in seeds.

In dicot storage tissues, which do not undergo cell division, DNA levels may or may not change after imbibition. In some, e.g., field pea *(Pisum arvense)*, DNA levels in the cotyledon remain constant during and after germination of the axis and then decline as senescence sets in. In others, e.g., peanut cotyledons, DNA levels may double up to the seventh to tenth day after imbibition and then decline again over several days. Part of the increase might be due to synthesis of mitochondrial or plastid DNA, although this probably does not account for more than 1–2% of the total. Thus most DNA synthesis is likely to be nuclear in origin: it could be due in part to amplification of genes for enzymes involved in reserve degradation, but there is no evidence at all to support this possibility.

During aging of dry seeds maintained in unfavorable storage conditions, or even for a long time in favorable conditions, DNA may develop lesions, which must be repaired during the early stages of germination if viability is to be maintained. This aspect of DNA synthesis is discussed in Chapter 3.

USEFUL LITERATURE REFERENCES

SECTION 4.1

Chen, S. C. C., and Varner, J. E., 1970, *Plant Physiol.* **46**:108–112 (respiration in wild oats).
Collins, D. M., and Wilson, A. T., 1975, *J. Exp. Bot.* **26**:737–740 (keto acid synthesis).
Duperon, R., 1955, *C. R. Acad. Sci. (Paris)* **241**:1817–1819 (oligosaccharide content of seeds).
Gould, S. E. B., and Rees, D. A., 1964, *J. Sci. Food Agr.* **16**:702–709 (oligosaccharides and respiration).
Kollöffel, C., 1967, *Acta Bot. Neerl.* **16**:111–122 (pea seed respiration).
Pazur, J. H., Shadaksharaswamy, M., and Meidell, G. E., 1962, *Arch. Biochem. Biophys.* **99**:78–85 (oligosaccharides in soybeans).

SECTION 4.2

Day, D. A., Arron, G. P., and Laties, G. G., 1980, in: *The Biochemistry of Plants,* Volume 2 (D. D. Davies, ed.), Academic Press, New York, pp. 197–241 (cyanide-sensitive and -resistant respiration: review).
Hourmant, A., and Pradet, A., 1981, *Plant Physiol.* **68**:631-635 (early oxidative phosphorylation).
James, T. W., and Spencer, M. S., 1979, *Plant Physiol.* **64**:431–434 (cyanide-insensitive respiration, peas).
Malhotra, S. S., Solomos, T., and Spencer, M., 1973, *Planta* **114**:169–184 (inhibitors and mitochondrial development).
Morohashi, Y., and Bewley, J. D., 1980, *Plant Physiol.* **66**:70–73 (pea mitochondria development).
Morohashi, Y., Bewley, J. D., and Yeung, E. C., 1981a, *J. Exp. Bot.* **32**:605–613 (peanut mitochondria development).

Morohashi, Y., Bewley, J. D., and Yeung, E. C., 1981b, *Plant Physiol.* **68**:318–323 (peanut mitochondria development).

Nakayama, N., Iwatsuki, N., and Asahi, T., 1978, *Plant Cell Physiol.* **19**:51–60 (mitochondrial degeneration and senescence).

Nakayama, N., Sugimoto, I., and Asahi, T., 1980, *Plant Physiol.* **65**:229–233 (integrity of mitochondrial inner membrane).

Nawa, Y., and Asahi, T., 1973, *Plant Physiol.* **51**:833–838 (pea mitochondria development).

Obendorf, R. L., and Marcus, A., 1974, *Plant Physiol.* **53**:779–781 (ATP synthesis in wheat).

Perl, M., 1980/81, *Isr. J. Bot.* **29**:307–311 (cytoplasmic ATP synthesis).

Pradet, A., Narayanan, A., and Vermeersch, J., 1968, *Bull. Soc. Franc. Physiol. Veget.* **14**:107–114 (adenosine phosphate levels).

Purvis, A. C., and Fites, R. C., 1979, *Bot. Gaz.* **140**:121–126 (glycolysis and PPP).

Rasi-Caldogno, F., and De Michelis, M. I., 1978, *Plant Physiol.* **61**:85–88 (AEC and metabolism).

Siedow, J. N., and Girvin, M. E., 1980, *Plant Physiol.* **65**:669–674 (alternative respiratory pathway).

Wilson, S. B., and Bonner, W. D., Jr., 1971, *Plant Physiol.* **48**:340–344 (respiration in peanut axes).

SECTION 4.3

Bertani, A., Brambilla, I., and Menegus, F., 1980, *J. Exp. Bot.* **31**:325–331 (anaerobiosis in rice seedlings).

Kollöffel, C., 1968, *Acta Bot. Neerl.* **17**:70–77 (ADH in peas).

Mocquot, B., Prat, C., Mouches, C., and Pradet, A., 1981, *Plant Physiol.* **68**:636–640 (anoxia and AEC in rice).

Pradet, A., and Prat, C., 1976, *Etudes de Biologie Végétale,* R. Jacques, Paris, pp. 561–574 (anoxia and AEC in rice).

Rumpho, M. E., and Kennedy, R. A., 1981, *Plant Physiol.* **68**:165–168 (anaerobic metabolism of oryzicola).

SECTION 4.4

Ajtkhozhin, M. A., Doschanov, Kh. J., and Akhanov, A. J., 1976, *FEBS Lett.* **66**:124–126 (informosomes in wheat embryos).

Anderson, J. W., and Fowden, L., 1970, *Biochem. J.* **119**:691–697 (tRNAs and tRNA synthetases).

Bewley, J. D., 1982, in: *Encyclopaedia of Plant Physiology,* Volume 14A (D. Boulter and B. Parthier, eds.), Springer, Berlin, Heidelberg, pp. 559–591 (protein and RNA synthesis in seeds: review).

Brooker, J. D., Tomaszewski, M., and Marcus., A., 1978, *Plant Physiol.* **61**:145–149 (wheat embryo mRNA).

Cheung, C. P., and Suhadolnik, R. J., 1978, *Nature* **271**:357–358 (ribonucleotide triphosphates in wheat).

Cuming, A. C., and Lane, B. G., 1979, *Eur. J. Biochem.* **99**:217–224 (mRNA changes during germination).

Delseny, M., Aspart, L., and Guitton, Y., 1977, *Planta* **135**:125–128 (loss of conserved mRNA during germination).

Delseny, M., Aspart, L., and Cooke, R., 1980/81, *Isr. J. Bot.* **29**:246–258 (RNA in radish: review).

Dziegielewski, T., Kedzierski, W., and Pawelkiewicz, J., 1979, *Biochim. Biophys. Acta* **564**:37–42 (tRNA repair).

Giesen, M., Roman, R., Seal, S. N., and Marcus, A., 1976, *J. Biol. Chem.* **251**:6075–6081 (initiation complex formation).

Gordon, M. E., and Payne, P. I., 1976, *Planta* **130**:269–273 (long-lived mRNAs).

Guilfoyle, T. J., and Jendrisak, J. J., 1978, *Biochemistry* **17**:1860–1866 (RNA polymerases).

Guilfoyle, T. J., and Malcolm, S., 1980, *Dev. Biol.* **78**:113–125 (RNA polymerases in soybean).

Hammett, J. R., and Katterman, F. R., 1975, *Biochemistry* **14**:4375–4379 (stored mRNAs in nucleus).

Huang, B.-F., Rodaway, S. J., Wood, A., and Marcus, A., 1980, *Plant Physiol.* **65**:1155–1159 (RNA synthesis in wheat and soybean).

Ingle, J., and Sinclair, J., 1972, *Nature* **235**:30–32 (rRNA gene amplification).

Marcus, A., Feeley, J., and Volcani, T., 1966, *Plant Physiol.* **41**:1167–1172 (early polysome formation in wheat).

Payne, P. I., Dobrzanska, M., Barlow, P. W., and Gordon, M. E., 1978, *J. Exp. Bot.* **29**:73–88 (conserved mRNAs: review).

Peumans, W. J., and Carlier, A. R., 1977, *Planta* **136**:195–201 (stored mRNP particles).

Sen, S., Payne, P. I., and Osborne, D. J., 1975, *Biochem. J.* **148**:381–387 (early RNA synthesis in rye).

Spiegel, S., and Marcus, A., 1975, *Nature* **256**:228–230 (protein synthesis without mRNA synthesis).

Spiegel, S., Obendorf, R. L., and Marcus, A., 1975, *Plant Physiol.* **56**:502–507 (early mRNA and rRNA synthesis in wheat).

Walbot, V., 1972, *Planta* **108**:161–171 (tRNA synthesis in *Phaseolus*).

SECTION 4.5

Bryant, T. R., 1969, *Caryologia* **22**:127–137 (DNA synthesis and mitosis in onion).

Conger, B. V., and Carabia, J. V., 1976, *Env. Exp. Bot.* **16**:171–175 (2C and 4C nuclear complements).

Hovemann, B., and Follmann, H., 1979, *Biochim. Biophys. Acta* **561**:42–52 (ribonucleoside reductase).

Mory, Y. Y., Chen, D., and Sarid, S., 1975, *Plant Physiol.* **55**:437–442 (DNA polymerase activity).

Wanka, F., Vasil, I. K., and Stern, H., 1964, *Biochim. Biophys. Acta* **85**:50–59 (thymidine kinase activity).

Chapter 5

Dormancy and the Control of Germination

5.1. INTRODUCTION

Whether or not a viable seed germinates and the time at which it does so depend on a number of factors, including those present in the seed's environment. First, the chemical environment must be right. Water must be available, oxygen may have to be present since the seed must respire, and noxious or inhibitory chemicals should be absent. The physical environment, too, must be favorable. The temperature must be suitable and so also, in many cases, must the light quality and quantity. But in many instances all these conditions may be satisfied and nevertheless the seed fails to germinate. The reason for this, as we have indicated in Chapter 1, is that there exists within the seed (or dispersal unit) itself some block(s) that must be removed or overcome before the germination process can proceed: such a seed is said to be *dormant*. To be released from dormancy, a seed must experience certain environmental factors or must undergo certain metabolic changes. Hence, the control of germination exists at two levels. One—dormancy—is due entirely to the state of the seed itself, and the second involves the operation of environmental factors on both dormancy and germination. We call these the *internal* and *external* controls, respectively. The relationship between dormancy and germination and the points at which control exists is shown in Figure 5.1.

5.2. INTERNAL CONTROLS

5.2.1. Dormancy—Its Biological Role

Since the function of a seed is to establish a new plant, it seems curious that dormancy—an intrinsic block to germination—should exist! But it may not necessarily be advantageous to a seed to germinate freely, and we shall see that dormancy, in fact, offers considerable benefits. One is that it is a means whereby distribution of germination in time can be achieved. This occurs in three ways.

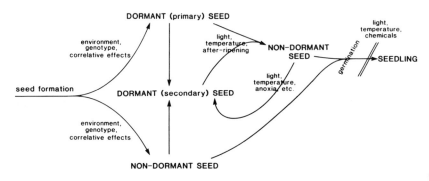

Figure 5.1. Dormancy and germination.

First, seeds are dispersed from the mother plant with different degrees of dormancy, a phenomenon known as polymorphism, heteromorphy, or heteroblasty. Frequently, the variation is reflected in the appearance of the seeds or dispersal units—color, size, and thickness of coat. In *Bidens bipinnata,* for example, the more dormant, outer cypselas of the inflorescence (a capitulum) are short, brown, and wrinkled, whereas the less dormant, inner ones are longer, black, and smooth. The green seeds produced by *Salsola volkensii* have virtually no dormancy, but the nongreen seeds have dormancy. And in *Chenopodium album* four types of seed can be found—brown or black ones each with reticulate or smooth coats: of these, the smooth, black seeds have the deepest dormancy. Differences such as these, in both appearance and dormancy, are displayed by seeds from the same mother plant (e.g., *Bidens pinnata*) or from different plants (e.g., *Chenopodium album*). In the former case, correlative effects (Section 5.4.3) operate to produce the variation, and in the second, both environmental and genetic causes can be traced. When there are polymorphic seeds, germination is spread temporally, with new seedlings emerging at irregular intervals thus reducing competition and increasing the likelihood that some individuals will survive. Such a temporal distribution clearly can have advantages with regard to the continuation and spread of the species.

Dormancy also gives distribution of germination in time through the dependence of dormancy breakage on some environmental factor which itself has a time distribution. For example, seeds are commonly released from dormancy by being chilled, sometimes for several weeks or months at temperatures of 1–5°C. Since such temperatures are found only during the winter, seeds that rely on such a means of dormancy breakage must await the passage of this cold season before they can germinate. The advantage of this strategy is that the young seedling emerges in the spring and establishes itself over the favorable succeeding months; emergence before winter would entail the risk of succumbing to the inclement conditions of that season. Seeds of many species enter a state of dormancy, called secondary dormancy, when they experience conditions unfavorable for germination such as relatively high or low temperatures, but in the succeeding weeks or months they slowly emerge from dormancy. Seeds forced into secondary dormancy by low temperatures (winter) end their dormancy at a different time of year from those made dormant by higher temperatures (summer), and thus the seedlings of the two types emerge at different seasons (see also Section 6.4).

Seed dormancy can also lead to a distribution of germination in space—another aspect of its biological importance. In many cases, dormancy is terminated by light, the effective wavelengths being in the red region of the spectrum. Those seeds which utilize this mechanism therefore do not germinate in light transmitted through green leaves, since it is poor in the red component, but do so only when unrestricted by leaf canopies. Hence, seedling emergence occurs in open, competition-free situations (Section 6.3). Moreover, germination of such kinds of seeds does not occur at depths in the soil to which suitable light cannot penetrate. This is important for small seeds whose food reserves can support only a very limited amount of seedling growth below ground (Section 6.2).

5.2.2. Categories of Dormancy

Dormancy is fundamentally the inability of the embryo to germinate because of some inherent inadequacy, but in many cases it is manifest only in the intact seed, and the isolated embryo can germinate normally. Here, the seed is dormant only because the tissues enclosing the embryo, generally referred to loosely as the seed coat (but often including endosperm, pericarp, or extrafloral organs), exert a constraint that the embryo cannot overcome. This type of dormancy is *coat-imposed dormancy.* Experimentally, the coats can often be replaced by other kinds of constraints, such as an osmoticum (to create water stress) or artificial wrappings (e.g., moist filter paper). For example, isolated embryos of lettuce germinate when placed on a substratum moistened with water, but they exhibit all the characteristics of dormant seeds when in contact with 0.3–0.4 M mannitol solution. In contrast to coat-imposed dormancy are those cases where the embryo itself is dormant. Removal of the coat does not permit such embryos to germinate normally, and so the block to germination is, in a sense, more profound than in seeds with coat-imposed dormancy. This category of dormancy—*embryo dormancy*—is common in woody species, especially in the Rosaceae, but is sometimes found in herbaceous plants such as some grasses (e.g., *Avena fatua,* wild oats). Examples of the two categories of dormancy are given in Table 5.1. Both types of dormancy exist simultaneously or successively in some species. In apple seeds, for example,

Table 5.1. Some Species Having Coat-Imposed and/or Embryo Dormancy

Coat-imposed dormancy[a]	Embryo dormancy
Avena fatua—some strains (palea, lemma, pericarp)	*Acer saccharum*
Hordeum spp. (palea, mainly pericarp)	*Avena fatua*—some strains
Betula pubescens (pericarp)	*Corylus avellana*
Peltandra virginica (pericarp)	*Fraxinus americana*
Acer pseudoplatanus (pericarp, testa)	*Hordeum* spp.
	Prunus persica
Phaseolus lunatus (testa)	*Pyrus communis*
Sinapis arvensis (testa)	*Pyrus malus*
	Sorbus aucuparia
Xanthium pennsylvanicum (testa)	*Syringa reflexa*
Lactuca sativa (endosperm)	*Taxus baccata*
Pyrus malus—some cvs. (endospermal membrane)	
Syringa spp. (endosperm)	

[a]Tissues responsible are in parentheses.

embryo dormancy predominates, but a contribution is made by the covering tissues, the endosperm and testa, and their removal reduces the amount of dormancy-breaking treatment (chilling) that is required. Mature sycamore dispersal units (actually fruits, not seeds) possess only coat-imposed dormancy, yet just before the end of their maturation on the mother plant they have embryo dormancy. And in the grasses *Aristada contorta* and *Bouteloua curtipendula,* dormancy of the seed in the newly dispersed units is so deep that removal of the covering hull has no effect, whereas some months later this treatment promotes their germination. The later dormancy is therefore coat imposed.

Seeds of several species display more complex patterns in which the parts of the embryonic axis differ in the depth of dormancy. In the so-called *epicotyl dormancy* (e.g., in *Paeonia* spp. and *Lilium* spp.), radicle emergence occurs readily but the epicotyl fails to grow. In *Trillium* spp. and *Caulophyllum thalictroides,* the radicle does have some dormancy but it is less deep than that of the epicotyl, and so the two organs differ in the degree of treatment (chilling) needed to break dormancy: such cases are said to exhibit *double dormancy.*

5.2.3. Mechanism of Dormancy

Embryo and coat-imposed dormancies share one common feature. In both, the embryo is unable to overcome the constraints imposed on it, in the former case by factors within the embryo itself and in the second by the enclosing tissues. So when we attempt to understand the mechanism of dormancy, we have to answer two questions, viz: (a) What is the mechanism of action of the constraints? (b) Why cannot the embryos overcome them? Let us begin with the first question in relation to embryo dormancy.

5.2.4. Embryo Dormancy—The Inherent Constraints

Unfortunately, few cases have been examined in detail, but in those that have, two factors appear to be involved: (1) the cotyledons and (2) germination inhibitors.

Amputation of the cotyledons often allows the embryonic axis of the dormant embryo to germinate and grow. In this way, dormancy is partially or completely broken in *Corylus avellana* (hazel) and *Euonymus europeus* (spindle tree) by excising one cotyledon, and in *Fraxinus excelsior* (ash) by cutting off two. Embryo dormancy in barley can be relieved by removal of the scutellum (which is probably a modified cotyledon), and dormancy of apple embryos is progressively reduced as increasing amounts of cotyledonary tissue are cut off (Fig. 5.2). Interesting evidence for the inhibitory role of the cotyledons is additionally provided by those species in which the isolated, dormant embryos succeed in making very sluggish growth, to form a dwarf plant—*physiological dwarfism*. If the cotyledons are excised at an early stage of dwarfism, a normal growth habit is resumed (e.g., in peach). Cotyledon excision also relieves epicotyl dormancy—in *Viburnum trilobum,* for example.

It appears, therefore, that some inhibitory effect emanates from the cotyledons to block the germination of the embryonic axis. This inhibitory influence also reveals itself in other ways, within the cotyledons themselves. Freshly harvested seeds of *Xanthium pennsylvanicum* (cocklebur) have embryo dormancy: if the cotyledons of these embryos are excised, they do not expand or form chlorophyll when placed on a moist substratum in the light, but after the loss of embryo dormancy the isolated cotyledons enlarge and become green. On the other hand, in some dormant embryos (e.g., apple, pear), the cotyledon in contact with the moist substratum does grow and develop chlorophyll in the light,

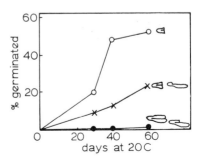

Figure 5.2. Effect of cotyledon removal on embryo dormancy in apple. Portions of cotyledon(s) were removed from isolated dormant embryos. The treated embryos were placed on moist cotton wool. After Thevenot and Côme (1973).

but the uppermost cotyledon, not in direct contact with the wet substratum, remains small and white. The beneficial effects of contact with a moist substratum suggest that the inhibitory influence can be leached from the cotyledon into the surrounding water, and implicit in this inference is that a chemical inhibitor is involved. Support for this possibility comes from observations on the effect of washing dormant embryos of *Taxus baccata* in a liquid medium. Two weeks of this treatment permit the embryos to germinate, though the same period of time spent on a medium solidified with agar (which, one imagines, would be less likely to encourage the leaching of inhibitor) is less effective (Fig. 5.3, A). What might the putative inhibitor be? During the 2 wk in the liquid medium, almost all of the free and bound forms of the inhibitor, ABA (Fig. 5.4), leave the embryo, a coincidence that prompts the suggestion that ABA, when remaining in the embryo, stops it from germinating. Supporting this contention (but not proving it!) is the finding that addition of ABA to the liquid medium very much reduces the efficacy of the washing treatment (Fig. 5.3, B).

It may be, therefore, that the cotyledons of dormant embryos exert their control by virtue of inhibitors within them. However, no conclusive evidence exists that this is the case, although dormant embryos of many species are known to contain inhibitors including ABA (see Table 5.3), much of which may be present in the cotyledons, as in apple. But as we shall see later, the evidence for the role of ABA in the imposition and maintenance of dormancy is equivocal on several grounds, and we are not yet sure what part it plays.

5.2.5. Coat-Imposed Dormancy—The Constraints

The following are possible effects of the tissues enclosing the embryo:

1. Interference with water uptake.
2. Mechanical restraint.
3. Interference with gas exchange.

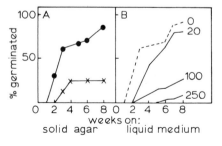

Figure 5.3. The effect of leaching on relief of embryo dormancy in *Taxus baccata*. In (A), embryos were placed on a liquid (●) or solid (X) agar medium, and germination over several weeks at 22°C was followed. In (B), embryos were on a liquid medium supplemented with ABA at the concentrations shown (in μg/l). After Le Page-Degivry (1973).

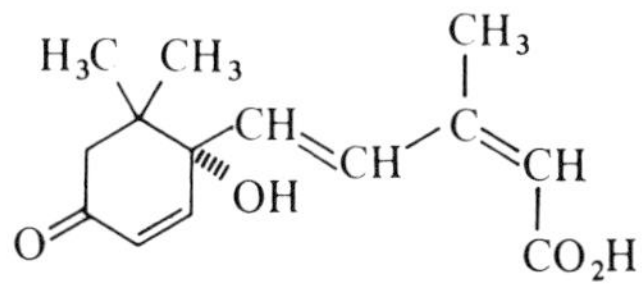

Figure 5.4. S-ABA (2-*cis,* 4-*trans*).

4. Prevention of the exit of inhibitors from the embryo.
5. Supply of inhibitors to the embryos.

Prevention of germination may be due to the action of one or more of these effects.

5.2.5.1. Interference with Water Uptake

This is a common effect especially in seeds of the Leguminosae, but it is also found in other families such as the Cannaceae, Convolvulaceae, Chenopodiaceae, and Malvaceae. Many species have seeds with extremely hard coats which, by preventing the entry of water, delay germination for many years. For example, about 20% of soaked *Robinia pseudoacacia* seeds remain ungerminated for 2 years because insufficient water reaches the embryo, and 1.5% are in this state for 20 years! It could be argued that these are not really cases of dormancy in the strict sense, because the embryos simply do not have one of the basic requirements for germination, i.e., water; nevertheless, hard-coated seeds, in which water entry is the factor limiting germination, are generally considered under the heading of dormancy.

The testa is generally responsible for impeding water uptake. *Melilotus alba* is an example which illustrates the salient anatomical features of this tissue (Fig. 5.5). Waterproofing can be conferred by several parts of the testa— the waxy cuticle, the suberin, and the thick-walled palisade and osteoscereid layers. All of these contribute to some extent, and, indeed, in some species of Leguminosae the waxy cuticle plays a major role. But careful experiments carried out on several species strongly suggest that the main barrier to water uptake is offered by the osteoscereids, because only when these cells are punctured do most seeds begin to imbibe water (Fig. 5.6). We must remember that the testa in the Leguminosae has the hilum, micropyle, and strophiole (Section 1.4.3). In species with impermeable coats, such as *Phaseolus lunatus,* the micropyle is occluded and the fissure in the hilum remains closed except under very low relative humidity. The strophiole stays intact unless the seed experiences certain conditions that cause the plug to be ejected, a point we will consider briefly in a later discussion. Thus, the testa is a very effective waterproofing tissue, which, because of this property, can delay germination of hard-coated seeds for considerable periods of time.

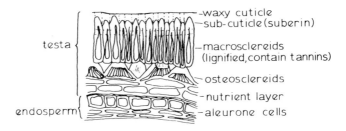

Figure 5.5. A section of the seed coat of *Meliotus alba*. After Hamly (1932).

5.2.5.2. Mechanical Restraint

The coats of many dispersal units are hard, tough tissues which may be expected to offer considerable resistance to the emergence of the embryo. The hard shells of different kinds of nuts are outstanding examples, and obviously if the embryos cannot generate enough force to penetrate these tissues, they

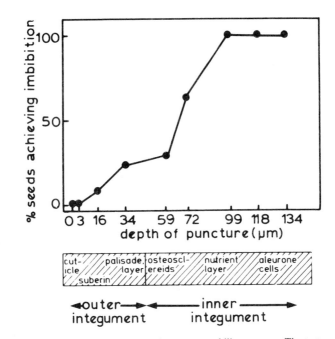

Figure 5.6. The effect of puncturing on seed coat permeability to water. The testae of *Coronilla varia* seeds were punctured to different depths with a fine needle and afterward placed on moist filter paper for imbibition. Adapted from McKee *et al.* (1977).

cannot properly germinate. Various tissues surrounding the embryo in other dispersal units are also extremely resistant. A rather good example is provided by the lettuce "seed" (actually an achenic fruit—a cypsela). Here, the enclosing tissue that is responsible for coat-imposed dormancy is the endosperm; so if the pericarp and testa are both removed, the embryo is still unable to germinate. The embryo is completely enclosed by the endosperm, which is two cells thick, each cell having relatively thick walls composed predominantly of mannans, not cellulose (i.e., a polymer of mannose, not glucose). That the endosperm is a tough, constraining tissue can be revealed by certain treatments with chlorine-generating compounds that do not stop the growth of the embryo but do prevent it from breaking through the endosperm. Because this tissue is so resistant the embryo grows inside it, developing a contorted radicle/hypocotyl that cannot penetrate the relatively thick, tough tissue surrounding it (Fig. 5.7). But nevertheless, in those species where the force needed to puncture the coat has been determined, there seems to be no clear correlation between the degree of dormancy and the resistance of the coat to puncturing; i.e., coats of dormant seeds are no tougher than those of nondormant ones. Of equal importance to the coat resistance, however, is the thrust generated by the embryo; clearly, the relationship between coat resistance and embryo "push" will determine whether or not the embryo can penetrate the enclosing tissues. Experiments on a few species, notably *Syringa* (lilac) and lettuce, in which

Figure 5.7. Resistance of lettuce endosperms. Lettuce seeds were treated with dichloroisocyanurate which did not stop the growth of the embryo but, for an as yet unknown reason, did prevent it from breaking the endosperm (e). This illustrates that the endosperm is a tough tissue that can even constrain a germinated embryo. By courtesy of A. Pavlista.

isolated embryos are placed in an osmoticum which acts as an external constraint, strongly suggest that before a dormancy-breaking treatment the embryo does not have enough growth potential to pierce the coat, whereas after the termination of dormancy it can generate the necessary force. In this connection we should note that embryos taken from nondormant *Xanthium* seeds generate more than double the growth thrust of embryos from dormant seeds. In summary, then, we can say that seed coats certainly can offer considerable mechanical resistance which the embryos in dormant seeds of certain species cannot overcome.

5.2.5.3. Interference with Gas Exchange

The several layers of tissue surrounding the embryo might limit the capacity for gaseous exchange by the embryo in two ways. First, entry of oxygen may be impeded; second, escape of carbon dioxide may be hindered. One important consequence could be the inhibition of respiration.

The possibility that the seed coat imposes dormancy by affecting gaseous exchange gains support from the fact that in many cases the inhibitory action of the tissues surrounding the embryo is much reduced simply by scratching or puncturing them. Thus, a pinprick through the endosperm near the radicle of the lettuce seed or through the pericarp over the embryo of the intact, dormant wheat grain can cause some germination. Numerous cases of such effects are known (Table 5.2), and it is difficult to understand how such moderate "surgical" operations can interfere appreciably with the mechanical resistance of the coat. Another finding that leads us to suspect that germination in the intact dispersal unit is held in check by insufficient oxygen is that dormancy is frequently overcome by oxygen-enriched atmospheres (Table 5.2). To summarize at this point, it can be suggested that the embryo in the intact, dormant dis-

Table 5.2. Some Treatments That Remove Coat-Imposed Dormancy[a]

Species	Tissue removal	Puncturing coat	High oxygen level
Acer pseudoplatanus	+ + + (testa)	+ + +	−
Avena fatua	+ + + (hull)	−	+ +
Betula pubescens	+ + + (pericarp, testa, endosperm)	+	+ +
Hordeum spp.	+ + + (hull)	+ +	+ +
Oryza sativa	+ + + (hull)	+ +	+ +
Phacelia tanacetifolia	+ + + (endosperm)	−	+ +
Triticum spp.	+ + + (pericarp, testa)	+	+ + +
Xanthium pennsylvanicum	+ + + (testa)	−	+ + +

[a] + + +, strong dormancy-relieving effect; + +, moderate effect; +, slight effect; −, no effect.

Figure 5.8. Gas uptake by components of the dispersal units of sugar
beet *(Beta vulgaris)*. Naked seeds
($\triangledown$), opercula (●), testas ($\triangle$), and
pericarps ($\square$) were placed sepa-
rately in Warburg flasks, wetted
(arrow), and the subsequent gas
exchange measured. There is initially a gas evolution (usually
occurring when dry seed parts are
moistened) followed by gas, probably oxygen, uptake. Redrawn from
Coumans *et al.* (1976).

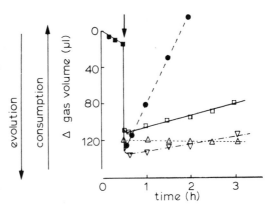

persal unit fails to germinate because of the restrictions by the enclosing tissues, particularly on oxygen uptake. Removal, abrasion, or puncturing of these
tissues gives the embryo access to oxygen and germination can proceed. To
assess the plausibility of this view we must have information relating to the
following three questions: (1) What is the permeability of seed coat tissues to
oxygen? (2) How much oxygen does the embryo need to support its respiration? (3) Does the impermeability of the coat lower the available oxygen to
below the necessary level for germination?

It has been found in several cases that the permeability of the seed coat
to oxygen is less than that of a layer of water of equivalent thickness. In *Sinapis
arvensis,* for example, it is lower by a factor of about 10^4 and in *Xanthium
pennsylvanicum* by about 10^2. The reasons for these differences are not clear
in all instances, but one possibility is that there is resistance to the entry of
oxygen offered by the layer of mucilage around many seeds (*Sinapis,* for example). Another way in which the coats act is by consuming oxygen themselves.
A good example of this is the dispersal unit of sugar beet *(Beta vulgaris)* in
which the ovary cap (the operculum) is the major barrier, for when this is
removed germination can occur. Though dead, the operculum when wetted in
air consumes relatively large quantities of gas, presumably oxygen (Fig. 5.8).
This is probably due to the enzymic oxidation of various chemical constituents,
an occurrence that is known in the testa of apple and other seeds, where various
phenolic compounds (e.g., phloridzin and chlorogenic acid) are implicated. The
dormancy-imposing hull of rice *(Oryza sativa)* also consumes oxygen. As dormancy slowly disappears during dry storage of the grains, so the oxygen-consuming capacity of the hull decreases: this correlation supports the possibility
that the hull imposes dormancy because it deprives the embryo of oxygen. Also
correlated with the level of dormancy in different cultivars of rice is the activity

of peroxidase in the hull—an enzyme that may form part of an oxygen-consuming complex.

The answer to the first of our three questions therefore supports the possibility that the coat prevents germination by limiting the embryo's oxygen supply. But when we come to answer the second question about how much oxygen an embryo needs, the situation becomes less clear. Investigations on several species have revealed that embryos are satisfied by extremely low partial pressures of oxygen. Isolated embryos of *Betula pubescens* (birch), *Sinapis arvensis,* and wheat can germinate even in nitrogen! Furthermore, the oxygen consumption of embryos of both *Sinapis arvensis* and *Xanthium pennsylvanicum* is considerably less than that permitted by the amount of oxygen that can diffuse through the testa, as we can see from the results shown in Figure 5.9. It is not the case, therefore, in answer to our third question, that the respiratory demand for oxygen by the embryo exceeds the possible supply. Taking the answers to our three questions into account, then, we are bound to conclude that coats do not seem to impose dormancy on the enclosed embryo simply by restricting the amount of oxygen available for respiration.

But we are still faced with the findings that both high partial pressures of oxygen and facilitated access to air (by pricking or scratching the seed coats) cause intact, dormant seeds to germinate. If the oxygen is not needed to support

Figure 5.9. Permeability of the coat and oxygen consumption in *Xanthium pennsylvanicum.* The burr of *Xanthium* has two seeds, an upper one and a lower one. Newly ripened seeds of both kinds are dormant (embryo dormancy). A period of afterripening causes loss of embryo dormancy in both kinds, but the upper seed retains a coat-imposed dormancy. In (A), the permeability of the coats and the oxygen uptake of both kinds of newly ripened seed are shown.

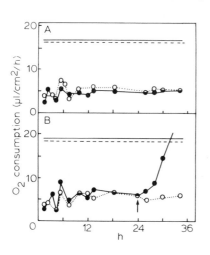

In (B), these features in afterripened seed are plotted. ——, lower seed coat permeability to oxygen; ----, upper seed coat permeability to oxygen; ···o···, upper seed oxygen uptake; —●—, lower seed oxygen uptake. Arrow, radicle emergence from nondormant (afterripened), lower seed. Note the following: (1) Permeabilities of upper and lower seed coats, of both dormant and nondormant seeds [compare (A) and (B)] are identical and higher than that needed to satisfy the oxygen uptake by the whole seed (i.e., presumably by the embryos inside). (2) Oxygen uptake by embryos of nondormant, afterripened seed is identical to that of dormant (coat-imposed) seed, as shown in (B). As expected, however, oxygen uptake increases when the radicle emerges from the nondormant seed. After Porter and Wareing (1974).

respiration, what *is* it for? Inhibitors have been invoked, in some cases, to answer this. In *Xanthium,* for example, growth inhibitors have been found in the embryo which, in the intact seed, are oxidized under high oxygen concentrations but not in air; hence, the effect of exposing intact seeds to oxygen is to remove these inhibitors. The inhibitors (which have not been identified chemically) can also diffuse out of the isolated embryo when it is set on a moist substratum, but they do not cross the seed coat of an intact seed. It is arguable, therefore, that in this case the major beneficial effect of removing the coat is to permit the escape of inhibitors and not to make more oxygen available. There is a twofold effect of the intact coat: primarily, it causes the retention of inhibitor, and secondarily, because it does act as a barrier to oxygen, it prevents the entry of sufficient oxygen from air to support the oxidation of the inhibitor. Inhibitors may be involved in a different way in *Sinapis arvensis.* Here, there is evidence that these substances are actually formed under conditions of low oxygen levels; thus, the effect of the coat is to lower the amount of oxygen reaching the embryo, and this, though it is still adequate to support respiration, encourages the production of inhibitor. Of course, the process occurs in isolated embryos held on moist filter paper under very low oxygen tensions, but in this situation the inhibitor can probably diffuse out. So here, too, there is a dual role for the coat, involving effects on gaseous exchange and on the exit of inhibitor.

Effects of oxygen and the seed coat in *Xanthium* may also be connected with the production and action of ethylene. Ethylene can break dormancy in some species of seed, including *Xanthium,* and can, in fact, be produced by the embryo itself. But relatively high oxygen tensions are needed for the operation of the so-called aerobic ethylene-producing system, a requirement that is not satisfied in the intact seed. Hence, the embryo may not be able to produce enough ethylene to relieve its dormancy. Which of the two sets of interpretation of coat and oxygen effects in *Xanthium* is more plausible (i.e., the "inhibitor hypothesis" or the "ethylene hypothesis") is not easy to judge at this stage.

5.2.5.4. Prevention of Exit of Inhibitors

We have already seen that the testa of *Xanthium* stops the loss of inhibitors from the embryo. Inhibitors of different chemical classes have been found in seeds of many species, some in the embryo and some in the coat itself (Table 5.3). If these inhibitors are retained by the imbibed seed, instead of being lost to the external medium, germination of the embryo may be blocked. We must be clear, though, that the discovery of an inhibitor in a seed does not necessarily mean that it functions in the dormancy mechanism. Important considerations that help us decide whether it does or not are (1) Where is the inhibitor

Table 5.3. Some Seeds Containing Germination Inhibitors

Species	Location	Inhibitor
Acer negundo	Pericarp	ABA
Avena fatua	Not determined	ABA
		Short-chain fatty acids
Beta vulgaris	Pericarp	Phenolic acids
		cis-Cyclohexene-1,2-dicarboximide
		Inorganic ions
Corylus avellana	Testa, Embryo	ABA
		ABA
Eleagnus angustifolia	Pericarp, Testa	Coumarin
	Embryo	Coumarin
		Coumarin
Fraxinus americana	Pericarp	ABA
	Embryo	ABA
Prunus domestica	Embryo	ABA
Rosa canina	Pericarp, Testa	ABA
		ABA
Taxus baccata	Embryo	ABA
Triticum spp.	Pericarp/testa	Catechin tannins

located? Is it available to the embryonic axis, the region where germination proceeds? (2) Even if it is present in the axis itself, is it sequestered in cell compartments or is it freely distributed? (3) Is it present at concentrations likely to be effective in maintaining dormancy? (4) Does it indeed have activity against the seed from which it is extracted or is it simply active in some general bioassay system such as the oat or wheat coleoptile? Unfortunately, in no case do we know the answers to all or even most of these questions. We can nevertheless assess whether the seed coat stops the escape of inhibitor by collecting diffusion products from intact and isolated embryos. The latter approach has been taken in the case of *Xanthium,* where it seems clear that the inhibitor readily diffuses out of the isolated embryo but not from the intact seed. In *Avena fatua* the hull is the part of coat that imposes dormancy, so the naked caryopsis germinates when set on moist filter paper. If only part of the hull is removed from one side, to leave the caryopsis in a boat-shaped portion, germination occurs only if the caryopsis is in contact with the wet paper and not if the unit is placed caryopsis uppermost with the hull in contact. Water uptake is unaffected by the intervening hull, and so it seems that outward movement of substances from the caryopsis might be impeded by the hull. Inhibitors in the caryopsis do not, in fact, leave the intact dispersal unit, but they do move out of a naked caryopsis. To test the effect of stopping outward diffusion in both *Xanthium* and *Avena fatua* the simple device has been used of allowing the isolated embryo or caryopsis to imbibe in a humid atmosphere, not in contact with liquid water: this does not reduce the total water uptake. Dormancy is retained after such treatment, whereas it is lost on moist filter paper. The interpretation is that an inhibitor cannot diffuse out in the humid atmosphere because no liquid is present.

5.2.5.5. Supply of Inhibitors to the Embryo

Examples of coats with inhibitors in them are given in Table 5.3. We should again note that their presence does not mean that they function in dormancy. In some cases, however, the coats do seem to have a powerful inhibitory effect, of a chemical rather than physical origin. Embryos of *Rosa* germinate when placed on a moist substratum, but if pieces of pericarp and testa are also present, many of the embryos fail to germinate; and dormancy is almost complete if each embryo is actually covered by a piece of pericarp. The inhibitor ABA is present in these coat tissues. In several cases, where repeated washing (leaching) of the seeds relieves dormancy, inhibitors are known to be removed.

5.2.6. Coat-Imposed Dormancy—A Summary

We have seen that there are several possible constraints imposed by the structures enclosing the embryo. In several species, e.g., in some of the Leguminosae, interference with water uptake is possibly the only factor involved. But in many species, the picture seems quite complex and it is not unlikely that more than one factor operates simultaneously. For example, it is clear that the coat must offer some mechanical resistance and that to penetrate it the embryo must exert a certain minimum force, or thrust. The ability to do so may be curtailed by the presence of inhibitors either held in the embryo by the impermeability of the coat or supplied to the embryo by the external layers. To overcome the inhibition, adequate oxygen levels must be available—these could be secured by elevating the external oxygen tensions or by puncturing the seed coat. Hence, in this scenario we can see how several of the possible constraints might conceivably act together to maintain the intact seed in a dormant state.

5.3. EMBRYONIC INADEQUACY—THE CAUSES

We now have to consider why the embryo in the intact seed cannot overcome the constraints to which it is subjected—the intrinsic ones and those originating in the coat. One possibility is that the embryo is metabolically deficient in some way. Let us now consider this.

Table 5.4. Gas Exchange in Dormant and Nondormant
Avena fatua[a]

	O_2 uptake (μl/h/ 10 embryos)	CO_2 evolution (μl/ h/10 embryos)
Dormant	12.07 ± 0.55	11.74 ± 0.63
Nondormant	12.16 ± 0.41	11.59 ± 1.14

[a]Determined during the first 10 h after the start of imbibition. After Simmonds and Simpson (1971).

5.3.1. Metabolism of Dormant Seeds

One problem here is that we have not yet been able to identify any particular aspect of metabolism that is peculiar to a germinable seed. Is there, for example, some metabolic pathway that occurs especially in a germinable seed and are there some special "germination enzymes" and "germination biochemical processes"? Since we do not yet know the answer to these questions, we cannot begin to investigate the fine differences between dormant and germinable seeds, and we must therefore deal only with generalized aspects of metabolism.

An approach frequently taken is to compare the metabolism of dormant seeds with that of afterripened, i.e., nondormant, seeds; where possible we will make such comparisons here. To ensure that the comparison is meaningful we must not contrast a dormant seed with one that has germinated, so in this discussion we will deal only with metabolism occurring in the early phases after imbibition, before radicle emergence from the nondormant seed. Beginning with respiration, dormant seeds are capable of carrying out this activity to the same extent as nondormant material. Dormant and afterripened seeds of *Xanthium* and *Avena fatua,* for example, show equivalent oxygen consumptions up to the time of emergence of the radicle from the germinable seed (Table 5.4). In some cases (e.g., barley and rice) dormant grains exhibit even higher oxygen consumption than the nondormant ones, but in these instances part of the consumption could be taking place in extraembryonic tissues. And hand in hand with the equal oxygen utilization by dormant and nondormant seeds, the ATP levels in both are comparable.

Another part of metabolism that may be particularly important for germination is nucleic acid and protein synthesis, for it is plausible that herein might lie the key differences between embryos in dormant and nondormant seeds. As we indicated previously, perhaps dormant seeds cannot make the right enzymes, because of an inability to transcribe certain genes. Again, however, dormant seeds seem to suffer no general deficiency in this respect. They are able to incorporate radioactive amino acids into protein at a rate no less than that of their fully germinable counterparts, and, similarly, RNA synthesis apparently proceeds normally in embryos of dormant seeds, from which active polysome preparations can also be made. A note of warning should be sounded here. The techniques used to study these metabolic processes have only covered overall macromolecular synthesis and have not yet been able to resolve fine differences that might arise, say, from the lack of synthesis of a small number of RNA species and proteins. Techniques are now available to approach these problems, and it is likely that we will see some important information emerging, to enable us to answer the question: are there special germination macromolecules that an embryo in a dormant seed cannot make?

5.3.2. Membranes and Dormancy

The state of cell membranes in relation to dormancy is receiving increasing attention. In some species, a striking feature of seed dormancy is its temperature dependence. In this so-called relative dormancy, the dormant condition is exhibited only at certain temperatures (Fig. 5.10). The noticeable characteristic of this is the abruptness with which the dormancy response to temperature occurs, suggestive of a sudden change in the cells. Events that are

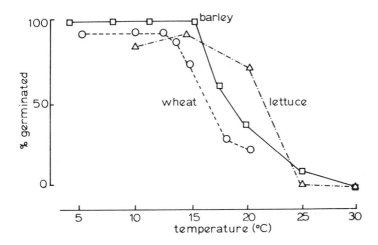

Figure 5.10. Relative seed dormancy in three species. In all three species there is no dormancy in seeds held at temperatures below approximately 15°C; i.e., almost all of the seeds germinate at these low temperatures. Dormancy is expressed as temperatures rise above approximately 15°C and is present in almost all of the barley grains and lettuce (cv. Grand Rapids) seeds at 25°C and in most of the wheat grains at 20°C. Data for barley after Roberts and Smith (1977) and for wheat and lettuce by Black.

known to occur in this abrupt manner in response to temperature are phase transitions of cell membranes; these undergo a sudden change from a crystalline or gel phase at lower temperatures to a fluid, liquid crystalline phase at higher temperatures. Many properties of the membrane (e.g., control of solute passage, activity of bound enzymes) alter when this happens, possibly with profound effects on the physiology of the cells, which could interfere with processes necessary for germination. Evidence that the temperature dependence of dormancy may in reality be connected with membrane changes come from two observations: (1) Leakage of amino acids from certain seeds occurs at the same temperature at which dormancy is expressed (Fig. 5.11); (2) a crude membrane fraction from temperature-sensitive seeds shows phase transitions, when tested by certain biophysical techniques (fluorescent probes), at the same temperature at which dormancy is expressed. Further support for a special role of cell membranes in the dormancy mechanism comes from the ability of certain chemicals to release seeds from dormancy. These chemicals are anesthetics—ethyl ether, chloroform, acetone and ethanol—which are known to act on cells by entering membranes, thus altering relationships between membrane components. It is of great interest that these chemicals, especially ethanol, are remarkably effective in terminating dormancy in seeds of several grass species, e.g., *Panicum* spp. and *Digitaria* spp.; propanol, too, has a similar effect on lettuce seeds. We have only just begun to suspect the particular participation of membranes in dormancy, but the evidence just outlined points to several lines of investigation worthy of further exploration.

The reader will have concluded that we cannot yet identify with certainty any area of metabolism or biochemistry of the dormant seed that accounts for the inadequacy of the embryo and its failure to overcome the constraints imposed on it. There is, however, an approach to the problem that we have not yet covered fully in our discussion: this is to elucidate the biochemical changes in the embryo that occur when dormancy is broken. We have already touched on this very briefly with respect to afterripening, but later we will examine it in more detail in connection with the action of chilling, light, and certain chemicals in the termination of dormancy (Section 5.5.1.9).

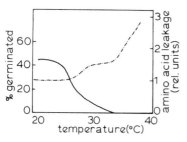

Figure 5.11. Temperature dependence of dormancy and amino acid leakage. In this batch of lettuce seeds (cv. Grand Rapids) there is a sharp drop in the percentage of germinated seeds in darkness (i.e., a rise in dormancy) beginning at 25–26°C (——). At these temperatures, amino acid leakage from the seeds shows a sharp rise (------) suggesting that a change in permeability of the plasma membrane of at least some embryo cells has occurred. Adapted from Hendricks and Taylorson (1979).

5.4. Development of Dormancy

A seed may become dormant at two stages in its life. The first is when it is still on the mother plant: when this happens the seed is dispersed in an already dormant state called *primary dormancy*. The second is generally when an imbibed, nondormant, mature seed encounters conditions that do not favor its germination, such as an unsuitable temperature, inhibitory light, or unfavorable atmosphere: the seed then enters a state of *secondary dormancy*. Sometimes a seed may pass from primary dormancy into secondary dormancy, a transition we can recognize because different factors are required to break the two kinds of dormancy. There is no evidence at present to suggest that the mechanisms of primary and secondary dormancy differ in any fundamental way.

5.4.1. When Does Primary Dormancy Occur?

The stage in the seed's development and maturation at which primary dormancy sets in varies from species to species. In some, such as certain embryo-dormant strains of *Avena fatua,* it is a very early event, detectable 10 days after fertilization, when the embryo is isolated and placed on a liquid medium. Other cereals—wheat is an example—also display at an early age what seems to be a kind of dormancy. When young (e.g., 20 days after fertilization) grains are tested, they are incapable of germination unless the pericarp is removed or pricked, or if they are treated with high oxygen concentrations— typical treatments to break dormancy. Thus, it seems that these grains may have a coat-imposed dormancy at quite an early age. On the other hand, dormancy may be delayed until a seed is almost fully mature. This happens in *Sida spinosa* and *Medicago lupulina* (Fig. 5.12), for example, and in both cases, changes in the seed coat (apparently precipitated by drying in *Sida*) seem to be responsible for initiating the dormant state. Because dormancy in *Medicago* is such a late event, seeds still on the mother plant frequently germinate when the plants fall and the fruiting heads touch the damp earth.

5.4.2. Genetic Control of Dormancy

Although environmental factors and correlative effects are important, the entry into dormancy is also under genetic control.

Pure lines of certain species have been isolated that show contrasting degrees of dormancy. There is a wide range of dormancy encountered in natural populations of wild oats *(Avena fatua)*, from types which have a short-

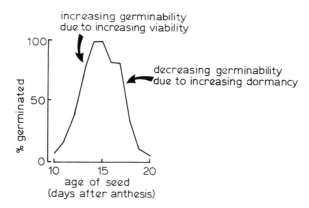

Figure 5.12. Onset of dormancy in *Medicago lupulina*. Germinability of seeds was determined at different ages (i.e., days after anthesis–dehiscence of the anther, which is very closely followed by pollination and fertilization). After approximately 15 days, germinability declines owing to the development of dormancy, which is complete in 20-day-old seeds. Adapted from Sidhu and Cavers (1977).

lived dormancy to those whose dormancy is very prolonged. From these a line with a deep embryo dormancy has been inbred. Pure lines have also been isolated of two groups whose grains respond to temperature in distinct ways, one showing little or no dormancy at temperatures in the range 4–32°C and the second exhibiting dormancy at intermediate temperatures. And in lettuce, different lines of the cultivar Grand Rapids exist, some showing the classical dormancy symptoms and others with no dormancy at all. We should recall, in this context, that dormancy can depend on an interaction between coat and embryo, and therefore three genetically distinct components are involved—the diploid embryo, with maternal and paternal genes, the diploid testa, pericarp, and hull, of maternal constitution only, and the triploid endosperm, bearing two sets of maternal genes. It may well be, then, that the genetics of coat-imposed dormancy are rather complex.

5.4.3. Correlative Effects in Dormancy

Though dormancy must have a genetic basis, there is obviously a great deal of plasticity in genotypic expression, as determined by correlative phenomena within the plant and by the environment. The phenomenon of polymorphism indicates the existence of the former, since genetically identical seeds or dispersal units might differ in their dormancy characteristics, depending on their position on the mother plant. In some species there are clearly influences of the seeds on each other. Spikelets of *Avena ludoviciana,* for example, produce a small distal caryopsis with a fairly deep dormancy and a large, proximal one which is less dormant. But if the latter is removed during grain development, the remaining distal caryopsis is much less dormant at maturity. This suggests that caryopses may compete for factors whose supply can determine the degree of dormancy. These factors might include hormones, but no information is available as to how these are distributed to different seeds in an inflorescence or to different inflorescences. However, the hormonal status of a seed, as established during its development and maturation, can certainly influence the degree of dormancy. For example, if gibberellic acid is fed to plants of *Avena fatua* and *A. ludoviciana* while the grains are developing, the mature grains have no dormancy, though the unfed, control grains are fairly deeply dormant. This artificial elevation of the growth regulator supply to the seeds might mirror the situation that could obtain in nature, where variations in hormonal flux from the mother plant to the seeds could be brought about by changes in maternal nutrition and by environmental factors.

5.4.4. Dormancy Induction by Inhibitors

The growth regulator (or hormone) that has received considerable attention with respect to the onset of dormancy is ABA. This substance is in developing seeds of many species, in many cases even where the mature seed is nondormant, such as the pea. So the existence of the inhibitor does not necessarily lead to the inception of dormancy, though this could be because the abscisic content in many cases falls considerably just before full maturity is reached. It has been claimed that in rice, however, the ABA is retained into grain maturity by those cultivars with dormancy, but not by those which are nondormant when mature. This seems to be an isolated example, however, and we more often find no correlation between the retention of ABA into maturity and the depth of dormancy: the nondormant Great Lakes cultivar of lettuce, for example, contains relatively high levels of ABA! The temporal relationship between entry into primary dormancy and the ABA content has been followed in detail in several species, including peach and apple, and we will use the latter to illustrate the findings (Fig. 5.13). No consistent relationship between the development of embryo dormancy and the ABA level in the embryonic axis is discernible. A rise in free ABA in the cotyledons does occur, however, at about the time when the embryos begin to move into dormancy. It is possible that this increase in ABA is responsible for the commencement of dormancy of the axis (recall the role of the cotyledons in embryo dormancy), but one would have to suggest that ABA is working through an intermediary, since there are no indications that the ABA level in the cotyledons is matched by a rise in the axis.

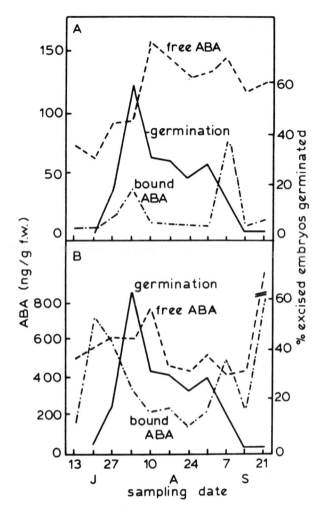

Figure 5.13. Embryo dormancy and levels of ABA in developing apple embryos. ABA (free, and bound to glucose) in the cotyledons (A) and embryonic axes (B) was determined at stages of embryo development during July (J), August (A), and September (S). Germinability of the embryos, isolated from the seeds, was also determined. Note that germinability declines (owing to the onset of dormancy) after August 3. After Balboa-Zavala and Dennis (1977).

We should be aware, also, that an involvement of ABA in dormancy inception does not necessarily require a continuous correlation between inhibitor content and the level of dormancy. It is conceivable, for example, that once the inhibitor has reached a certain threshold level, events are slowly set in train that lead to dormancy. Alternatively, sensitivity to ABA may change and only when the embryo becomes responsive does dormancy begin (this could happen, for example, on August 3 in Fig. 5.13!). Supporting this possibility is the finding, not in seeds but in the aquatic plant *Spirodela,* that induction of dormant buds (turions) by added ABA can occur only during a fairly short time period, when the plant is sensitive to the inhibitor. At present, however, we must keep an open mind as to the role of ABA in dormancy induction in seeds: future investigations will, we hope, throw light on the possibilities mentioned.

5.4.5. The Environment in Dormancy Inception

Although there is uncertainty as to the nature of the internal controls that force a seed into dormancy, there is no doubt that environmental factors play an important part. The effect of these factors explains why dormancy varies with provenance. Certain species used commonly by researchers on seed physiology (e.g., the Grand Rapids cultivar of lettuce) are notorious in this respect, and one cannot be certain that a particular batch will have any dormancy at all: this depends on where it was grown and what environmental factors were operative at the time. What are these factors which can have such a profound effect on the development of dormancy? The edaphic factor is likely to be important, but relatively little information is available about this. Much more is known about the effects of temperature and of light—its quantity, daily distribution, and spectral quality.

A theory of dormancy inception has been put forward by Vegis which attributes potent effects to relatively high temperatures and partial anaerobiosis—the latter enforced by the tissues covering the embryo. As a result, the seed moves in time through predormancy, dormancy, and then after maturity into postdormancy. Each phase is characterized by a response of the seed to temperature: in the first, as the seed approaches dormancy, there is a narrowing of the temperature range over which germination can occur until, in full dormancy, it might not occur at any temperature: finally, the temperature range widens as full dormancy declines during the postdormant phase. Although the theory is attractive in some respects, especially that it draws attention to change in response to temperature as dormancy lessens, parts of it seem to have some inherent difficulties. There are cases, for example, where development of dormancy is promoted not by high temperatures, but by relatively low ones, as in *Rosa* spp. and in grasses such as *Avena fatua* and the cereals, wheat, and barley. It is hard to understand, also, how polymorphism can be accommodated within the theory, or how various daylength effects might operate. There are nevertheless, some cases in which dormancy is induced by the elevated temperatures. *Syringa vulgaris,* for example, can be made to produce dormant seeds by holding the plant at relatively high temperatures (18–24°C) during the last 2 wk of seed maturation, a treatment that appears to make the endosperm (the tissue imposing dormancy) tougher.

Photoperiodic effects in the inception of seed dormancy are well known in several species, and *Chenopodium* spp. provide good examples. *C. album* (Fig. 5.14) has deeply dormant seeds when the fecund plants are held under long days, but nondormant seeds under short days. Not only is the dormancy pattern affected by daylength but also the structure of the seeds, for those maturing in long days are smaller and thicker coated than those in short days. Dormancy induced by photoperiod is not always associated with coat thickness, however, since a short-term dormancy of seeds with thin coats is brought about by long photoperiods given for just a few days after the end of flowering. An effect of photoperiodic conditions on coat structure is also seen in seeds of several other species. When grown in long days (the last 8 days of maturation are most important), seeds of *Ononis sicula* are relatively large with thick, yellow coats, are poorly permeable to water, and show low germinability. After short days, they are, in contrast, smaller, green, and thin coated and, because they readily take up water, have high levels of germination.

Light, quite apart from the photoperiod, has an important role in dormancy induction in several species. Seeds of *Arabidopsis thaliana* maturing in white fluorescent light have little dormancy when harvested, whereas those which have experienced incandescent light remain deeply dormant for at least several months (Fig. 5.15). The explanation for this effect is that white fluorescent light is relatively rich in the red wavelengths whereas incandescent light has a relatively high component of far-red light. Under the former illumination conditions there is a higher concentration of the active Pfr form of phytochrome (Section 5.5.1.4) set up in seeds than there is under the latter type of light. Seeds with a high Pfr content can germinate in darkness while those with a low Pfr concentration remain dormant. Treatment with far-red light alone, of course, has the same effect: cucumber seeds *(Cucumis sativus)* are made dormant when fruits are irradiated with far-red light. It appears that

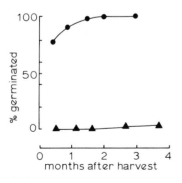

Figure 5.14. Germination of *Chenopodium album* seeds that developed under different photoperiods. *Chenopodium album* plants with developing seeds were held under short days or long days. After harvesting, germination was tested at intervals. Note that seeds developing and maturing under short days (●) have high germination percentages whereas those from long-day treated plants (▲) have a dormancy that lasts for at least 3½ mo. All germination tests were carried out in darkness. Adapted from Karssen (1970).

Figure 5.15. Light quality during seed maturation and subsequent dormancy. Seed-bearing plants of *Arabidopsis thaliana* were kept in white incandescent or white fluorescent light during seed maturation. Seed germination was subsequently tested in darkness. Seeds maturing in white fluorescent light (●) have little dormancy compared with those maturing in incandescent light, which are deeply dormant for more than 100 days. (■). Adapted from Hayes and Klein (1974).

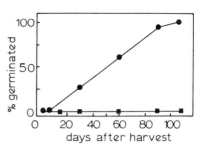

this phenomenon also occurs in nature, where the source of far-red light is light filtered through green tissues (Section 6.3). Chlorophyll absorbs red light (peak ca. 660 nm) but not wavelengths longer than about 710 nm; hence the transmitted light, being rich in the far-red component, serves to lower the amount of Pfr. Seeds of many species mature while the surrounding tissues (fruit or seed coat) are still green, and thus the embryos are in a far-red rich environment: such seeds are dormant, whereas those whose covering tissues have little chlorophyll are nondormant when mature (Fig. 5.16).

Figure 5.16. Extraembryonic chlorophyll content and subsequent seed dormancy. Chlorophyll content of the tissues surrounding the maturing embryos was determined at the midpoint of seed drying. Mature seeds were collected and their germination in the light and dark was tested. Germination in the dark as a percentage of that in the light is shown for each species. Note that all the species that have little or no requirement for light (i.e., no light-sensitive dormancy)—Hc, Ae, Hm, Lc, Sn, Sd—have a low chlorophyll content in the extraembryonic tissues. Of the light-requiring seeds, the majority (Dm, Ma, Dp, Tp, Me, Ss, Cb) have moderate to high extraembryonic chlorophyll. Ae, *Arrhenatherum eliatus;* Ah, *Arabis hirsuta;* Ao, *Anthoxanthum odoratum;* Cb, *Capsella bursa-pastoris;* Dm, *Draba muralis;* Dp, *Digitalis purpurea;* Hc, *Helianthemum chamaecistus;* Hm, *Hordeum murinum;* Lc, *Lotus cornicula-*

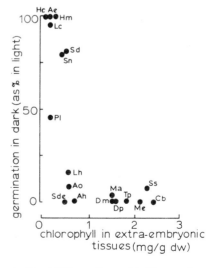

tus; Lh, *Leontodon hispidus;* Ma, *Myosotis arvensis;* Me, *Millium effusum;* Pl, *Plantago lanceolata;* Sd, *Silene dioica;* Sde, *Sieglingia decumbens;* Sn, *Silene nutans;* Ss, *Senecio squalidus;* Tp, *Tragopogon pratensis.* After Cresswell and Grime (1981).

5.4.6. Development of Hard Coats

We have already mentioned that photoperiodic conditions can affect seed coat thickness and color. The hardness of seed coats also depends on the rate and degree of drying, and while drying in air is proceeding, phenolic substances are oxidized to dark-colored compounds that may contribute to coat impermeability. In some leguminous seeds, drying is controlled by the hilum. When the relative humidity is low, the cells bordering the hilar fissure shrink owing to loss of water, thus opening the tissue and allowing even more drying to occur. Under damp conditions, however, the cells expand to close the fissure. Drying of the seed is thus regulated according to the prevailing environmental conditions.

5.4.7. Secondary Dormancy

Secondary dormancy develops in already dispersed, mature seeds in response to unfavorable germination conditions (see also Section 6.7). Some examples are given in Table 5.5. The mechanism of induction is unknown.

Table 5.5. Factors Inducing Secondary Dormancy

Inducing factor	Example
Anaerobic conditions	*Xanthium pennsylvanicum*
Darkness (skotodormancy)	*Lactuca sativa*
	Lamium amplexicaule
	Phleum pratense
Prolonged white light (photodormancy)	*Lactuca sativa*
	Nemophila insignis
	Phacelia tanacetifolia
Prolonged far-red light	*Arabis hirsuta*
	Amaranthus caudatus
	Lactuca sativa
Temperatures above maximum for germination	*Ambrosia trifida*
	Chenopodium bonus-henricus
	Taraxacum megalorhizon
Temperatures below minimum for germination	*Phacelia dubia*
	Taraxacum megalorhizon
	Torilis japonica
Water stress	*Lactuca sativa*

5.5. THE EXTERNAL CONTROLS

We will divide our discussion of the external factors controlling germination into two parts. First, we will see how external factors are involved in the termination of dormancy, and second, we will examine the ways in which germination of a nondormant seed is regulated by the environment.

5.5.1. The Release from Dormancy

In the following account the termination of dormancy by several factors will be considered. For the purposes of this discussion we must treat these factors separately and examine some of the characteristics of their effects. It must be appreciated, however, that in nature, a seed is not subject to the influence of just one factor but to several simultaneously. Most species of seed are affected by more than one factor. For example, dormancy of many light-requiring seeds is also broken in darkness by chilling, by alternating temperatures, by exposure to nitrate in the soil, and by afterripening in the "dry" or near-dry state. The termination of dormancy in the field is not likely to be the perrogative of just one factor in the seed's environment but will be influenced by several, and in some circumstances, seeds of the same species might have their dormancy ended by different agencies. With these qualifications in mind we can now consider the factors that operate in dormancy breakage.

5.5.1.1. Afterripening

Dormant seeds, when "dry," slowly lose their dormancy by the process of afterripening, perhaps requiring as little time as a few weeks (e.g., barley) or as long as 60 months (e.g., *Rumex crispus*). "Dry" seeds, we should note, can have up to 18–20% water content. In ways that are not all understood, dry seeds change, and these changes somehow lead to the termination of dormancy. The efficacy of afterripening depends on the environmental conditions—moisture, temperature, and oxygen. Since afterripening generally occurs in dry seeds, or more accurately in seeds at below a certain water content, it may be prevented in the presence of water. Indeed, at intermediate moisture contents not only might afterripening fail but the seed may also lose viability, and at higher water contents secondary dormancy may set in. It seems, nevertheless, that a minimum water content is required for afterripening, and if seeds become too dry (e.g., 5% water content), the process is delayed. The rate of afterripening depends on the temperature, as is shown in Table 5.6, a property that is sometimes exploited to accelerate the loss of dormancy in agriculturally

Table 5.6. Effect of Temperature on Afterripening Rate in
Rice *(Oryza sativa)*[a]

Temperature (°C)	Time for 50% loss of dormancy (days)
27	50
32	30
37	15
42	8
47	5

[a]Based on data of Roberts (1965).

important species such as barley and wheat. Afterripening is delayed when the oxygen tension is low and accelerated when it is high. But here we mean concentrations of oxygen of near zero to 100%, a range that dry seeds in nature would never encounter, although it might be employed experimentally in the laboratory.

5.5.1.2. Low Temperatures (Chilling)

A high proportion of species—probably the majority of nontropical species—can be released from dormancy when, in the hydrated condition, they experience relatively low temperatures, generally in the range 1–10°C, but in some cases as high as 15°C (See Table 5.7). The operation of this kind of control in nature is obvious: dormancy of the hydrated seed is slowly broken over the winter and, as mentioned before, this is presumably a means of preventing germination until after the winter has passed (see also Section 6.4). Chilling of seeds to break dormancy is a long-standing practice in horticulture and forestry and is generally referred to as stratification, because the seeds are sometimes arranged in layers (i.e., stratified) in moist substrata.

Chilling is effective in seeds with embryo, coat-imposed, primary, relative, and secondary dormancy. We can recognize, in apple for example, how the different components of dormancy are differentially broken by chilling. Figure 5.17 shows that chilling for 60 days suffices to remove the embryo dormancy; the presence of the endosperm increases the required time to about 80 days, whereas the whole seed needs even longer than this. In general, woody species require fairly extensive treatment times, sometimes as much as 180 days *(Crataegus mollis)*, but usually 60–90 days are satisfactory. In contrast, dormancy in some herbaceous species is broken by just a few days of low temperature (7 and 14, respectively, in *Poa annua* and *Delphinium ambiguum*) and only 12 h in wheat!

Recorded optimum temperatures are generally in the region of 5°C, but this figure may be misleading if the situation exemplified by *Rumex obtusifol-*

Table 5.7. Termination of Dormancy by Various Factors

	Factor[a]			
Afterripening	Alternating temperatures	Chilling	Light	Species
+		+		*Acer pseudoplatanus*
+		+		*Ambrosia trifida*
+	+		+	*Agrostis tenuis*
+		+		*Avena fatua*
+		+	+	*Betula pubescens*
+			+	*Chenopodium album*
+		+		*Corylus avellana*
+		+		*Delphinium ambiguum*
+		+		*Hordeum* spp.
+			+	*Kalanchoë blossfeldiana*
+		+	+	*Lactuca sativa* (some cvs.)
	+	+	+	*Lepidium virginicum*
	+		+	*Lythrum salicaria*
	+		+	*Nicotiana tabacum*
		+	+	*Pinus sylvestris*
+		+		*Prunus domestica*
		+	+	*Poa annua*
+		+	(+)?	*Pyrus malus*
+	+	+	+	*Rumex obtusifolius*
+		+	+	*Senecio vulgaris*
+		+		*Triticum aestivum*

[a] +, effective in dormancy breaking

ius is widespread. Provided the seeds are illuminated to prevent the onset of secondary dormancy, they are released from dormancy almost as effectively by 1.5°, 10°, and 15°C, at least with 2-wk treatment times. As the treatment time lengthens, secondary dormancy sets in, the temperatures of 10° and 15°C cease to be as efficacious, and it then appears that 1.5°C is the most suitable temperature (Fig. 5.18). Temperatures as high as 10°C also serve to promote dormancy breakage in lettuce and wheat—a point to which we will return

Figure 5.17. Termination of apple seed dormancy by chilling. Imbibed seeds were kept at 5°C and periodically removed for testing. The percentage germination of the following was determined: (a) isolated embryos; (b) seeds with testa removed but endosperm intact; (c) intact, whole seeds. Note that embryo dormancy is terminated by 40–50 days' chilling, endosperm-imposed dormancy by 70–80 days' chilling, and whole-seed dormancy by more than 80 days' chilling. After Visser (1956).

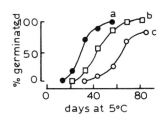

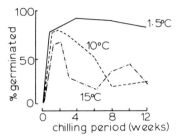

Figure 5.18. Temperature–time relationships for the termination of dormancy in *Rumex obtusifolius*. Imbibed seeds were held in the light at three temperatures for times up to 12 wk. They were then transferred to 25°C for 4 wk, after which the percentage of germinated seeds was determined. After Totterdell and Roberts (1979).

when we consider the chilling mechanism. To return briefly to secondary dormancy and chilling: the onset of this type of dormancy could explain why an interruption of chilling by high temperatures seems to cancel out the previous low-temperature experience and why the seed then has to undergo the full chilling treatment, without interruption, once more; i.e., short periods of chilling are not cumulative. Secondary dormancy might set in during the high-temperature interpolation.

What changes does chilling bring about in the seed to terminate its dormancy? The first problem we have in answering this question is to understand how the low temperature is sensed. Several possibilities exist. First, low temperatures might act simply to lower the rate of enzymic reactions taking place in the seed. It would be expected that all the metabolic processes are retarded by the chilling treatment, but it is not easy to see how this would have a positive effect in removing dormancy unless some of the affected processes are ones stopping the germination mechanism; i.e., they could be inhibitory, dormancy-imposing processes: it is then conceivable that if these are arrested, certain germination steps could therefore proceed, albeit slowly. On the other hand, the low temperature might have differential effects, perhaps because of differences in the activation energies of separate reactions. This would lead to some processes being less affected than others by the chilling treatment, and in this way the germination reactions could be favored. Other effects of low temperature might include differential changes in enzyme concentration or in enzyme production. It must be admitted, however, that there is no convincing experimental evidence that any of these possible sensing mechanisms actually occurs.

One particular feature of the chilling syndrome nevertheless gives us a clue and suggests exciting avenues for exploration. From some cases that have recently been studied in detail, we see that termination of dormancy can be achieved by temperatures as high as 12–15°C, as in wheat and *Rumex obtusifolius*. And in lettuce, where chilling enhances the dormancy-breaking effect of light, temperatures up to about 10°C suffice. When these temperatures are exceeded, there is a fairly abrupt decline in effectiveness, seen very strikingly in *Rumex,* and rather less so in lettuce (Fig. 5.19). If we look at this in another way, it seems that as the temperature is decreased, a value is reached when temperatures sharply become efficacious in chilling, i.e., somewhere between 20°C and 15°C in *Rumex,* about 12°C in wheat, and about 13°C in lettuce. Now we have already mentioned an event in cells that occurs at certain critical temperatures—the phase transition of cell membranes (Section 5.3.2). The characteristics of the temperature dependence for chilling would be consistent with the occurrence of such transitions. One might speculate, therefore, that a membrane could be the sensor for the low-temperature breakage of dormancy, and at a certain critical temperature (let us say around 15°C in the case of chilling), the membrane becomes crystalline (a gel), a change that would interfere with its functioning. Homeostatic mechanisms involving membranes are well known in many organisms—bacteria, algae, protozoa, and higher animals and plants—which, in response to temperatures lower than a critical value, adjust the composition of the membrane to maintain a certain degree of fluidity. This acclimation—sometimes called homeoviscous adaptation—involves alterations in membrane components, partly by increasing the proportion of unsaturated fatty acids in the phospholipids. A membrane (or membranes) could thus alter in response to the low temperature, and such a change might

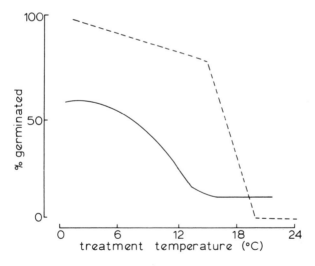

Figure 5.19. Temperature requirements for termination of dormancy. Imbibed seeds of *Rumex obtusifolius* were held at different temperatures for 4 wk in dim, diffuse light, after which they were transferred to 25°C for germination (----). Based on data from Totterdell and Roberts (1979). Imbibed lettuce seeds (cv. Grand Rapids) were held at different temperatures for 6 h, then given far-red light for a few minutes, to generate low levels of Pfr (see Section 5.5.1.4). They were then transferred to 20°C in darkness for 48 h after which germinated seeds were counted (——). After Van Der Woude and Toole (1980).

be an essential prelude to the dormancy-breaking mechanism, which then proceeds when the seeds later experience higher temperatures.

5.5.1.3. Other Effects of Temperature on Dormancy

In the field, dormant seeds are commonly subjected to fluctuating temperatures, for example, low night temperatures and high daytime temperatures. Such temperature fluctuations, or temperature alternations, are frequently effective in dormancy breakage, in cases such as *Bidens tripartitus,
Nicotiana tabacum,* and *Rumex* spp., which all have coat-imposed dormancy.
The effect of alternating temperatures on *Rumex obtusifolius* is shown in Fig.
5.20. At constant temperatures, the seeds remain dormant—this is shown by
the valley running across the diagonal of the figure. Dormancy is broken when
different temperatures are combined, with the maximum effect at the greatest
temperature differentials (i.e., amplitudes). So to be effective, the temperature
alternation must have a certain minimum amplitude, and in some species this
need be only a few degrees. Other parameters are also important. Without
going into detail, we can recognize that the temperatures of the pair must be
above and below certain values, that the duration of exposure to each is important, and that the number of cycles of fluctuating temperature can be decisive.

In a few species, relatively high temperatures can break, or assist in breaking, dormancy. Seeds of *Hyacinthoides non-scripta* require several weeks at
26–31°C followed by a germination phase at 11°C. Several species that are
chilling sensitive need a period at relatively high temperature before the cold.
Fraxinus spp., for example, require a few weeks at about 20°C prior to chilling
at 1–7°C. Softening of the seed coat of woody species might occur at the higher
temperature.

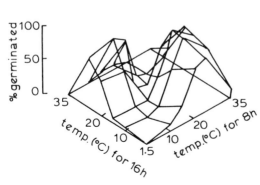

Figure 5.20. Breaking dormancy in
Rumex obtusifolius by alternating
temperatures. Imbibed seeds were
held in darkness for 28 days at different temperature combinations—
16 h at one temperature followed by
8 h at another. Germinated seeds
were counted after 28 days. Note
that high germination percentages
(i.e., termination of dormancy)
occur in the following temperature
combinations: (a) 16 h at 25–35°C,
8 h at 1.5–15°C; (b) 8 h at 25–
35°C, 16 h at 1.5–20°C. After Totterdell and Roberts (1980).

Table 5.8. Illumination Conditions Required for the Breaking of Dormancy

Illumination conditions	Examples
Seconds or minutes	*Agrostis tenuis*
	Chenopodium album
	Lactuca sativa cv. Grand Rapids
	Nicotiana tabacum
Several hours (or intermittent)	*Hyptis suaveolens*
	Lythrum salicaria
Days (or intermittent)	*Epilobium cephalostigma*
	Kalanchoë blossfeldiana
Long days	*Begonia evansiana*
	Betula pubescens (at 15°C)
	Chenopodium botrys (at 30°C)
Short days	*Chenopodium botrys* (>20 °C)
	Tsuga canadensis
	Betula pubescens (>15°C)

5.5.1.4. Light

Light is an extremely important factor for releasing seeds from dormancy (see examples in Table 5.7). Almost all light-requiring seeds have coat-imposed dormancy. Seeds of many species are affected by exposure to white light for just a few minutes or seconds (e.g., lettuce), whereas others require intermittent illumination (e.g., *Kalanchoë blossfeldiana*). Photoperiodic effects also exist, so that some species require exposure to long days and others to short days (Table 5.8). The light requirement frequently depends on the temperature. Grand Rapids lettuce, for example, generally is dormant in darkness only above about 23°C, so below this value seeds germinate without illumination. Seeds of some species of *Betula,* on the other hand, are dormant in darkness at lower temperatures (e.g., 15°C) but not at 25–30°C. Sensitivity to light in many species is enhanced by chilling, but in some (e.g., *Betula maximowicziana*) light terminates dormancy only in seeds that have previously been chilled. Various temperature alternations and temperature shifts also interact with light, as we shall see later in this section.

5.5.1.4.a. Action Spectra and Phytochrome.

In nature, white light (i.e., sunlight) breaks dormancy, but we know that the wavelengths in the orange/red region of the spectrum are responsible. In 1954, a detailed action spectrum for the breaking of dormancy in the Grand Rapids cultivar of lettuce, obtained by Borthwick, Hendricks, and their colleagues, revealed that major activity is at 660 nm (Fig. 5.21, A). Prior to this, inhibitory parts of the spectrum were also known, an especially potent wave band being far-red light, i.e., wave-

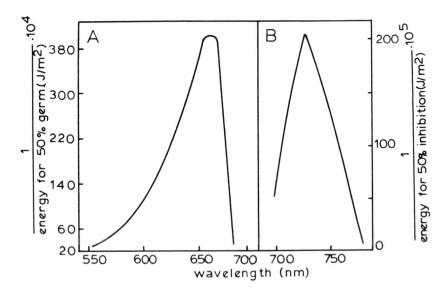

Figure 5.21. Action spectra for effects on seed dormancy in lettuce cv. Grand Rapids. In (A), seeds were imbibed for 16 h in darkness and then irradiated with various energy levels of light at wavelengths between 560 nm and 690 nm. They were then returned to darkness and the percentage of germinated seeds was later counted. The energy at each wavelength required to produce 50% germinated seeds (i.e., to break dormancy in 50% of the seeds) was determined. In (B), energies at different wavelengths of far-red light effective in reversing the red-light termination of dormancy were determined. Redrawn from Borthwick *et al.* (1954).

lengths longer than about 700 nm; the action spectrum showed that 730 nm is the wavelength of maximum activity (Fig. 5.21, B). At about the same time as the action spectrum was discovered, Borthwick, Hendricks, Parker, E. H. Toole, and V. K. Toole showed that the red and far-red light are mutually antagonistic. This was done by exposing lettuce seeds to a sequence of red and far-red irradiations: only when the last exposure in the sequence was to red light was dormancy terminated (see Table 5.9). This established the fact of photoreversibility; i.e., the two wavelengths 660 nm and 730 nm are able to reverse each other's effect. Light is, of course, absorbed by molecules of a pigment, and the one participating in the breaking of dormancy came to be called *phytochrome*. Phytochrome therefore exists in two forms. One is present in unirradiated, dormant seeds; it absorbs red light (peak at 660 nm) and is therefore designated as Pr. This form of the phytochrome obviously cannot break dormancy (if it could, the seeds would not be dormant!), but when activated by 660-nm light, it is changed into an active, dormancy-breaking form. But this active form absorbs far-red light (730 nm)—hence it is designated as Pfr—and having done so it reverts to Pr as follows:

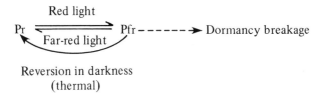

Ignoring for the moment the dark-reversion path (we will consider this soon), it is easy to see how the reversibility of dormancy breakage works: if Pfr is left in the seed at the end of the irradiation sequence, dormancy is terminated, but if Pr is left, dormancy is retained. Note that photoreversion only stops the breaking of dormancy if Pfr has been given insufficient time in which to act. If dosage with far-red light is delayed for a few hours, Pfr can then operate, and even if photoreversion occurs later, dormancy nevertheless has been terminated. The time needed for Pfr to act, after which photoreversion is inconsequential, is called the *escape time* (see Fig. 5.33).

The energies needed to carry out these photoconversions are relatively small. Saturating dose of red light in lettuce seeds is about $10 \ Jm^{-2}$—an amount given by about 0.2 s of direct, summer sunlight. Up to saturation value, the effect is directly proportional to the total amount of energy, irrespective of how the energy is delivered, i.e., by a lower fluence rate (irradiance) for a longer time or by a higher fluence rate for a shorter time. As long as the prod-

Table 5.9. Phytochrome Photoreversibility and the Breaking of Dormancy[a]

Irradiation sequence	Germinated (%)
None (darkness)	4
R	98
FR	3
R, FR	2
R, FR, R	97
R, FR, R, FR	0
R, FR, R, FR, R	95

[a]Seeds of the Grand Rapids cultivar of lettuce were imbibed in darkness and then exposed to red light (640–680 nm) for 1.5 min and far-red light (> 710 nm) for 4 min in the sequence shown. After irradiation, they were returned to darkness for 24 h before germinated seeds were counted.

ucts, fluence rate $\times$ time, are equal, the same effect (e.g., percentage of seeds breaking dormancy) is achieved. This is to say that the phytochrome system shows *reciprocity*. In a more complex response type, e.g., where repeated doses of red light are required, reciprocity is shown for each exposure, and not for the total amount of light involved. The dosage of far-red light needed for photoreversion is higher; in lettuce, for example, approximately 600 Jm^{-2} of light at 730 nm causes about 50% reversion of the effect of saturating red light. In the laboratory, higher doses of far-red are usually secured by higher fluence rates and/or longer irradiation times.

Turning back to the scheme for pigment photoconversion given previously, we must now explain the dark-reversion component. This was discovered when it was found that when lettuce seeds are transferred to a relatively high temperature for a few hours immediately after exposure to red light, they fail to germinate (Fig. 5.22): they come to require a second dose of red light. Thus, the high temperature causes a slow loss of active phytochrome by a reversion of Pfr to Pr—so-called *dark reversion.*

5.5.1.4b. Photoequilibria. The properties of phytochrome that we have so far discussed—the Pr and Pfr forms, their peak absorptions at 660 nm and 730 nm, respectively, the photoreversibility, and the dark reversion—were all suspected simply on the basis of experimental work on the physiology of light action in lettuce seed dormancy. Complete confirmation of these points came when phytochrome was isolated from plants—not from seeds, but from dark-grown (i.e., nongreen) tissues such as oat or rye coleoptiles. Phytochrome is a blue chromoprotein, the chromophore being an open-chain tetrapyrrole not unlike the phycocyanins of the blue–green algae. In solution, it shows photoreversibility with red and far-red light, and the absorption spectrum of the two forms can be determined (Fig. 5.23, A). Peak absorption of Pr is at 660 nm and of Pfr at 730 nm—just as predicted from the experiments with lettuce

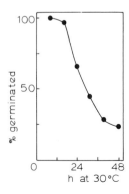

Figure 5.22. Inhibitory effect of high postillumination temperature. Lettuce seeds (cv. Grand Rapids) were exposed to a saturating red irradiation at 20°C and then kept in darkness at 30°C for periods up to 48 h. They were then returned to 20°C, still in darkness. The percentage of seeds that later germinated was determined. Note that more than 18 h at 30°C causes some reversion of the red-light potentiation of germination. After Borthwick *et al.* (1954).

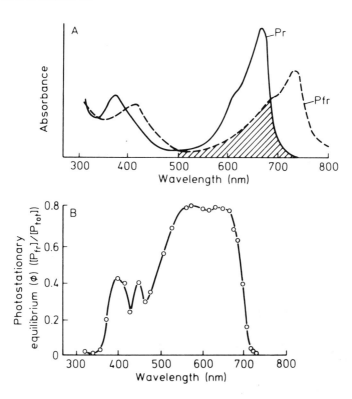

Figure 5.23. Absorption spectra and photostationary equilibria of phytochrome. Absorption spectra were determined using solutions of Pr and Pfr (A). Photostationary equilibria were measured in hypocotyls of dark-grown *Sinapis alba* (white mustard) seedlings (B). (A) After Hartmann (1966). (B) After Hanke *et al.* (1969).

seeds! An important point to note is the overlap in absorbance of the Pr and Pfr. Because both forms absorb over the spectrum from about 300 nm to about 730 nm, irradiation with monochromatic light in this range sets up an equilibrium mixture of Pr and Pfr—the photoequilibrium or photostationary state Pfr/Ptotal or ϕ (Fig. 5.23, B). At 660 nm, for example, Pr is photoconverted to Pfr, but Pfr also absorbs at this wavelength and hence some Pfr molecules are phototransformed back to Pr. A mixture of approximately 75% Pfr and 25% Pr is thus established (Pfr/Ptot, ϕ = approximately 0.75). Even at 730 nm, where Pfr absorbs most strongly, there is also some absorption by Pr— here the ϕ value is 0.02, i.e., 2% Pfr. Inspection of the absorption spectrum shows that no wave band region can produce 100% Pfr; on the other hand, photoreversion to give almost 100% Pr can be achieved by irradiation at 740–800 nm. It will be appreciated that irradiation with mixed wavelengths also

**Table 5.10. Photoequilibrium Values of
Phytochrome (Pfr/P$_{total}$) Required for
Dormancy-Breakage^a**

Species	Pfr/P$_{total}$
Amaranthus retroflexus	0.001
Amaranthus caudatus	0.02
Wittrockia superba	0.02
Sinapis arvensis	0.05
Cucumis sativus	0.1–0.15
Chenopodium album	0.3
Lactuca sativa	0.59

^{a}Values given are sufficient to break dormancy in most seeds
of the population.

establishes a certain ϕ value. In midday sunlight ϕ is about 0.55, in white incandescent light it is about 0.45, and in white fluorescent light about 0.65. Now the effectiveness of phytochrome in terminating dormancy is determined by the ϕ value set up in the seeds, the required value depending on the species, as can be seen in Table 5.10. Lettuce is clearly satisfied by the ϕ value brought about by sunlight, and *Wittrockia superba* can even be stimulated to germinate by broad-band far-red light ($\phi = > 0.02$).

What is the function of this apparently complex photoreceptive system? The importance of control by the ϕ value lies in the fact that light-requiring seeds can use phytochrome to detect different light qualities. In nature, the quality varies when the proportions of red and far-red light change. This happens to some extent according to the time of day, but more importantly it occurs when light is transmitted through green leaves, whose chlorophyll absorbs red light but allows far-red light to pass, or through the soil, which also transmits far-red light. Thus, a seed under a leaf canopy is in far-red light, which sets up a ϕ value too low to satisfy most light-requiring seeds. In a sense, phytochrome is the device used by the seed to detect where it is, especially in relation to other plants. Phytochrome is also involved in the photoinhibition of germination, when light of rather high fluence rates given for relatively long periods of time prevents even nondormant seeds from germinating. We will return to these aspects of the ecophysiology of dormancy in Chapter 6.

5.5.1.4c. Where Is Phytochrome Located? Selective irradiation (using narrow beams of light) of different parts of light-requiring seeds—the radicle/hypocotyl and the cotyledons—shows that dormancy is terminated only when the former, i.e., the embryonic axis, is illuminated. This suggests that phytochrome is located here, and indeed, the technique of *in vivo* spectrophotometry, by which phytochrome can be measured in intact plant tissues, confirms its presence in the axis and its very low concentration in the cotyledons.

5.5.1.4d. Photoconversions of Phytochrome. To appreciate fully the role of phytochrome in dormancy and germination we should understand something of the conversion pathway of the pigment. Photoconversion does not occur directly between Pr and Pfr but through several intermediates, as shown in Fig. 5.24. Some of the steps can occur only in highly hydrated seeds, whereas others can proceed in seeds at a very low moisture content. Thus, "dry" seeds have some light sensitivity, but all the photoconversions cannot take place. While the seed is developing on the mother plant, phytochrome conversions proceed normally, until the maturing seed begins to dry. Some of the intermediates (meta Rb and possibly meta Fa) are then trapped, so the dry seed has consigned to it some Pfr, meta Rb and meta Fa, and some Pr, in proportions depending on the quality of light reaching the embryo and on the relative sensitivity of intermediate steps to different hydration levels. On rehydration, meta Rb can pass to Pfr in darkness, and if there is much Pfr already present, the total might be sufficient to promote germination. Such a seed would be non-

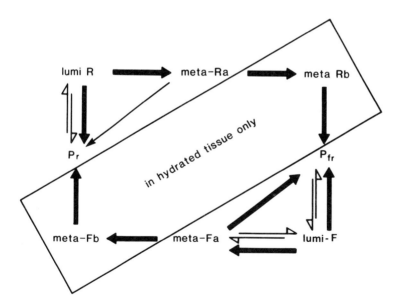

Figure 5.24. Transformations of phytochrome. Conversion of Pr to Pfr and Pfr to Pr is in each case through three intermediates, involving light-driven steps (thin arrows) and nonphotochemical steps (thick arrows). Adapted from Kendrick and Spruit (1977).

dormant. But if sufficient Pfr, or intermediates that can produce it, is not present in the seed, it matures into a dormant, light-requiring seed.

5.5.1.4e. Phytochrome—The Sensor. The formation of Pfr alone does not break dormancy. Rather, Pfr sets in train some events that culminate in the termination of dormancy, as manifest by germination of the seed. To put this in another way, we can suggest that there is some transduction mechanism that intervenes between Pfr formation and the final display—germination. Let us now consider what the earliest events in this transduction process might be.

There is good evidence from several physiological systems other than seeds that phytochrome acts on cell membranes. Fast changes in the flux of ions and water occur in some cells, for example. Several pieces of evidence, mainly relating to effects of temperature, suggest that in seeds also there may be an association of Pfr with membranes. Many dormant seeds—*Amaranthus retroflexus* is a good example—are stimulated to germinate when they experience a shift in temperature, perhaps for only 1–2 h. In *Amaranthus,* Pfr cannot act to break dormancy if the temperature is too high, say 40°C. But a drop in temperature to 26°C for about 60 min *after* (but *not* before) Pfr is generated by light enables Pfr to operate, even if the seeds are transferred back to 40°C, so that dormancy is terminated. When this is investigated carefully, it becomes clear that there is a critical temperature below which Pfr action is initiated— it is almost exactly 32°C (Fig. 5.25). So if the seed falls below 32°C for a relatively short time, the Pfr that is present is able to initiate events leading to dormancy breakage, irrespective of later temperatures. The critical nature of the temperature dependence is highly suggestive of a cell membrane transition, and the implication is that a membrane must be in the correct condition for Pfr to interact with it; in this case, it seems that the membrane must be in the gel state. Crude membrane fractions extracted from the seed show temperature-controlled transitions at this temperature. The enhancement of Pfr action

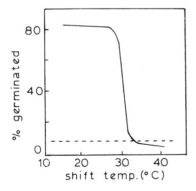

Figure 5.25. Effect of a temperature shift on phytochrome action in *Amaranthus retroflexus.* Imbibed seeds were held in darkness at 40°C except for a shift to lower temperatures for 2 h. Using far-red light, a small amount of Pfr was generated before (——) or after (----) the shift. Note that a temperature of 32°C or below for 2 h in the presence of Pfr breaks the dormancy. Seeds not given the temperature shift (i.e., at 40°C throughout) do not respond to Pfr. Adapted form Hendricks and Taylorson (1978).

by the chilling of lettuce seeds prior to illumination might also depend on membrane effects. Here, too, it seems that there may be a critical temperature (Fig. 5.19), though the effect is not as sharp as in *Amaranthus.*

This is attractive, preliminary evidence suggesting that newly formed Pfr in seeds associates with a cell membrane. The process clearly occurs fairly quickly, in 60 min at most, and we can envisage it as the primary action of Pfr. Once the association with the membrane has taken place, certain membrane-dependent events might follow, such as alterations in ion movement or possibly activation of membrane-bound enzymes: unfortunately, there is no information to tell us what events do occur and the matter therefore remains speculative. Presumably, however, these events eventually allow processes to occur that terminate in emergence of the radicle from the seed, which occurs by cell elongation. Hence, the ultimate effect of phytochrome may be to control the capacity for cell enlargement, a point which is discussed further in Section 5.5.1.10.

5.5.1.5. Seeds with Impermeable Coats

The environment is important in softening hard coats that are impermeable to water. Microbial attack is thought to be important, as well as abrasion by soil particles. Of particular interest is the effect of high temperatures on some leguminous seeds. During exposure to heat, cracks appear in the seed coat of some species, especially in the region of the micropyle. A spectacular response is seen in some seeds, such as *Albizzia lophantha,* in which the strophiolar plug is audibly ejected from the seed as high temperature is reached, leaving a strophiolar crater through which water can enter. Such effects of high temperature are thought to be important in pyric species whose seedlings emerge as a consequence of forest fires.

5.5.1.6. Breaking of Dormancy by Chemicals

A selected list of chemicals that can break dormancy is given in Table 5.11. Only a few of these are likely to be encountered by seeds in their natural environment, but they are nevertheless of great interest because they may help us understand the mechanism of dormancy breakage. The possible action of many of these substances is discussed elsewhere in this chapter—anesthetics in relation to membranes (Section 5.3.2), respiratory inhibitors, nitrate, nitrite, methylene blue in relation to the pentose phosphate pathway, and oxygen in connection with various oxidation processes (Section 5.5.1.9). Something needs to be said about the growth regulators, however, since their effects have suggested that their naturally occurring counterparts participate in the dormancy mechanism. The growth regulators, gibberellin (usually gibberellic acid GA_3, GA_4, and GA_7), cytokinin (usually kinetin, benzyladenine), and ethylene var-

Table 5.11. Some Chemicals That Break Seed
Dormancy

Class	Example
Respiratory inhibitors	
Cyanide	*Lactuca sativa*
Azide	*Hordeum distichum*
Iodoacetate	*Hordeum distichum*
Dinitrophenol	*Lactuca sativa*
Sulfhydryl compounds	
Dithiothreitol	*Hordeum distichum*
2-Mercaptoethanol	*Hordeum distichum*
Oxidants	
Hypochlorite	*Avena fatua*
Oxygen	*Xanthium pennsylvanicum*
Nitrogenous compounds	
Nitrate	*Lactuca sativa*
Nitrite	*Hordeum distichum*
Thiourea	*Lactuca sativa*
Growth regulators	
Gibberellins	*Lactuca sativa*
Cytokinins	*Lactuca sativa*
Ethylene	*Chenopodium album*
Various	
Ethanol	*Panicum capillare*
Methylene blue	*Hordeum distichum*
Ethyl ether	*Panicum capillare*
Fusicoccin	*Lactuca sativa*

iously affect seed dormancy (Table 5.12). Seeds that normally require chilling or light or afterripening exhibit striking responses to these substances, which often also accelerate germination of nondormant seeds. It should be said, however, that the success of these regulators is mixed, and many species of seed do not respond at all.

The growth regulators with the widest spectrum of activity are the gibberellins. As an example, we can quote their effect on dormant lettuce seeds, which respond to a concentration of around 10^{-4}M GA_3. Cytokinins are less widely effective, and even when they do act, they often induce abnormal germination—in lettuce, for example, the cotyledons tend to emerge from the seed before the radicle. Ethylene is also only limitedly effective, and many species remain unaffected. Frequently, the growth regulators interact with other factors or among themselves. Kinetin, for example, promotes normal germination of dormant lettuce in combination with low levels of light, and ethylene stimulates *Chenopodium album* most effectively in the presence of light and gibberellin. Numerous studies have demonstrated interactions among the growth

Table 5.12. Breaking of Dormancy by Some Growth Regulators

Species	Dormancy broken by	Effect of growth regulator[a]		
		GA[b]	CK[c]	Ethylene
Apium graveolens	Light	+	−	+
Avena fatua	Afterripening, chilling	+	−	+
Corylus avellana	Chilling	+	−	−
Hordeum distichum	Afterripening, chilling	+	−	+
Lactuca sativa	Light	+	±	−
Lamium amplexicaule	Light	+	nr	nr
Pyrus malus	Chilling	−	+	nr
Ruellia humilis	Chilling	+	nr	nr
Xanthium pennsylvanicum	Afterripening	+	−	+

[a]+, regulator effective; −, regulator ineffective; ±, moderately effective; nr, not recorded.
[b]GA, gibberellin (usually gibberellic acid GA$_3$, GA$_4$, GA$_7$).
[c]CK, cytokinin (usually kinetin, benzyladenine).

regulators. One example will suffice for our present purposes (Fig. 5.26). It should be noted that in lettuce, kinetin can overcome the inhibitory effect of ABA. At low concentrations of ABA, GA$_3$ is also effective in this respect.

5.5.1.7. The Mechanism of Dormancy Release

What changes are brought about in dormant seeds by the action of the various factors we have just considered? Are the changes different, according to the particular factor, or do they all have the same fundamental effect? In short, is there a universal dormancy-breaking mechanism? We should be clear at the outset that whether we are dealing with embryo dormancy or coat-

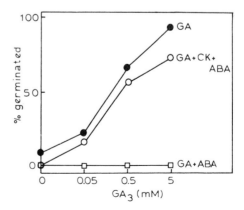

Figure 5.26. Interactions between growth regulators in the breaking of dormancy. Lettuce seeds (cv. Grand Rapids) were placed in darkness on filter papers wetted with solutions of gibberellic acid (GA), GA plus ABA, and GA plus ABA plus the cytokinin kinetin (CK). GA alone breaks dormancy, ABA stops GA action, and the cytokinin overcomes ABA. After Thomas (1977).

imposed dormancy, it is the embryo that is affected by the active factor—and possibly only the axis, since its growth culminates the germination process. Apart from the hard coats of impermeable seeds, there is no evidence that any major change in the tissues enclosing the embryo accompanies the termination of dormancy. The mechanical resistance remains unaltered, as does the permeability to oxygen and carbon dioxide.

A concept that has attracted a great deal of attention is the hormonal theory of dormancy. This attributes the control of dormancy to various growth regulators or hormones—inhibitors, such as ABA, and promoters, such as gibberellins, cytokinins, and ethylene. According to the theory, dormancy is maintained (and possibly even induced) by inhibitors, and it can end only when the inhibitor is removed or when promoters overcome it. The theory owes its inception primarily to known effects of applied growth regulators on dormancy, some of which, as we saw earlier, cause a dormant seed to germinate whereas others inhibit germination of a nondormant seed. A second concept is that important metabolic changes occur as a consequence of the action of the dormancy-breaking factor. One such change is thought to involve the synthesis of RNA and protein, and another, the operation of the pentose phosphate pathway. We will discuss these, and others, in the following sections.

5.5.1.8. Hormones in Dormancy Breakage

At one time it was thought that chilling acts by causing a drop in ABA content. Although the level of ABA of seeds does indeed decrease dramatically during the chilling treatment, it also does so when seeds are held imbibed at nonchilling temperatures, so the effect is not confined to the chilling experience. Another finding that makes it certain that a fall in ABA alone is insufficient to break dormancy is that the inhibitor disappears long before the effective duration of cold has been experienced: in apple seeds, for example, ABA disappears after 3 wk of chilling, when the seeds are still dormant! Although it can be concluded that a reduction in ABA cannot, on its own, account for dormancy breakage, the loss of the inhibitor might nevertheless be essential to it, but we do not know if this is really the case or not. Other hormonal changes might also be implicated in the termination of dormancy. Transient increases in various gibberellins and cytokinins take place in seeds of several species during chilling, but the timing of these events does not seem related in any clear way to the course of dormancy breakage, unless a sequential appearance of the promotive hormones is necessary (Fig. 5.27).

One case is known—hazel *(Corylus avellana)*—where chilling has a marked effect on the capacity for gibberellin biosynthesis. Gibberellin production takes place when the chilled seed subsequently experiences higher temperatures and leads to substantial increases in GA_1 and GA_9 (Table 5.13).

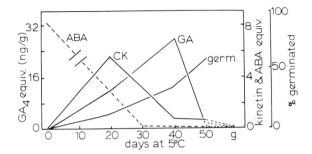

Figure 5.27. Changes in endogenous growth regulators during chilling of sugar maple *(Acer saccharum)*. ABA, cytokinins (CK), and gibberellins (GA) were extracted from seeds that had been chilled at 5°C for different durations. ABA was determined after gas liquid chromatography. CK was assayed using the soybean callus bioassay, and the results are expressed in terms of equivalent amounts of kinetin having the same activity. GA was assayed by the lettuce hypocotyl assay; results are expressed in terms of equivalent amounts of GA_3 known to have the same activity. The effect of different periods of chilling on the breaking of dormancy (germ.) was also determined. Adapted from Webb *et al.* (1973).

Significantly, the application of 2-chloroethyltrimethyl ammonium chloride (a substance that blocks gibberellin biosynthesis) to seeds just after the cold treatment stops germination, and so it seems that the chilling-induced buildup of gibberellin is essential for dormancy breakage to culminate in germination. Dormant, nonchilled hazel seeds germinate after treatment with exogenous gibberellic acid; presumably, the endogenous gibberellin which appears after chilling is also effective in promoting germination—certainly the concentration formed is high enough to be physiologically active. The termination of hazel seed dormancy by chilling therefore involves an enhancement of the capacity subsequently to make gibberellin, and not an increase in the promoter during the chilling treatment itself. The reason for the enhancement is unknown. It is important to discover if seeds of other species behave in the same way as hazel, but this has not yet been reported.

**Table 5.13. Enhancement by Chilling of Gibberellin-Biosynthetic Capacity in
Corylus avellana Embryos[a]**

	GA content (nmol/seed)	
Temperature–treatment	GA_1	GA_9
None	1.02	< 0.01
42 days at 5°C	0.12	< 0.01
42 days at 5°C, then 8 days at 20°C	4.92	3.06

[a]Based on Williams *et al.* (1974).

Transient increases in gibberellins and cytokinins have been found to occur shortly after illumination of some seeds, such as pine *(Pinus sylvestris)* and *Rumex* spp., and in lettuce, fairly high increases in GA_9 content have been reported. Whether the hormones play any role in dormancy breakage is unclear, however. Some doubt arises from the fact that the increases occur quickly, well within the escape time for Pfr (Section 5.5.1.4a). This means that even though the hormones may be formed, further Pfr action is nevertheless still needed for dormancy to be broken, suggesting that termination of dormancy cannot depend solely on the Pfr-induced formation of endogenous gibberellin or cytokinin.

Ethylene is another hormone that might feature in the dormancy-breaking mechanism. The production of ethylene by *Xanthium* and peanut embryos is thought to be important for the loss of dormancy. Imbibed afterripened peanut seeds, for example, eventually evolve more ethylene than dormant seeds, though it is difficult to decide whether the substance is produced simply as the result of germination or is really the cause of it.

5.5.1.9. Dormancy Breakage and Metabolism

Metabolic activity that has been investigated in relation to the breaking of dormancy by light, chilling, afterripening, and growth regulators such as gibberellic acid includes nucleic acid and protein synthesis and aspects of respiratory metabolism—gas exchange, ATP production, the pentose phosphate pathway, and the so-called alternative (i.e., cyanide-insensitive) pathway. The rationale is that dormant seeds may be metabolically deficient in some respect and that in the breaking of dormancy the deficiencies are made good.

There is no evidence that synthesis of RNA or protein is promoted by light or chilling. Dormant, light-requiring seeds are able to carry out macromolecular syntheses in darkness, and abundant polyribosomes are produced; there is some slight evidence, however, that polyribosome numbers in lettuce and pine spp. increase after irradiation, but the effect awaits confirmation. These observations, however, do not rule out the possibility that the termination of dormancy involves the production of discrete RNA (and hence protein) types— the techniques that have been used to date would certainly fail to detect small quantitative changes. Application of gibberellin to hazel seeds causes an increase in total RNA of the embryonic axes, but the effect, prior to radicle emergence itself, is small. RNA synthesis, as detected by the incorporation of radioactive precursors, apparently is promoted, and on this basis it has been suggested that the growth regulator increases DNA template availability and RNA polymerase activity; i.e., gibberellin derepresses certain genes. However, there is only limited evidence that RNA synthesis is enhanced, and no evidence that there is synthesis of RNA(s) essential for germination. Some stimulation by gibberellin of polyribosome formation and protein synthesis occurs in lettuce and charlock, but whether the protein is essentially linked with germination is unknown. Similarly, cytokinins increase RNA and protein synthesis, especially when they overcome the inhibitory effects of ABA.

Respiration, as measured by gaseous exchange, on the whole seems unresponsive to light or chilling. There are no convincing changes in oxygen consumption or production of ATP in illuminated or chilled seeds. Chilling seems to affect phosphate metabolism in embryonic axes of cherry. Here, organic phosphates (nucleotides) accumulate during cold stratification, as opposed to inorganic phosphates during the warmth. It is possible that in this species the change in phosphate metabolism brought about by cold treatment could make more precursors available for nucleic acid synthesis, but this has not yet been shown to be the case.

The concept that the pentose phosphate pathway plays a unique role in dormancy breakage arises largely from studies of the effect of certain chemicals on dormancy. Dormancy of several different kinds of seeds, including lettuce, rice, and barley, is broken by the application of inhibitors of respiration. Active substances include those which inhibit terminal oxidation and the tricarboxylic acid cycle in the mitochondria (e.g., cyanide and malonate) and some which inhibit glycolysis (e.g., fluoride). Electron acceptors such as nitrate, nitrite, and methylene blue can also break dormancy. High oxygen concentrations are also active. An explanation for the effects of these substances is as follows: The consumption of oxygen by conventional respiration deprives other processes that also require oxygen. When conventional respiration is blocked by the inhibitors, oxygen then becomes available for the other processes; alternatively, the oxygen could be made available by elevating the external oxygen concentration. The pentose phosphate pathway has been suggested to be the important "other" process, and it may require oxygen for the oxidation of reduced NADP (NADPH $+$ H$^+$) which the pathway generates. Oxidation of NADPH can also be brought about by the electron acceptors methylene blue, nitrate, and nitrite, in which case oxygen is not needed (Fig. 5.28).

Although the special significance and function of the PPP has hever been explained, the hypothesis has nevertheless been thoroughly investigated in several ways. First, the contribution of the pathway to glucose oxidation has been measured in several kinds of dormant seeds and in seeds whose dormancy is being broken. The results are mixed, some of them showing a marked participation of the PPP during dormancy breakage, whereas others are less convincing. Moreover, some criticisms have been leveled at the techniques used in these determinations, and it has been suggested that they might give misleading results. A second approach is to examine the activity of two key enzymes in the pathway, glucose-6-phosphate dehydrogenase and 6-phosphogluconate dehydrogenase. In wild oats and barley, the extracted enzymes from nondormant grains are no more active than those from dormant material (Fig. 5.29). Some enhancement of activity seems to occur in hazel cotyledons during chilling, which might be significant since the cotyledons are involved in the regulation of embryo dormancy. Third, levels of reduced and oxidized NADP and NAD have been studied, but these are completely unrelated to the dormancy status of seeds. Finally, although the suggested role for oxygen is for the oxidation of NADPH, such an oxidation system seems to be unknown in plants. The evidence for the participation of the PPP in dormancy breakage is therefore, at best, equivocal. Information is accumulating in the research literature that casts great doubt on the validity of the hypothesis that the termination of dormancy requires a switch in respiratory metabolism toward increased PPP activity.

Notwithstanding the doubt surrounding the PPP, an "alternative" respiratory pathway (one that is insensitive to cyanide and therefore does not use the normal terminal oxidation mechanism—see Section 4.2.2) has been sug-

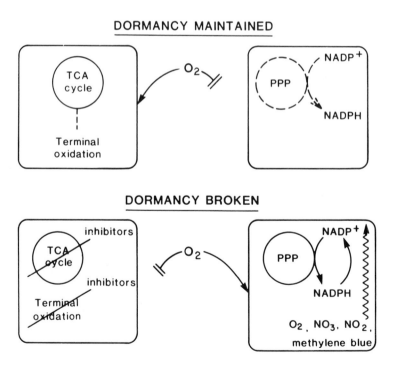

Figure 5.28. The pentose phosphate pathway and dormancy breakage. See text for explanation.

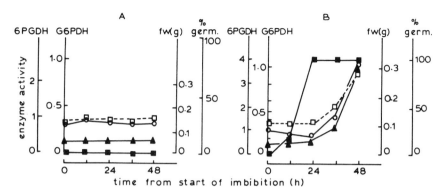

time from start of imbibition (h)

Figure 5.29. Enzymes of the pentose phosphate pathway and dormancy in wild oats *(Avena fatua)*. For this experiment, two strains of wild oats were used—a highly dormant line (A) and a nondormant line (B). In both, the activity was determined of two enzymes in the pentose phosphate pathway, glucose-6-phosphate dehydrogenase (G6PDH) and 6-phosphogluconate dehydrogenase (6PGDH). Units of enzyme activity refer to the amount of reduced nicotinamide dinucleotide phosphate produced in the reaction, measured in absorbance units. Activity was determined at time intervals over 48 h, during which the percentage of germinated seeds and the change in fresh weight in grams (a measure of seedling growth) were followed. □---□, 6PGDH; o—o, G6PDH; ■—■, % germinated seeds; ▲—▲, fresh weight (fw). Note the absence of significant differences in enzyme activity between nondormant and dormant grains until after germination of the former has occurred. This shows that dormancy is not associated with lower enzyme activity. Adapted from Upadhyaya *et al.* (1981).

gested to be important in the ending of dormancy. Little is known about this pathway, though, and its special role in dormancy breakage is not yet clear. In the case of *Xanthium,* however, it seems that chilling temperatures enhance subsequent electron flow through the cyanide-insensitive pathway, an effect that appears to increase the ATP level and the total adenylate pool (but not the energy charge). It has been suggested that such changes promote subsequent germination, but for reasons that remain to be elucidated.

5.5.1.10. Dormancy Breakage and Cell Enlargement

The release from dormancy, in the end, means that the embryo is able to develop enough growth thrust to overcome the constraints upon it. Measurements on lettuce embryos demonstrate that one dormancy-breaking factor—light—has this effect. The lettuce is a seed whose dormancy is coat-imposed (actually the endosperm is the effective tissue). Therefore, isolated embryos have no apparent dormancy and germinate in darkness. But their dormant behavior is restored if they are placed in solutions of high osmotic strength (mannitol or polyethylene glycol), when they become light requiring once more. Embryos that have received no light require a concentration equivalent to 0.1 molal mannitol to make them dormant (i.e., to prevent their germination and growth). When embryos are illuminated, the necessary concentration to stop growth is about 0.35 molal mannitol (Fig. 5.30). Now, the arrest of embryo growth occurs when the external osmotic potential is lower than the embryo's water potential. Clearly, since a higher osmotic strength (i.e., lower osmotic potential) is needed to inhibit growth of irradiated embryos, these embryos must have a lower water potential than the unirradiated ones; in short, light has brought about a lowering of embryonic water potential. It follows, then, that embryos exposed to light are more able to take up water, i.e., to generate a growth thrust. Is this ultimately the effect of light in dormancy breakage? And if so, how is it achieved? The water potential of the embryo (ψ_{emb}) is comprised of its pressure or turgor potential (ψ_p) and its osmotic potential (ψ_π) (remember that ψ_π is a negative quantity).

$$\psi_{emb} = \psi_p + \psi_\pi$$

Thus, a low ψ_{emb} could be achieved by decreasing ψ_p or by making ψ_π more negative. There is no evidence for a substantial change in ψ_π as a result of illumination, and so a fall in ψ_p must occur. Now this means that the resistance of the cell walls is lowered; i.e., they become softer. Softening of cell walls is known to occur during growth of plant cells, so perhaps it is not surprising that it seems to take place in the lettuce embryo capable of growth. But the implication is that here, light (i.e., Pfr) ultimately is responsible. Cell wall softening is thought to occur as a consequence of the secretion of hydrogen ions by the protoplast into the wall. From measurements of the effects of red-light-promoted lettuce embryos on the acidity of the external liquid, it does seem that Pfr can stimulate the hydrogen-ion-secreting system (Fig. 5.31). We might conjecture, therefore, that whatever the preceding effects of Pfr are in the dor-

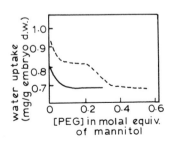

[PEG] in molal equiv. of mannitol

Figure 5.30. Effect of light on the water potential of lettuce embryos. Lettuce embryos were isolated from red-light-treated (----) and dark-treated (——) seeds (cv. Grand Rapids). They were then placed in darkness on polyethylene glycol solutions (PEG) of different osmotic strengths, expressed as molal equivalents of mannitol. Water uptake after 15–16 h was determined by measuring the change in weight. Water uptake in dark-treated embryos is stopped by concentrations higher than 0.1 molal equivalents, whereas in light-treated embryos, water uptake continues in solutions up to approximately 0.35 molal equivalent. Adapted from Nabors and Lang (1971).

Figure 5.31. Effect of Pfr on hydrogen ion secretion. Embryos of lettuce (cv. Grand Rapids) were isolated from red-light-treated (r) and far-red-treated (fr) seeds and placed in a medium of 1 mM phosphate buffer for 24 h. Over this period, the growth of the embryonic axes and the pH of the external medium were determined. Note the greater acidification of the medium by red-light-treated (i.e., containing Pfr) embryos. After Carpita *et al.* (1979).

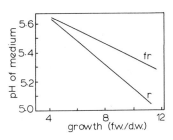

mancy-terminating mechanism, the final result might be the activation of a proton pump, leading to softer cell walls and thereby the capacity for radicle extension. That chemical stimulation of germination might act via activation of a proton pump is less clear-cut, and some reservations are considered in Section 3.6.

5.5.2. Environmental Control of Germination

Several factors in the environment—water, oxygen, light, temperature, and chemicals—determine whether or not germination occurs and the rate at which it does so. These controls apply to seeds whose dormancy has been broken, as well as to those which never had any dormancy. As far as the dormant seed is concerned, controlling factors that act over a fairly long duration might have effects on dormancy and germination which merge, in some cases antagonistically. For example, exposure of a dormant seed to an extended period of cold may serve to break its dormancy, but it will not germinate because the temperature is too low, and it could even be forced into a secondary dormancy. Similarly, a light-requiring seed under long-term illumination with white light could initially have its dormancy broken by the light but then have its radicle extension inhibited by the same irradiation, and again, it too might enter secondary dormancy.

We will not, in this section, consider environmentally limited water, which is covered briefly in Section 3.5.1, and some effects of chemicals have been mentioned earlier. Attention will be confined to light and temperature.

5.5.2.1. Effects of Light

Seed germination in many species is inhibited by continuous white light. Well-known examples are *Nemophila insignis, Phacelia tanacetifolia, Amaranthus caudatus,* and several cultivars of lettuce. Such seeds are normally dark germinators: they are frequently referred to as light-inhibited or negatively photoblastic seeds. But even some seeds whose dormancy is broken by light can also be inhibited by prolonged exposure, especially if the fluence rate is high. Dual effects of this kind are neatly illustrated by dormant *Oryzopsis miliacea*, which germinates in response to a few minutes of light but does not do so under continuous exposure (Fig. 5.32). Even the classical light-requiring seed— Grand Rapids lettuce—behaves like this under certain circumstances.

We have emphasized that white light inhibits when exposure is prolonged; hence the phenomenon is time dependent. In some cases, however, inhibition is brought about by intermittent light of a few hours each day. Since the proportion of inhibited seeds increases with the duration of each light period, a quasiphotoperiodic effect results, when most of the seeds can germinate under short days but few can do so in long days of, say, about 20 h light per day. *Nemophila insignis* has this kind of behavior. Photoinhibition is also fluence-rate dependent, the degree of inhibition generally increasing linearly with the logarithm of the fluence rate. There is no reciprocity, which distinguishes the response from the phytochrome effects discussed earlier (Section 5.5.1.4a). The degree of inhibition also may depend on temperature, but no general rule can be formulated. *Nemophila insignis*, for example, at moderate fluence rates is inhibited only at temperatures above 21°C; *Amaranthus* spp., however, are photoinhibited only at lower temperatures! It is interesting that for photoinhibition to occur, the embryo must be under constraint. If the embryo of *Nemophila insignis* is isolated from the enclosing tissues, or even if just the radicle

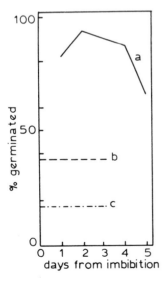

Figure 5.32. Dual effects of light on seeds of *Oryzopsis miliacea*. Imbibed seeds were held in darkness (b) or continuous white fluorescent light (c) or given a few minutes of such light at different times (a). The percentage of germinated seeds was determined at 8 days after the start of the experiment. After Koller and Negbi (1959).

tip is freed, light ceases to inhibit. On the other hand, application of a constraint to a seed not normally affected by light can make it inhibitable. This happens in cucumber and radish seeds when they are placed on osmotica (about 0.5 M mannitol)—only a few of them germinate in the light whereas all do so in darkness, although on water they are almost light insensitive.

It seems that if a seed is to be inhibited it must receive light at a critical stage in its germination. The most sensitive time for a population of *Nemophila* is 40–60 h after the start of imbibition, which is the period when the radicles would otherwise start to emerge. Hence, seeds are susceptible to photoinhibition just prior to and during the time when the radicle cells are about to elongate: the inhibitory effect of light is therefore probably on cell elongation.

Photoinhibition of seed germination by prolonged illumination is a manifestation of the high irradiance reaction (HIR) of photomorphogenesis. It has received most experimental attention in seeds in connection with the action of extended periods of far-red light, which act after the escape time has passed and therefore do not simply reverse Pfr (Fig. 5.33, A). The effectiveness of the inhibitory far-red light depends on its duration and fluence rate (Fig. 5.33, A, B). It is not appropriate here to enter into a detailed discussion of the mechanism of the HIR, but it will suffice to say that the prolonged far-red effect undoubtedly involves the operation of phytochrome in a special way. Most important, it requires a high cycling rate of the pigment, i.e., a high rate of the photoconversions Pr $\rightleftharpoons$ Pfr. This gives us a clue, for high cycling rates also occur under white light, the rates obviously being dependent on the fluence rate. This, then, accounts partly for the inhibitory effect of white light, though the mechanism is unknown. But white light also inhibits because of its blue component. Prolonged, pure blue light—again it is fluence-rate dependent—inhibits germination, the sensitive stage again being around the time of incipient radicle emergence. Part of the effect of blue light may be due to Pr $\rightleftharpoons$ Pfr cycling, since phytochrome absorbs in the blue region of the spectrum (Fig. 5.23), but probably a substantial part operates through another pigment system—the special blue-light receptor, or cryptochrome. Phytochrome cycling and cryptochrome, therefore, together account for photoinhibition by white light.

5.5.2.2. Inhibition by Short Periods of Far-Red Light

Many dark-germinating seeds are inhibited by far-red light given for just a few minutes or intermittently. This happens in seeds of certain cultivars of tomato and lettuce and in cucumber, for example. In these cases, far-red light is not acting through the HIR but is effective through phytochrome in the low-energy mode, which is why inhibition can be reversed by just a few minutes of

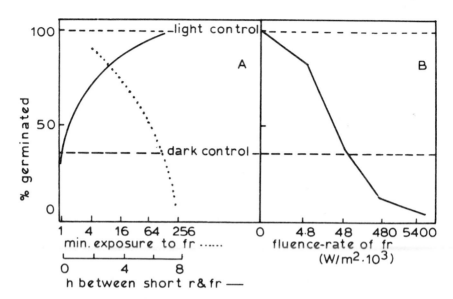

Figure 5.33. Effects of far-red light on lettuce seeds (cv. Grand Rapids). (A) Imbibed seeds were exposed to 1 min red light; this caused 100% of the seeds to germinate (----, light control). Other seeds were exposed to 1 min red light and then, at different intervals, to 4 min far-red light. The solid line shows the effect of these treatments on the subsequent germination. This curve shows the escape time, i.e., the time taken for the dormancy-breaking process to escape from the effects of reversal of Pfr. Reversal of Pfr after 8 h has no effect on subsequent germinability, and reversal after 4 h has only a relatively small effect. Seeds were also exposed to 1 min red light and then allowed to spend 4 h in darkness, after which they were irradiated with different durations of far-red light (· · · · ·). Even though seeds have almost completely escaped from the effect of a short far-red exposure, they can nevertheless be inhibited by prolonged exposure (e.g., completely inhibited by 256 min far-red light). (B) Seeds were given the following treatment: 1 min red light, 4 h darkness, 256 min far-red light at different fluence rates. Note the increasing inhibition as the fluence rate rises. Adapted from Mohr and Appuhn (1963).

red light. The short periods of far-red light therefore convert these nondormant seeds into dormant, light-requiring ones. The explanation for the phenomenon is that the dark-germinating seeds initially have no dormancy because they mature on the mother plant with sufficient Pfr already in them, or because they contain phytochrome intermediates (such as meta Rb, Section 5.5.1.4d) that generate Pfr when the seed is hydrated. Those seeds which have Pfr, but no intermediates, are inhibited by a single dose of far-red light (e.g., some cultivars of tomato). In those with intermediates which slowly generate Pfr, intermittent far-red light is required to drive the developing Pfr back to Pr (e.g., some cultivars of lettuce). This process might happen in the field when a seed that contains Pfr hydrates in light rich in far-red, such as under leaf canopies.

5.5.2.3. Effects of Temperature

It is worth emphasizing that when considering the effects of temperature on germination we must be careful to exclude dormancy. For example, when seeds with relative dormancy are tested at different temperatures, germination is found to occur only over a certain range (see Fig. 5.10). But this is not strictly the temperature range for germination—it is the temperature range over which there is no dormancy. When dormancy of such a seed is removed, germination then occurs over a substantially wider range of temperatures (Fig. 5.34). Similarly, dormant seeds with a requirement for chilling may actually succeed in germinating at the same low temperature needed for the termination of dormancy, albeit sluggishly: they cannot germinate at higher temperatures (since they are dormant) so they appear to have a very narrow temperature range for germination. In this section, then, we omit complications due to dormancy and consider only the control by temperature of germination in nondormant seeds.

Temperature affects both the capacity for germination and the rate of germination. Seeds have the capacity to germinate over a defined range, characteristic for each species (Fig. 5.35); hence, there are clear minimum and maximum temperatures for germination, and between them a broad range over which germination of all seeds can be attained. Species can have widely different temperature minima and maxima: a few examples are shown in Table 5.14. It is sometimes useful for experimental and descriptive purposes not to state maximum or minimum temperatures, but to define the temperature at which 50% germination occurs. These are referred to as the GT_{50}, or the temperature cutoff points. Interestingly the GT_{50} of several species is affected by growth regulators. The cytokinin kinetin raises the upper GT_{50} for lettuce seeds in the light from 31° to 40°C, for example.

Although all seeds of a species may germinate over a fairly wide range of temperatures, the time needed for the maximum germination percentage to be

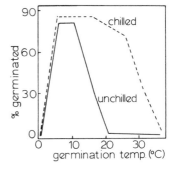

Figure 5.34. Temperature effects on dormancy and germination. Unchilled seeds of *Delphinium ambiguum* were set to germinate at different temperatures. A similar experiment was carried out with seeds that had been chilled at 6°C for 2 wk to remove dormancy. All the unchilled seeds are dormant at 20°C and above. The chilled seeds have no dormancy and their temperature range for germination extends to well above 30°C. Data adapted from Ezumah (Ph.D. thesis, 1980).

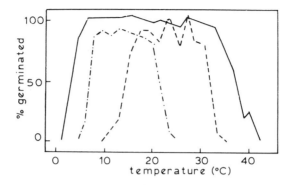

Figure 5.35. Temperature ranges for germination. Nondormant seeds of four different species were set to germinate over the range 1–43°C. Note that each species has a distinct range, which differs so far as the maximum and minimum temperatures for germination are concerned. ——, *Gypsophila perfoliata;* ·······, *Allium porrum;* ----, *Lychnis flox-cucili.* Adapted from Thompson (1973).

reached varies with the temperature; that is to say, the rate of germination is temperature dependent, the maximum rate occurring over a range of just 1–2°C—in the example shown, at about 30°C (Fig. 5.36). Because of the high rate of germination at certain temperatures, a seed population whose germination total is determined early in a test appears to have a narrow optimum temperature (Fig. 5.37), but this, of course, is only the optimum after a limited time period. The Q_{10} for germination can be estimated from germination rates. Since the rate curves generally consist of two linear portions (Fig. 5.36), the Q_{10} values on both the rising and falling parts of the curve are not constant. In *Dolichos,* for example, as temperatures increase, the Q_{10} between 10° and 20°C is approximately 5, while that between 15° and 25°C is 2.5. Because of the differences in rate, seeds in a population germinate at different times. A

Table 5.14. Temperature Maximum and Minimum for Seed Germination of Some Species

Species	Minimum (°C)	Maximum (°C)
Allium porrum	7	23
Apium graveolens (cv. Golden self-blanching)	10	15
Brassica oleracea (var. *gemmifera*)	4	42
Dolichos biflorus	6	42
Gypsophila perfoliata	2	40
Lychnis flos-coculi	9	35
Lycopersicon esculentum cv. Top cross	12	36
Silene gallica	2	32

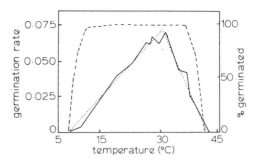

Figure 5.36. Temperature and the rate of germination. Seeds of *Dolichos biflorus* were allowed to germinate at different temperatures. Percentages of germinated seeds were recorded (----) and the rate of germination (1/hours taken to reach maximum percentage) was determined (——). The regression line for the germination rate is shown (· · · · ·). Adapted from Labouriau and Pacheco (1979).

few succeed in germinating quickly even at low temperatures, but the majority need more time. Thus, to a certain extent, this heterogeneous response to temperature acts like dormancy, by distributing germination in time.

Rate data for germination can be treated mathematically in the same manner as an enzyme reaction, and activation energies can be calculated. The values obtained are those characteristic of protein denaturation, and so it is suggested that as the temperature rises, changes in protein conformation occur which actually promote the germination process, but further conformational changes occur which are deleterious to it as temperatures become too high. Because of the lack of agreement between effects of temperature on respiration and on germination, an increase in respiratory metabolism is not thought to account for the increase in germination rate as the temperature rises to the optimum. There are indications, however, that changes in membrane states could be involved.

5.5.2.4. Water Stress

Water stress can reduce both the rate and percentage of germination. The range of response among species is wide, from the very sensitive (e.g., soybean)

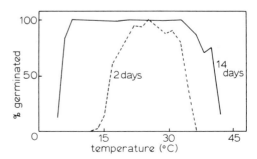

Figure 5.37. Time and the temperature optima for germination. Brussels sprout seeds (*Brassica oleracea* var. *gemmifera*) were set to germinate over the temperature range 4–42°C. Germination percentages were recorded after 2 and 14 days. The highest rates of germination are at temperatures around 25°C, but all the seeds are capable of germinating at temperatures from about 10° to about 35°C. Adapted from Thompson (1973).

Table 5.15. Effect of Water Stress on Germinability and Subsequent Radicle Growth

Species	Radicle length (mm)[a]			
	0 bar	−3 bar	−6 bar	−10 bar
Dandelion	13	15	6	0
Bitter sneezeweed	4	1	1	0
Hemp sesbania	10	7	0	0
Jimson weed	33	22	13	0
Soybean	68	12	0	0
Prickly sida	32	38	33	16
Pearl millet	124	125	94	102

[a]Germinated for 96 h at 29°C

to the resistant (e.g., pearl millet) (Table 5.15). Resistant seeds may have an ecological advantage in that they can establish plants in areas in which drought-sensitive seeds cannot do so.

Water stress is sometimes used in the laboratory, and to some extent in horticulture, to delay germination. Seeds treated with osmotica carry out many of the germination processes, but radicle growth cannot occur: when introduced to water, these seeds, which are nearly all at the same germination stage, show rapid, almost synchronous radicle emergence (Fig. 5.38). Advantages of this treatment—called osmotic priming—are that treated seeds take less time to germinate, and they produce a crop of uniform age (See Chapter 9).

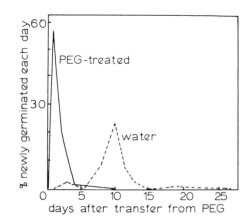

Figure 5.38. Synchronizing radicle emergence by osmotic treatments. Onion seeds were treated with a solution of polyethylene glycol (PEG) at −10 bar for 23 days at 10°C. They were then transferred to water for germination (——). Controls (i.e., seeds not given a prior PEG treatment) are shown (-----). Note that nearly 80% of the PEG-treated seeds germinate in the first 2 days on water. After Heydecker et al. (1973).

5.5.2.5. Oxygen and Carbon Dioxide

Seed germination in the majority of species is delayed by oxygen concentrations less than that in air. A few exceptions occur, and seeds of certain aquatic plants (for example *Typha latifolia*) are actually inhibited by air. High carbon dioxide concentrations (e.g., 4%) can prevent germination of several species (e.g., *Capsella bursa pastoris*). In the terrestrial habitat, concentrations of oxygen low enough to affect germination adversely are rarely found. The air in soil, for example, contains about 19% oxygen and no more than 1% carbon dioxide; the latter is not likely to inhibit germination.

USEFUL LITERATURE REFERENCES

SECTION 5.2

Ballard, L. A. T., 1973, *Seed Sci. Technol.* **1**:285–303 (seed coat effects).

Barton, L. V., 1965, in: *Encyclopedia of Plant Physiology,* Volume 15/2 (W. Ruhland, ed.), Springer, Berlin, pp. 909–924 (general review on dormancy types).

Barton, L. V., 1965, in: *Encyclopedia of Plant Physiology,* Volume 15/2 (W. Ruhland, ed.), Springer, Berlin, pp. 727–745 (coat-imposed dormancy).

Bewley, J. D., and Black, M., 1982, *Physiology and Biochemistry of Seeds,* Volume 2, Springer, Berlin (all aspects of dormancy).

Cavers, P. B., and Harper, J. L., 1966, *J. Ecol.* **54**:367–382 (polymorphism in *Rumex*).

Côme, D., and Thevenot, C., 1982, in: *The Physiology of Biochemistry of Seed Development, Dormancy and Germination* (A. A. Khan, ed.), Elsevier Biomedical Press, Amsterdam, pp. 271–298 (embryo dormancy).

Coumans, M., Côme, D., and Gaspar, T., 1976, *Bot. Gaz.* **137**:274–278 (oxygen consumption by beet seed coats).

Datta, S. C., Evenari, M., and Gutterman, Y., 1970, *Isr. J. Bot.* **19**:463–483 (polymorphism in *Aegilops*).

Edwards, M. M., 1969, *J. Exp. Bot.* **20**:876–894 (oxygen and coat effects in *Sinapis arvensis*).

Esashi, Y., and Leopold, A. C., 1968, *Plant Physiol.* **43**:871–876 (coat strength and embryo thrust).

Esashi, Y., Watanabe, K., Ohhara, Y., and Katoh, H., 1976, *Aust. J. Plant Physiol.* **3**:701–710 (ethylene and oxygen in *Xanthium*).

Hamly, D. H., 1932, *Bot. Gaz.* **93**:345–375 (seed coat structure).

Le Page-Degivry, M.-T., 1973, *Biol. Plant.* **15**:264–269 (ABA and embryo dormancy).

McKee, G. W., Pfeiffer, R. A., and Mohsenin, N. N., 1977, *Agron. J.* **69**:53–58 (seed coat impermeability to water).

Porter, N. G., and Wareing, P. F., 1974, *J. Exp. Bot.* **25**:583–594 (coat permeability to oxygen).

Rolston, M. P., 1978, *Bot. Rev.* **44**:365–396 (coat impermeability to water).

Taylorson, R. B., and Hendricks, S. B., 1977, *Ann. Rev. Pl. Physiol.* **28**:331–354 (review on dormancy).

Thevenot, C., and Côme, D., 1973, *C. R. Acad. Sci.* Ser. D **277**:1873–1876 (cotyledons and embryo dormancy).

Wareing, P. F., 1965, in: *Encyclopedia of Plant Physiology,* Volume 15/2 (W. Ruhland, ed.) Springer, Berlin, pp. 909–924 (germination inhibitors).

SECTION 5.3

Bewley, J. D., and Black, M., 1982, *Physiology and Biochemistry of Seeds,* Volume 2, Springer, Berlin (metabolism, control of dormancy).

Hendricks, S. B., and Taylorson, R. B., 1979, *Proc. Natl. Acad. Sci.* USA **76**:778–781 (membranes, dormancy, and germination).

Roberts, E. H., and Smith, R. D., 1977, in: *The Physiology and Biochemistry of Seed Dormancy and Germination* (A. A. Khan, ed.), North Holland, Amsterdam, pp. 385–411 (pentose phosphate pathway and germination).

Simmonds, J. A., and Simpson, G. M.,1971, *Can. J. Bot.* **49**:1833–1840 (respiration of dormant wild oats).

SECTION 5.4

Balboa-Zavala, O., and Dennis, F. G., 1977, *J. Am. Soc. Hort. Sci.* **102**:633–637 (ABA and the onset of dormancy in apple).

Cresswell, E. G., and Grime, J. P., 1981, *Nature (London)* **291**:583–585 (green enclosing tissue and the onset of dormancy).

Gutterman, Y., 1980/81, *Isr. J. Bot.* **29**:105–117 (onset of dormancy).

Hayes, R. G., and Klein, W. H., 1974, *Plant Cell Physiol.* **15**:643–563 (spectral quality of light and onset of dormancy).

Karssen, C. M., 1970, *Acta Bot. Neerl.* **19**:81–94 (photoperiodic induction of dormancy in *Chenopodium*).

Karssen, C. M., 1980/81, *Isr. J. Bot.* **29**:45–64 (secondary dormancy).

Sawhney, R., and Naylor, J. M., 1979, *Can. J. Bot.* **57**:59–63 (genetic aspects of dormancy in *Avena fatua*).

Sidhu, S. S., and Cavers, P. B., 1977, *Bot. Gaz.* **138**:174–182 (onset of dormancy in *Medicago*).

SECTION 5.5

Bartley, M. R., and Frankland, B., 1982, *Nature (London)* **300**:750–752 (high-irradiance inhibition of germination).

Bewley, J. D., 1979, in: *The Plant Seed* (I. Rubenstein, R. L. Phillips, C. B. Green, and B. E. Gengenbach, eds.), Academic Press, New York, pp. 219–239 (hormonal and chemical effects on dormancy).

Black, M., 1980/81, *Isr. J. Bot.* **29**:181–192 (hormones and dormancy).

Borthwick, H. A., Hendricks, S. B., Toole, E. H., and Toole, V. K., 1954, *Bot. Gaz.* **115**:205–225 (action spectrum for breaking of dormancy in lettuce).

Carpita, N. C., Nabors, M. W., Ross, C. W., and Petretic, N. L., 1979, *Planta* **144**:225–233 (phytochrome and cell elongation).

Esashi, Y., Ishihara, N., Saijoh, K., and Saitoh, M., 1983, *Plant Cell Environ.* **6**:47–54 (cyanide-resistant respiration and dormancy).

Hanke, J., Hartmann, K. M., and Mohr, H., 1969, *Planta* **86**:235–241 (phytochrome photoequilibria).

Hartmann, K. M., 1966, *Photochem. Photobiol.* **5**:349–354 (absorption spectrum of phytochrome).

Hendricks, S. B., and Taylorson, R. B., 1978, *Plant Physiol.* **61**:17–19 (temperature shifts and phytochrome).

Heydecker, W., Higgins, J., and Gulliver, R. L., 1973, *Nature* **246**:42–44 (advancing germination with PEG).

Kendrick, R. E., 1976, *Sci. Prog. (London)* **63**:347–367 (light and germination).

Kendrick, R. E., and Frankland, B., 1982, *Phytochrome and Plant Growth.* Arnold, London (a general account of phytochrome).

Kendrick, R. E., and Spruit, C. J. P., 1977, *Photochem. Photobiol.* **26**:201–204 (phytochrome intermediates).

Koller, D., and Negbi, M., 1959, *Ecology* **40**:20–36 (dual effect of light on generation).

Labouriau, L. G., and Pacheco, A., 1979, *Bol. Venez. Cienc. Nat.* **34**:73–112 (temperature and germination).

Mohr, H., and Appuhn, U., 1963, *Planta* **60**:274–288 (high-irradiance far-red effects).

Nabors, M. W., and Lang, A., 1971, *Planta* **101**:1–25 (light and embryo water relations).

Roberts, E. H., 1965, *J. Exp. Bot.* **16**:341–349 (temperature and afterripening).

Roberts, E. H., and Smith, R. D., 1977, in: *The Physiology and Biochemistry of Seed Dormancy and Germination* (A. A. Khan, ed.), North Holland, Amsterdam, pp. 385–411 (pentose phosphate pathway and germination).

Rollin, P., 1972, in: *Phytochrome* (K. Mitrakos and W. Shropshire, eds.), Academic Press, London, pp. 229–256 (light and seed germination).

Stokes, P., 1965, in: *Encyclopedia of Plant Physiology,* Volume 15/2 (W. Ruhland, ed.), Springer, Berlin, pp. 746–803 (temperature and seed dormancy).

Taylorson, R. B., 1982, in: *The Physiology and Biochemistry of Seed Development, Dormancy and Germination* (A. A. Khan, ed.), Elsevier Biomedical Press, Amsterdam, pp. 323–346 (light and other factors in seed germination).

Thomas, T. H., 1977, in: *The Physiology and Biochemistry of Seed Dormancy and Germination* (A. A. Khan, ed.), North Holland, Amsterdam, pp. 111–114 (growth regulators and germination).

Thompson, P. A., 1973, in: *Seed Ecology* (W. Heydecker, ed.), Butterworths, London, pp. 31–58 (temperature and germination).

Toole, V. K., 1973, *Seed Sci. Technol.* **1**:339–396 (light, temperature and other factors in germination).

Totterdell, S., and Roberts, E. H., 1979, *Plant Cell Environ.* **2**:131–137 (chilling of Rumex).

Totterdell, S., and Roberts, E. H., 1980, *Plant Cell Environ.* **3**:3–12 (alternating temperatures and germination).

Upadhyaya, M. K., Simpson, G. M., and Naylor, J. M., 1981, *Can. J. Bot.* **59**:1640–1646 (pentose phosphate pathway enzymes and dormancy).

Van Der Woude, W. J., and Toole, V. K., 1980, *Plant Physiol.* **66**:220–224 (chilling and phytochrome action).

Vincent, E. M., and Roberts, E. H., 1977, *Seed Sci. Technol.* **5**:659–670 (interacting factors and dormancy).

Visser, T., 1956, *Proc. K. Ned. Akad. Wet. C* **59**:314–324 (chilling and apple seed dormancy).

Webb, D. P., Van Staden, J., and Wareing, P. F., 1973, *J. Exp. Bot.* **24**:105–116 (chilling and hormonal changes).

Williams, P. M., Bradbeer, J. W., Gaskin, P., and MacMillan, J., 1974, *Planta* **117**:101–108 (chilling and gibberellins in hazel).

Chapter 6

Some Ecophysiological Aspects of Germination

6.1. INTRODUCTION

We pointed out in the previous chapter that the biological importance of dormancy must be seen in relation to the ecology of germination—when and where a seed germinates. Interactions between the dormancy-releasing agents—light, temperature, and afterripening—and the sensitivity of germination to light, temperature, and, say, water stress, are responsible for determining if a seed will germinate in a particular situation and season. The germination and dormancy mechanisms are therefore of great adaptive importance in ensuring that seedling emergence occurs at the most advantageous time and place. In this chapter we will consider some examples to illustrate the ecological significance of these control processes.

6.2. SEED BURIAL

6.2.1. The Seed Bank

A great quantity of seeds is buried in the soil—the seed bank. One determination for the top layer of cultivated land put the number of buried seeds therein at tens of millions per hectare. The distinction has been drawn between transient and persistent seed banks, the former consisting of seeds that mostly are viable in the soil for no more than 1 yr, and the latter having a significant proportion of the seeds remaining viable for many years. In Fig. 6.1, which illustrates the different types of seed bank, we can see the seasonal acquisition of germinability which gives each group of plants its characteristic emergence pattern. These seasonal patterns are largely controlled by the seeds' response to environmental factors, particularly light, temperature, and moisture. Also involved are seasonal environmental factors that govern the onset of dormancy. It is worth noting, however, that the environment may not have exclusive control, for in a limited number of species (e.g., *Mesembryanthemum* spp.) the

238

Chapter 6

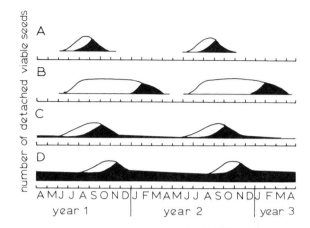

Figure 6.1. Seed banks common in soils of temperate regions. ■, seeds that can germinate immediately when tested in the laboratory (i.e., nondormant seeds). □, viable seeds that are dormant.
(A) Seeds of annual and perennial grasses of dry or disturbed habitats (e.g., *Lolium perenne*).
(B) Seeds that germinate in early spring, colonizing gaps in the vegetation. They include annual
and perennial herbs, trees, and shrubs (e.g., *Impatiens glandulifera, Acer pseudoplatanus*). (C)
Seeds of winter annuals, which germinate mainly in the fall. A small, persistent seed bank of
nondormant seeds is maintained (e.g., *Erophila verna*). (D) Seeds of annual and perennial herbs
and shrubs, which have large, persistent seed banks (e.g., *Stellaria media*). After Grime (1979).

seeds seem to have an inbuilt rhythm of germinability that governs seedling
emergence.

6.2.2. Light and Seed Burial

Clearly, germination of buried seeds is held in check, sometimes for many
years, perhaps until some disturbance occurs. If cultivated soil is turned over,
or if natural disturbance takes place as in a river bank, many of the seeds
germinate and a flush of seedlings results. One investigation reported that
weeds such as *Sinapis arvensis, Polygonum aviculare, Veronica persica,* and
others increased from about 35 to 780 seedlings per square meter when the
ground was dug. A factor that has major responsibility for this effect is light,
for when steps are taken to exclude light when the earth is disturbed, very few
seedlings appear: and as little as 0.5 sec exposure can cause many seeds to
germinate and product seedlings. But many of the species that appear as a
consequence of soil disturbance and the accompanying illumination do not dis-
perse light-requiring seeds. And so the seeds of these species (such as *Hera-*

cleum sphondylium, Senecio vulgaris, and *Spergula arvensis*) must develop a light requirement after their dispersal, i.e., during burial. This is probably because burial induces secondary dormancy by mechanisms that are still unexplained. Light sensitivity can also be lost during burial, as in *Barbarea vulgaris* and *Chenopodium bonus-henricus.* Seeds of both these species are light requiring when freshly dispersed, become resistant to light after a period of burial, and then acquire light sensitivity once more. Throughout this sequence all the seeds are completely dormant in darkness (Fig. 6.2). The development of a type of secondary dormancy called skotodormancy could account for the loss of photosensitivity; this is a phenomenon that occurs in light-requiring seeds when they are held in darkness for extended periods of time. Why photosensitivity is regained is unclear, but it might be a consequence of chilling; in Chapter 5 we mentioned that prior treatment with low temperatures can confer photosensitivity on seeds of several species. The onset of secondary dormancy in buried seeds can be an important contribution to the germination strategy of a species. We will discuss some examples later in our considerations of the role of temperature in the temporal distribution of germination.

Burial is also important in protecting seeds from the inhibitory effects of light (Section 5.5.2.1). For example, several cultivars of lettuce seeds are unable to germinate in full sunlight but can do so when buried. It has been observed that the light-inhibited *Nemophila menziesii* germinates in cracks in the earth where the seeds are shaded from direct sunlight, whereas seeds a few centimeters distant, exposed on the soil surface, are inhibited.

The transmittance of light in soil is clearly of great importance in light-controlled dormancy and germination. How effective is the soil as a light filter and how deep must seeds be buried to escape from the effects of light? Less than 2% of the light passes through 2 mm of sand, and the light that is transmitted consists only of wavelengths longer than about 700 nm (Fig. 6.3). A wide part of the spectrum can pass through clay loams, but if the grain size is small (when they will be closely packed), a layer 1.1 mm thick is virtually opaque. Soil moisture content affects transmission, increasing it somewhat in the clay loam but decreasing it in larger-grained sand. Lettuce seeds remain dormant under 6 mm of sand even when the surface is illuminated, but beneath

Figure 6.2. Changes in light sensitivity of buried seeds of *Barbarea vulgaris.* Seeds were buried at a depth of 2.5 cm and samples were taken periodically for germination tests in darkness (●) or after red light (o). After Taylorson (1972).

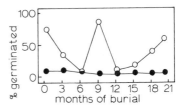

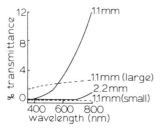

Figure 6.3. Transmission of light through soils. Two soils were used, sand (——) and a clay loam (----), at thicknesses indicated on the curves and, in the case of clay, at large (approximately 0.9 mm) and small (approximately 0.45 mm) grain size. Adapted from Woolley and Stoller (1978).

a thinner layer (2 mm), about 24% of the seeds eventually germinate when given 1 h of illumination, and 72% are promoted by 10 h of light. These values are a good illustration of the importance of burial depth to a light-requiring seed. Inhibition of germination by other soils may require greater depths, however. About 8 mm of loam are needed to completely suppress germination of *Sinapis arvensis* (Fig. 6.4). The quality of transmitted light is, of course, as important as the quantity. Since the longer wavelengths penetrate earth more easily, a buried seed experiences a relatively high proportion of far-red light (i.e., > 700 nm), which obviously affects the phytochrome photoequilibrium. Light passing through 1 mm of dry, sandy soil sets up a photoequilibrium of about 0.45, which should be sufficient to break dormancy in a high percentage of light-requiring seeds. That it does not necessarily do so may be because at the very low fluence rates in the soil the rate at which Pfr is formed is extremely low, and the rate of thermal reversion may succeed in removing a large proportion of the Pfr. The germination of dormant, light-requiring seed is therefore restricted to the uppermost layers of soil where the light stimulus for dormancy breakage can operate. This is extremely important for small seeds with limited food reserves, because if germination occurs at too great a depth the seedling's reserves may be exhausted before it is able to reach daylight and begin to photosynthesize.

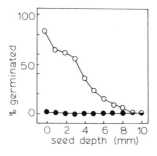

Figure 6.4. Effect of burial on germination of *Plantago major*. Seeds were buried at different depths in wet sunlit soil (o) or in wet soil held in darkness (●). The percentage of germinated seeds at each depth was recorded. Adapted from Frankland and Poo (1980).

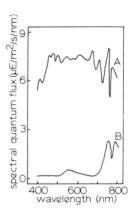

Figure 6.5. Spectral energy distribution of sunlight and vegetational shade light. (A) Direct sunlight at 11.00 h. (B) Shade light (under sugar beet plants). The ratio of red to far-red light (660/730 nm) in A is 1.18 and in B is 0.12. After Holmes and McCartney (1975).

6.3. GERMINATION UNDER LEAF SHADE

Chloroplasts absorb light most strongly at approximately 675 nm and allow wavelengths longer than 720 nm to be transmitted completely. Because of this, sunlight passing through green leaves has a very low red/far-red ratio (Fig. 6.5). The phytochrome photoequilibrium established by such light can be as low as 0.15, if the light has passed through a sufficient number of leaves. Most light-requiring, dormant seeds will not germinate under such conditions. Indeed, if they are exposed to this light for many hours over several days they will probably be forced into secondary dormancy. Nondormant seeds are also inhibited by canopy light (Table 6.1) for two reasons: first, the Pfr that some contain is reversed to Pr, and second, if the far-red rich light is at a high enough fluence rate, it will inhibit through the high-irradiance reaction (Section 5.5.2.1). These seeds therefore become dormant as a result of irradiation with canopy light and will germinate later only when a dormancy-releasing factor, such as chilling or light, has been experienced. Seeds of *Bidens pilosa,* for example, when forced into dormancy by canopy light, become light-requiring

Table 6.1. Some Nondormant Seeds Whose Germination is Inhibited by Light Filtered through Green Leaves

Achillea millefolium	*Leontodon hispidus*
Amaranthus caudataus	*Melandrium album*
Bidens pilosa	*Ruellia tuberosa*
Bromus tectorum	*Rumex acetosa*
Cucumis sativus	*Tragopogon orientalis*
Lactuca sativa	*Veronica arvensis*

and subsequently germinate when exposed to light of suitable quality to establish a relatively high photoequilibrium value of phytochrome: direct sunlight would, of course, suffice. This mechanism accounts for the paucity of seedlings on forest floors and the flush that follows the appearance of a gap in the leaf canopy, brought about when individual trees die and fall or when tree clearance occurs. That is, the seed bank in forest soils may be rich but germination is held in check, to a large extent by the far-red environment. One example will illustrate this. In forests of Brazil, seeds of *Cecropia glaziovi* are distributed by the bats and other small mammals; yet seedlings are not found there. These emerge only after tree clearance. In the field, vegetational shade is provided by grasses and herbs, of course, as well as by trees. Plants of *Arenaria* spp., *Veronica* spp., and *Cerastium* spp. establish themselves on vegetation-free ant mounds but not in the surrounding pasture. Seeds of these species are inhibited by light filtered through green leaves, which probably explains their difficulty in germinating among grasses and herbs: they germinate well on bare earth. Some species are, of course, able to colonize shaded sites. It is known that seeds of some of these species, e.g., *Centaurium erythrea,* are certainly less sensitive to far-red canopy light.

As mentioned, the spectral energy distribution of canopy light is determined by the number of leaves through which the light passes. And hence, the effects on germination will vary according to the density of leaf cover. As far as we know, no investigations on this have been carried out in field situations, but it has been simulated under experimental conditions using seeds of *Plantago major.* When placed under increasing leaf cover, provided by *Sinapis alba* plants, germination of these light-requiring seeds is increasingly inhibited as the cover becomes more dense (measured as the leaf area index) and the phytochrome photoequilibrium beneath the leaves decreases (Fig. 6.6). Variations in the density of a leaf canopy occur in the field, of course, as buds break and leaves expand and later senesce and fall. Thus, light-sensitive seeds among vegetation are exposed to changing light environments throughout the growing season, with concomitant effects on their dormancy breakage and germination.

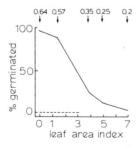

Figure 6.6. Germination of light-requiring seeds under leaf cover. Seeds of *Plantago major* were sown beneath plants of *Sinapis alba* (white mustard) at different densities, as indicated by the leaf area index (i.e., unit leaf area per unit land area). $Pfr/Ptotal$ (ϕ) values (at soil level) at different densities are indicated at the top of the graph. -----, germination in darkness. From data in Frankland and Poo (1980).

We must appreciate, however, that other environmental factors, such as temperature, are also influential, so that even when shade ends, seed germination will not necessarily proceed.

6.4. TEMPERATURE

Temperature acts to regulate germination in the field in three ways: (a) by determining the capacity and rate of germination, (b) by removing primary and/or secondary dormancy, and (c) by inducing secondary dormancy. Unfortunately, it is not always possible to distinguish from the available experimental work which of these is involved, but where we are able to do so we will make the distinction.

In Chapter 5, we saw how species have a particular temperature range for germination, and how temperature also determines germination rate. This sensitivity to temperature limits germination of some species to particular times of the year. *Amaranthus retroflexus* prefers relatively high temperatures for germination (approximately $> 25°C$), whereas *Chenopodium album* and *Ambrosia artemsiifolia* are both satisfied by lower temperatures ($> 10°C$ and $> 15°C$, respectively). Of these three, then, *Amaranthus* germinates only in late spring and early summer in Kentucky, but the other two can germinate in early spring. The *Ambrosia* seeds that do not succeed in germinating in early spring enter secondary dormancy and must await chilling to be released from this. The combination of a low-temperature requirement for germination and the induced secondary dormancy thus serves to restrict *Ambrosia* germination to early spring. *Nemophila menziesii* is another interesting example that illustrates the ecological importance of the germination temperature. Although these seeds are light-inhibited dark germinators, they are nevertheless incapable of germinating even in darkness at temperatures above a certain value, and freshly mature seeds are inhibited by a lower temperature than are older seeds (Fig. 6.7). In California, the seeds are dispersed in springtime (the plant is a winter annual) when the mean daily temperature is about $15°C$, and so, even though moisture is available, they do not germinate. To do so, the seeds must await the passage of time, which widens their temperature tolerance, and then they can germinate in September. This strategy prevents germination in spring, after which the young seedlings would be exposed to the hot, dry summer, and restricts it to late summer, just prior to the wet fall and winter. Summer and winter annuals of the flora of the Mojave and Colorado deserts also use their germination temperature requirement to achieve seasonal emergence—the winter annuals germinating only at lower temperatures (e.g., $10°C$) and the summer ones only at higher values (e.g., $26-30°C$).

Seeds whose dormancy is broken by chilling germinate when temperatures

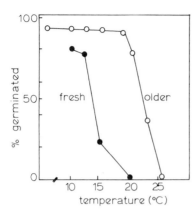

Figure 6.7. Sensitivity to temperature of germination in *Nemophila menziesii*. The percentage of seeds germinating at different temperatures is shown for freshly mature seeds (●) and seeds several months old (○). After Cruden (1974).

begin to rise in early spring, which has dual advantages. First, seedling emergence immediately before the inclement winter is prevented. Second, seedlings can establish themselves at a favorable time when there is no shading by leaves. Plants inhabiting meadows, for example, benefit from this mechanism because vegetation density in such habitats is lowest in early spring. Species inhabiting deciduous woodlands, including the nascent tree seedlings themselves, also derive benefit from germinating before the tree leaf canopy appears.

The bluebell *(Hyancinthoides non-scripta)*—a herbaceous plant of woodlands—uses another strategy involving temperature to ensure germination before the leaves are out. Dormancy in these seeds is best terminated by relatively high temperatures (26–31 °C) for a few weeks. Germination cannot occur immediately but it does take place at an optimum of approximately 11 °C. Many of the seeds shed in midsummer are warmed up in the soil and germinate later, in the fall. Seedlings in southern English woodlands therefore emerge in late fall, overwinter, and establish vigorous growth in spring before the development of the leaf canopy.

Afterripening, as a dormancy-relieving mechanism, operates under field conditions, and here, too, the importance of temperature can be discerned. For example, dormant seeds of the winter annuals *Sedum pulchellum* and *Valerianella umbilicata* require a fairly high temperature for afterripening, so that after being shed in winter they must experience the hot, dry summer before losing their dormancy: germination is therefore restricted to the fall and avoids the unfavorable midsummer droughts.

Dormancy of many species in the field is broken by alternating temperatures. As far as buried seeds are concerned, it seems that the vegetation cover is important, for it can influence the amplitude of the temperature fluctuation. This effect is seen rather well in an experiment in which the diurnal temperature fluctuations in soil under different sizes of gaps in the overlying vegetation

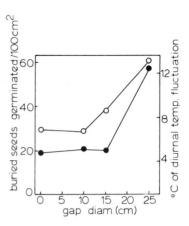

Figure 6.8. Effect of vegetation gap size on soil temperature fluctuation and seed germination. Gaps of different sizes were made in the grass of a pasture in northern England. The mean daily temperature fluctuation at a depth of 1 cm during May was recorded (○), together with the percentage of germination (●) of a naturally buried seed population of *Holcus lanatus*. After Grime (1979).

was determined, together with the germination of buried *Holcus lanatus* seeds (Fig. 6.8). Gap size clearly influences the amplitude of the temperature fluctuation, which in turn appears to correlate with the germination percentages. The temperature amplitude that a soil can achieve after vegetation clearance varies according to the depth at which it is measured (Fig. 6.9). It has been suggested that the dampening of the amplitude can be used by dormant seeds to sense the depth at which they are buried. Clearance of vegetation can therefore have important effects in addition to an alteration of the light environment in the top layer of the soil, and thus buried seeds can use a mechanism other than the phytochrome system to give them positional information.

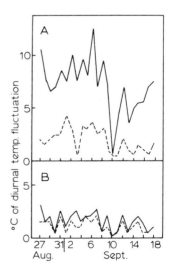

Figure 6.9. The effect of vegetation clearance on soil temperatures. The daily temperature fluctuation was recorded in soil at a depth of 1 cm (A) and 5 cm (B). The amplitude of fluctuation in gaps of 20-cm diameter cleared of overlying grass (——) and in uncleared soil (----) is shown. After Grime (1979).

6.5. WATER

Seeds of many species are nondormant when shed from the mother plant and can germinate over a wide temperature range. It seems that in these cases the major, perhaps the only, determinant of germination is the availability of water. Clearly, if the seeds are nondormant and germinate when sufficient moisture is present, there is no reservoir of seeds left in the soil; i.e., there is no persistent seed bank [(A) in Fig. 6.1]. Seeds of this type are characteristic of the common grasses of European meadows and pastures, such as *Festuca* spp., *Lolium* spp., *Bromus* spp., and *Dactylis glomerata.*

It will be appreciated that in desert habitats germination should be rigorously linked to water supply if the newly emerged seedlings are going to survive. This means that seedlings should not emerge when there is just enough water to support germination but rather when the soil has enough water in it to sustain subsequent plant growth. Interesting findings on the germination physiology of certain desert plants suggest how this might occur. Many seed species in desert soils germinate only after heavy rain (e.g., 12–15 mm), but not after light rain, even when the latter is composed of several, intermittent falls. Moreover, the precipitation (say 15 mm) is more effective when spread over 10 h than when received over 1 h. The explanation for this is that the seeds (e.g., of *Euphorbia* spp. and *Pectis papposa*) contain water-soluble germination inhibitors, which must be leached out before germination can commence. Thus, the seeds have a device that acts like a rain gauge and ensures that germination takes place only when the earth has received sufficient rain to support the establishment and growth of the seedling. Another way by which insufficient rainfall can be detected is through the seeds' sensitivity to salt solutions. Germination of *Baeria chrysostoma* is reported to be inhibited by a solution having an osmotic potential of -1 bar approximately, and it therefore occurs only when the salt in the top layer of earth has been diluted to an osmotic potential below this value. Desert seeds with very hard coats, such as *Cercidium aculeatum,* can germinate only after the seeds have been carried along by heavy rain in the company of abrasive sand and small stones, which scarify the testa. This species, in company with several other desert inhabitants, is found only in the former courses of such torrents.

6.6. INTERACTIONS

From time to time we have indicated that various factors operate together to control germination in the field. Perhaps a few more specific examples would be appropriate at this juncture. Chilling, light, and alternating temperatures appear to interact in *Aster pilosus,* for example, where at maturity seed dor-

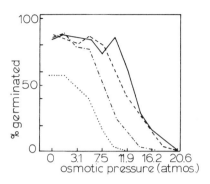

Figure 6.10. Interacting factors controlling germination of *Nemophila menziesii.* Seeds were held on various osmotic concentrations in darkness (——), in darkness at temperatures at incipient inhibition (4 h at 23°C, 20 h at 10°C) (----), in the light (4 h per day) (·······), or in the light plus inhibitory temperature (······). Maximum germination percentages are shown. Adapted from Cruden (1974).

mancy can be broken only by light, combined with high, alternating temperature. This cannot occur when the seeds are dispersed in the fall because the temperatures are not high enough. The temperature requirement is lowered by chilling, however, so that the seeds can then subsequently germinate in the alternating temperatures of the fairly cool spring of Kentucky. Germination before the very cold winter is thus prevented, and in the spring a vegetative rosette can be established before the summer drought begins. Temperature, light, and available water all regulate germination in *Nemophila menziesii.* We can see from Fig. 6.10 that high temperature and light, which are both inhibitory, would permit some germination as long as water stress was not serious. Thus, some *Nemophila* seeds might germinate in summer and fall in years when water is adequate, but not otherwise.

6.7. SECONDARY DORMANCY

Secondary dormancy is of great importance in germination ecophysiology, since it develops in many species exposed to unfavorable conditions. The residual level of dormant seeds in the persistent seed bank (Fig. 6.1) includes seeds that failed to germinate and entered secondary dormancy. Some of these seeds succeed in germinating at a later stage, and those which do not enter dormancy once more. This partially accounts for the rhythm of withdrawal from the seed bank indicated in Fig. 6.1. Seeds of winter and summer annuals are forced into dormancy by different factors. Winter annual seeds, such as those of *Veronica hederofolia* and *Phacelia dubia,* become dormant because of the low winter temperatures they experience after dispersal. They slowly emerge from dormancy over the summer, partly by afterripening, and germinate in the fall. Any seed that does not germinate goes back into secondary dormancy and follows the sequence once more. The rhythm of their germination (Fig. 6.11, A) therefore shows peaks in the fall seasons. The summer annuals (e.g., *Polygonum*

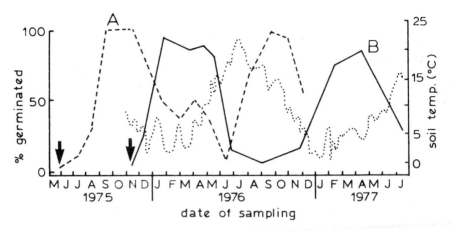

Figure 6.11. Patterns of seed dormancy during burial. The patterns of germination and dormancy of *Veronica hederofolia* (A) and *Polygonum persicaria* (B) are shown. Seeds of both species were buried in sandy loam at 5 cm *(Veronica)* or 10 cm *(Polygonum)*. Germination at alternating temperatures in the light was tested at intervals. The dotted line shows the morning soil temperature at the site of burial of *Polygonum*. N.B. *Veronica hederofolia* is a winter annual; *Polygonum persicaria,* a summer annual. Note that the two species enter a deep secondary dormancy at different times of the year, in response to low and high temperatures, respectively. After Roberts and Lockett (1978) and Karssen (1980/81).

persicaria), on the other hand, develop secondary dormancy in response to the high temperatures of summer. It is broken by chilling, at least in many of the seeds, so that germination occurs in springtime (Fig. 6.11, B). Again, the unsuccessful seeds are forced back into secondary dormancy, and reenter the bank of dormant seeds in the soil. Some of them germinate after the following winter. Their germination rhythm therefore peaks in springtime (Fig. 6.11, B), so the seeds of the winter and summer annuals are out of phase. These examples are a striking illustration of how the germination physiology has important ecological, adaptive significance.

6.8. GERMINATION, PLANT DISTRIBUTION, AND PLANT ORIGIN

In Chapter 5, some examples were given of species differences in the temperature requirements for germination. Such differences are important in determining the distribution of plants, for they obviously limit germination to regions that have suitable temperatures. It follows, also, that indigenous species of a particular region show characteristic temperature requirements, since they are adapted to the temperature conditions prevailing in the environment. To illustrate this we can take one family, the Caryophyllaceae, and see how a few members differ. One way of doing this is to define the "germination character" of a species by determining the time taken at each temperature to reach 50% of the maximum germination level. The curve so derived therefore expresses rate data and also indicates the cut-off temperatures for germination. Germination characters of three caryophyllaceous species are shown in Fig. 6.12, each typifying geographical origin. The species from the continental grassland (steppe) *(Petrorhagia prolifera)* shows quick germination at the favorable temperatures (12–40°C), with high minimum (ca. 8°C) and maximum (ca. 42°C) temperatures. The Mediterranean species *(Silene echinata)* also germinates fairly rapidly at median temperature but, in contrast to the grassland type, has rather low minimum (< 5°C) and maximum (ca. 25°C) requirements. Finally, the European woodland species *(Silene dioica)* has relatively slow germination at median temperatures, with a high minimum (10–15°C) and moderate maximum (ca. 35°C) temperature requirement. The Mediterranean character, belonging to the winter annuals, has been interpreted as favoring fall germination of the shed seed, in anticipation of the winter growing season. Seeds of the European woodland species, with a median temperature range, if shed in summer, would germinate at once, whereas fall-shed seed would have to await the following spring. Here, germination only in spring/summer is encouraged. The grassland species have an opportunist character and are able to germinate over a wide range of temperatures; they would germinate when the seeds are shed in mid- to late summer. Of course, the same species may be found in a wide variety of climatic regions: here, the germination behavior may differ according to provenance. This is seen in the case of *Tsuga canadensis* where distinct ecological types can be recognized—the seed coming from northerly latitudes (Quebec) showing distinguishable temperature requirements from those from southerly regions (Tennessee) (Fig. 6.13). The germination characters of representatives of a single species from different regions are very revealing (Fig. 6.14). Seeds of *Silene vulgaris* from three origins show the typical germination characters of the regional types (Mediterranean, continental grassland, and European woodland—cf. Fig. 6.12), so clearly, adap-

Figure 6.12. Germination "character" curves. Non-dormant seeds of three species were held at different temperatures, and the number of days taken to reach 50% germinated seeds was determined. a, *Silene echinata* (Mediterranean); b, *Petrorhagia prolifera* (continental grassland); c, *Silene dioica* (European woodland). Adapted from Thompson (1973b).

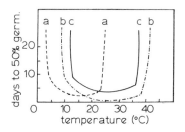

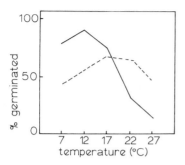

Figure 6.13. Germination of *Tsuga canadensis* seeds of two provenances. Seeds collected from Quebec, Canada (——), and Eastern Tennessee, U.S.A. (----), were cold stratified to remove dormancy. Germination tests were carried out over a range of temperatures: maximum germination percentages are shown. Adapted from Stearns and Olson (1958).

tation to the local conditions occurs within a species. On the other hand, some species have remained remarkably stable as far as their germination pattern is concerned, even in widely distributed populations. Figure 6.15 shows germination characters of seeds of *Agrostemma githago* (the weed corncockle) from four widely separated origins in Europe. Despite the geographical origins the characters are strikingly close and bear the greatest resemblance to the Mediterranean Caryophyllaceae shown in Fig. 6.12 (i.e., *Silene echinata*). It would seem, therefore, that *A. githago* has a Mediterranean origin and has retained this character in its spread through Europe, which accompanied the cultivation of wheat, i.e., over the last 3500–4500 years. Seeds of this species are difficult to separate from wheat grains, and so they have been sown and harvested along with the crop, generally in spring and late summer, respectively. Thus, although their germination is adapted to a Mediterranean climate, the seeds would not have been subjected to the same selective pressures that a feral population would experience. Interestingly, there is evidence of the decline of the species in Europe in the last 100 years, probably as seed-cleaning methods improved.

Finally, we should note that the germination character can give a clue as to the geographical origin of crop plants. When we examine the germination

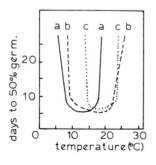

Figure 6.14. Germination "character" curves of *Silene vulgaris* seeds from different regions. Seeds of *Silene vulgaris* from three geographical localities were tested at a range of temperatures. The number of days for the germinated seeds to reach 50% was determined at each temperature. a, seeds from Portugal; b, seeds from Czechoslovakia; c, seeds from England. After Thompson (1973a).

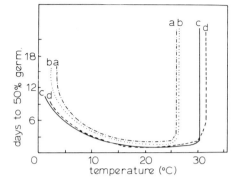

Figure 6.15. Germination "character" curves for *Agrostemma githago* seeds from different geographical localities. a, from Switzerland; b, from Poland; c, from Greece; d, from Germany. Adapted from Thompson (1973a).

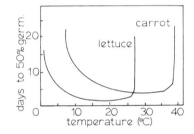

Figure 6.16. Germination "character" curves of lettuce and carrot seeds. Note the similarity of the curve for lettuce to the Mediterranean type (curve a), and that for carrot to the continental type (curve b) in Fig. 6.12. After Thompson (1973a).

characters of lettuce and carrot seeds (Fig. 6.16), we see similarities to the Mediterranean and continental types, respectively (see Fig. 6.12), which suggest their origins. Throughout a long history of cultivation, the characters of these species must have remained unchanged, though they do, of course, limit the climatic range over which cultivation can be practiced.

USEFUL LITERATURE REFERENCES

SECTION 6.2

Bewley, J. D., and Black, M., 1982, *Physiology and Biochemistry of Seeds,* Volume 2, Springer, Berlin, Chapter 6 (ecophysiological aspects of seeds).

Frankland, B., and Poo, W. K., 1980, in: *Photoreceptors and Plant Development* (J. DeGreef, ed.), Proc. Ann. Europ. Symp. Plant Photomorphogenesis, University Press, Antwerp, pp. 357–366.

Grime, J. P., 1979, *Plant Strategies and Vegetation Processes,* Wiley, New York (seed banks, dormancy in the field, temperature effects).

Harper, J. L., 1977, *Population Biology of Plants,* Academic Press, London, Chapters 2–5 (seed ecology).

Taylorson, R. B., 1972, *Weed Sci.* **20:**417–422 (light sensitivity of buried seeds).

Wesson, G., and Wareing, P. F., 1969, *J. Exp. Bot.* **20:**414–425 (buried seeds and light).

Woolley, J. T., and Stoller, E. W., 1978, *Plant Physiol.* **61:**597–600 (light transmittance of soil).

SECTION 6.3

Frankland, B., 1981, in: *Plants and the Daylight Spectrum* (H. Smith, ed.), Academic Press, New York, pp. 187–204 (shade and germination).

Frankland, B., and Poo, W. K., 1980, in: *Photoreceptors and Plant Development* (J. DeGreef, ed.), Proc. Ann. Europ. Symp. Plant Photomorphogenesis, University Press, Antwerp, pp. 357–366 (shade, light, and seed burial).

Holmes, M. G., and McCartney, H. A., 1975, in: *Light and Plant Development* (H. Smith, ed.), Butterworths, London, pp. 446–476 (shade light quality).

SECTIONS 6.4 AND 6.6

Baskin, J. M., and Baskin, C. C., 1977, *Oecologia* **30**:377–382 (temperature in germination ecology).

Cruden, R. W., 1974, *Ecology* **55**:1295–1305 (ecophysiology of *Nemophila* germination).

Grime, J. P., 1979, *Plant Strategies and Vegetation Processes,* Wiley, New York.

Karssen, C. M., 1980/81, *Isr. J. Bot.* **29**:65–73 (dormancy and seed burial).

SECTION 6.7

Baskin, J. M., and Baskin, C. C., 1980, *Ecology* **61**:475–480 (temperature and secondary dormancy).

Karssen, C. M., 1980/81, *Isr. J. Bot.* **29**:45–64 (secondary dormancy and the environment).

Roberts, A., and Lockett, P. M., 1978, *Weed Res.* **18**:41–48 (seed dormancy and burial).

SECTION 6.8

Stearns, F., and Olson, J., 1958, *Am. J. Bot.* **45**:53–58 (seed provenance and germination).

Thompson, P. A., 1973a, in: *Seed Ecology* (W. Heydecker, ed.), Butterworths, London, pp. 31–58 (geographical adaptation of seeds).

Thompson, P. A., 1973b, *Ann. Bot. (London)* **37**:133–154 (temperature and germination character).

Chapter 7

Mobilization of Stored Seed Reserves

The mobilization of the stored reserves in the storage organs invariably commences after radicle elongation; i.e., it is a postgerminative event. In the growing regions (i.e., axis) some mobilization can occur before germination is completed; here the reserves are generally present in minor amounts, although the products of their hydrolysis might be important for early seedling establishment.

As the high-molecular-weight reserves contained within the storage organs of the seed are mobilized, they are converted into a readily transportable form and at their sites of requirement (usually the most rapidly metabolizing and growing tissues) are utilized for energy-producing and synthetic events. Reliance on the stored reserves diminishes as the seedling emerges above the soil and becomes photosynthetically active (i.e., autotrophic). For the purpose of clarity we have divided this chapter into sections, each of which covers the mobilization of one major type of storage reserve. It must be remembered, however, that storage organs often contain substantial quantities of two or more major reserves (Table 1.2) and that hydrolysis and utilization of these can occur concurrently.

7.1. STORED CARBOHYDRATE CATABOLISM

Starch is the most common reserve carbohydrate in seeds, and hence most attention will be paid to this. Other stored carbohydrates of note are the hemicelluloses (such as galactomannans—see Chapter 1), which are to be found particularly as cell wall components of the endosperm of various legumes.

7.1.1. Pathways of Starch Catabolism

There are two catabolic pathways of starch: one hydrolytic and the other phosphorolytic.

The amylose and amylopectin in the native starch grain are first hydrolyzed by α-amylase, which breaks the α-1,4 glycosidic links between the glucose residues randomly throughout the chains. The released oligosaccharides are further hydrolyzed by α-amylase until glucose and maltose are produced.

$$\text{Amylose} \xrightarrow{\alpha\text{-Amylase}} \text{Glucose} + \text{maltose}$$

But α-amylase cannot hydrolyze the α-1,6 branch points of amylopectin, and hence highly branched cores of glucose units, called limit dextrins, are produced.

$$\text{Amylopectin} \xrightarrow{\alpha\text{-Amylase}} \text{Glucose} + \text{maltose} + \text{limit dextrin}$$

The small branches must be released by an enzyme specific for the α-1,6 link (debranching enzyme) before being hydrolyzed to the monomer. Limit dextrinase is specific for the small limit dextrins, and it cannot release branches from high-molecular-weight amylopectin; R-enzyme can attack both.

$$\text{Limit dextrins} \xrightarrow[\text{Limit dextrinase}]{\text{Debranching (R-)enzyme}} \text{Glucose oligomers} \xrightarrow{\alpha\text{-Amylase}} \text{Maltose} + \text{glucose}$$

α-Glucosidase

Glucose

Although β-amylase cannot hydrolyze native starch grains, it can cleave away successive maltose units from the nonreducing end of large oligomers released by prior α-amylolytic attack. Again, amylopectin cannot be completely hydrolyzed, and the involvement of a debranching enzyme is essential.

$$\text{Amylose} \xrightarrow{\beta\text{-Amylase}} \text{Maltose}$$

$$\text{Amylopectin} \xrightarrow{\beta\text{-Amylase}} \text{Maltose} + \text{limit dextrin}$$

The disaccharide maltose, produced by α- and β-amylase action, is converted by α-glucosidase to two glucose molecules.

$$\text{Maltose} \xrightarrow[\text{(Maltase)}]{\alpha\text{-Glucosidase}} \text{Glucose}$$

Starch phosphorylase releases glucose-1-phosphate by incorporating a phosphate moiety, rather than water, across the α-1,4 linkage between the penultimate and last glucose at the nonreducing end of the polysaccharide chain. Complete phosphorolysis of amylose is theoretically possible, and amylopectin can be degraded to within two to three glucose residues of an α-1,6 branch linkage.

$$\text{Amylose/amylopectin} + \text{Pi} \xrightarrow{\text{Starch phosphorylase}} \text{Glucose-1-P} + \text{limit dextrin}$$

The relative activities of the two amylases and phosphorylase vary between different seed species: β-amylase tends to be more active in germinated rice than in some other cereals, and although phosphorylase activity is low or negligible in cereals, it is appreciable in legumes. Examples are given in Table 7.1.

7.1.2. Synthesis of Sucrose

The products of starch (and triglyceride) catabolism are transported as sucrose into the root and shoot of the developing seedling. Glucose-1-phosphate (Glc-1-P) released by phosphorolysis can be used directly as a substrate for sucrose synthesis, but free glucose released by amylolysis must first be phosphorylated to glucose-6-phosphate (Glc-6-P) and then isomerized to Glc-1-P. This is combined with a uridine nucleotide (UTP) to yield the nucleotide sugar uridine diphosphoglucose (UDPGlc), which in turn transfers glucose to free fructose or to fructose-6-phosphate.

$$\text{Glc-1-P} + \text{UTP} \xrightarrow{\text{UDPGlc pyrophosphorylase}} \text{UDPGlc} + \text{PPi (pyrophosphate)}$$

$$\text{UDPGlc} + \text{fructose} \underset{\text{Sucrose synthetase}}{\overset{}{\rightleftharpoons}} \text{Sucrose} + \text{UDP}$$

Table 7.1. Activities of Starch-Hydrolyzing Enzymes during the Period of Rapid Starch Breakdown in Two Cereals and a Legume[a]

Enzyme	Barley	Rice	Pea
α-Amylase	34.4	31.8	19
β-Amylase	11.4	120	Very low
Starch phosphorylase	Negligible	0.09	14.6

[a]The values are in milligrams starch hydrolyzed per hour per seed. Based on Ap Rees (1974).

or

$$\text{UDPGlc} + \text{fructose-6-P} \underset{\longleftarrow}{\overset{\text{Sucrose-6-P synthetase}}{\longrightarrow}} \text{Sucrose-6-P} + \text{UDP}$$

The phosphate moiety is cleaved from sucrose-6-P by sucrose phosphatase. In the seedling tissues, sucrose can be hydrolyzed to free glucose and fructose by a β-fructofuranosidase (e.g., sucrase, invertase).

7.2. MOBILIZATION OF STORED CARBOHYDRATE RESERVES IN CEREALS

7.2.1. The Embryo Reserves

Although the endosperm is the major source of carbohydrate reserves in cereals, some low-molecular-weight sugars are stored in much lower quantities within the embryo. These probably provide an early source of respirable substrate during germination and early seedling growth, until the hydrolytic products of starch are made available from the endosperm. Sucrose and raffinose (galactosyl sucrose) are present in the ratio of about 3:1 in dry grains of Proctor barley, and both decline to imperceptible levels within the first 1–2 days from the start of imbibition. A replenishment of carbohydrate from the endosperm, as sucrose, begins on the third day.

7.2.2. The Endosperm Reserves

The first change during starch hydrolysis that the barley endosperm appears to undergo is the digestion of the intermediate layer adjacent to, and in between, the cells of the absorptive epithelium of the scutellum (Fig. 7.1). The enzymes mooted to be responsible for digestion of the cell walls of the intermediate layer are endo-β-glucanases, a class of enzymes capable of degrading hemicelluloses (glucans containing β-1,3 and β-1,4 links). In some cereal grains (e.g., those of the oat family, but not barley or wheat), the scutellum and its epithelial cells then elongate (Fig. 7.2, A and B), thus presenting a larger surface area for absorption of the hydrolytic products of the endosperm reserves.

The initial production of α-amylase occurs in the region of the scutellum, in some species of grain perhaps in the epithelial layer (e.g., rice), or entire scutellum (e.g., sorghum), and in others in the few aleurone cells that penetrate the peripheral regions of the scutellum (e.g., barley). In barley, the α-amylase

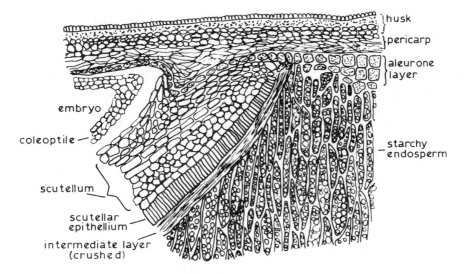

husk
pericarp
aleurone layer
embryo
coleoptile
starchy endosperm
scutellum
scutellar epithellium
intermediate layer (crushed)

Figure 7.1. Detailed structure of the barley grain in the region of the scutellum. Based on Briggs (1973).

is then released into the starchy endosperm and diffuses away from the scutellum (Fig. 7.3, A–C). Later, the enzyme is synthesized within the aleurone layer (Fig. 7.3 D) and is secreted inward to complete the hydrolysis of starch reserves. Although α-amylase from the scutellum may be important at early stages of mobilization, it is likely that later most hydrolysis is effected by enzyme from the aleurone layer. The relative contributions to total starch hydrolysis by the α-amylases from these two sources probably varies greatly between cereal species. The utilization of endosperm reserves in some cereals (or cultivars) is controlled by the embryo and mediated by the production of the plant hormone gibberellin (see Chapter 8 for details). Gibberellin is synthesized by the embryo during and after germination and is released through the scutellum into the starchy endosperm. It is known that the hormone diffuses to the aleurone layer and there sets off a sequence of events that culminate in the synthesis and release of α-amylase. It has not been clearly established whether the α-amylase released initially from the scutellum is also under hormonal control.

β-Amylase is involved also in the digestion of barley starch, but this enzyme is not *de novo* synthesized, nor is it released from either the scutellum or the aleurone layer. Instead, it is present in the starchy endosperm of the mature ungerminated grain (Fig. 7.3 E) where, presumably, it becomes active (or activated) after initial digestion of the starch grains by α-amylase. In wheat

Figure 7.2. (A) Scanning electron micrograph of the epithelial cells of the scutellum of germinated *Avena fatua* showing the extent of their elongation. (B) shows greater detail. Bar in (A) is 0.1 mm, and in (B) 0.05 mm. The scutellum itself grows through the starchy endosperm and

by 8 days from imbibition may extend from the proximal (embryo) end to the distal end. Courtesy of Dr. J. Sargent and Dr. M. Negbi.

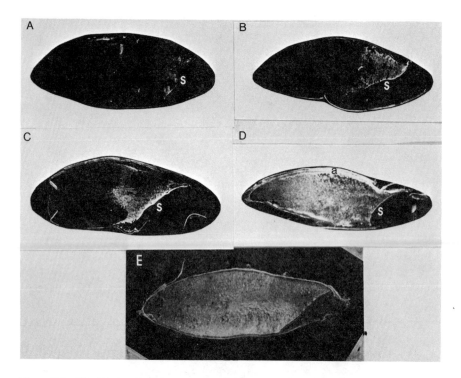

Figure 7.3. (A)–(D) Immunofluorescent localization of α-amylase in barley grains at various times after the start of imbibition. (A) After 30 h (fluorescence near scutellum). (B) After 54 h (fluorescence spreading from scutellum). (C) After 78 h (fluorescence spreading from scutellum; little near aleurone layer). (D) After 7 days (fluorescence throughout endosperm; more near aleurone layer). Rabbit antibodies specific to α-amylase were made (monospecific antibodies). Grains were germinated and at various times thereafter fixed and sectioned. The antibodies were then added, which tightly bound to the α-amylase in the sections. After this, a fluorescent probe attached to a rabbit antiserum protein was introduced to the sections (this bound specifically to the rabbit antibody: α-amylase complex) and was detected and photographed using a fluorescence microscope. (E) Immunofluorescent localization of β-amylase in mature dry barley grains using monospecific β-amylase antibodies. This enzyme is found throughout the starchy endosperm of the dry grain. The fluorescence in the outer coat region on all sections is an artifact of the technique. s, scutellum; a, aleurone layer. Photographs kindly provided by Dr. G. C. Gibbons. Details in Gibbons (1979).

kernels, a large proportion of the β-amylase is present in an inactive, latent form bound by disulfide linkages to protein (glutenin) bodies, or their remnants, in the starchy endosperm. Extracts of the enzyme can be activated through release from its bound form, either by digestion of the glutenin with proteolytic enzymes or by disrupting the binding linkages with sulfhydryl-containing compounds. In the grain, gibberellins stimulate the activation of β-amylase, probably by first increasing proteinase synthesis and release from the aleurone layer, which in turn effects the release of the bound enzyme. In rice grains, there is initially some *de novo* synthesis of β-amylase in the scutellum and release therefrom. Later there is activation of a latent form of this enzyme, which is bound to starch grains, within the endosperm.

Enzymic solubilization of the endosperm cell walls accompanies digestion of the starch reserves in many cereals (sorghum is one exception). In wheat grains, the cell walls contain about 15% protein and 75% polysaccharide, of which 85% is arabinoxylan. In the endosperm cell walls of barley about 74% of the carbohydrate component is a mixed-linkage (β-1,3 and β-1,4) glucan, with a lesser amount of arabinoxylan. Compared to primary cell walls there is much less cellulose and pectin and thus they are simpler in composition; consequently, their mobilization may be effected more easily. Cell wall dissolution may precede starch hydrolysis in any given cell, and thus the synthesis and release of arabinoxylanases (pentosanases) and glucanases (β-glucan solubilase, endo-β-glucanase) may, in turn, precede those of α-amylase. Initially then, in barley at least, the cell-wall-hydrolyzing enzymes may be released from the scutellum, and later from the aleurone layer. In wheat, release of pentosanases from this latter tissue results first in the hydrolysis of endosperm walls in the subaleurone layer, and then a wave of degradation progresses toward the center of the endosperm (Fig. 7.4), the disappearance of cell walls accompanying an increase in the formation of water-soluble arabinoxylans.

Complete amylolysis of the amylose component of starch can be achieved by α- and β-amylases, but digestion of amylopectin ceases with the production of limit dextrins. R-enzyme from barley malt can debranch amylopectin (although only after solubilization by α-amylase) and limit dextrin, whereas barley and oat limit dextrinases are specific for the smaller molecule. Little is known about the site and mode of production of the two debranching enzymes, but in barley it appears that at least the limit dextrinase is synthesized *de novo* in the aleurone layer at the time of starch breakdown, rather than being activated from a latent form. In rice grains, on the other hand, debranching enzyme is synthesized in an active form during the ripening stage of development, is present in an inactive and insoluble form in the dry seed, and is subsequently solubilized and activated after germination.

The major product of α- and β-amylolysis is maltose, which is hydrolyzed by α-glucosidase (maltase), an enzyme present in the embryo and aleurone layer of ungerminated barley. The enzyme increases in activity at both sites after germination although only that released from the aleurone layer is involved in the mobilization of endosperm reserves. That in the axis participates in local sugar metabolism.

A notable variation from this general pattern of events is found in

Figure 7.4. Scanning electron micrograph of a median cross section of a wheat grain after 2 days from imbibition to show dissolution of the cells of the endosperm in the subaleurone region (s). The aleurone layer (a) remains adhered to the grain coat. From Fincher and Stone (1974).

sorghum grains. Here, α- and β-amylase and α-glucosidase are present in the dry grain, sequestered in organelles (lytic bodies), along with other hydrolytic enzymes, e.g., proteinase, phosphatase, and nuclease. Presumably these enzymes are released at the appropriate time after germination to effect hydrolysis of the stored reserves.

7.2.3. The Fate of the Products of Starch Hydrolysis

The growing embryonic axis and the endosperm are separated by the shield-shaped scutellum (Fig. 7.1)—the modified single cotyledon of the Gramineae. The products of starch digestion (mostly glucose and maltose) are absorbed by the scutellum and converted there to sucrose. This is transported to the axis (perhaps through the phloem of the vascular system connecting this with the scutellum) and is utilized by the growing root and shoot tissues. Glucose is absorbed by the scutellum both passively and by active transport: maltose is probably hydrolyzed very quickly by α-glucosidase contained within the free space of the scutellar cells. Some of the absorbed glucose is converted to fructose-6-phosphate, but neither of these sugars accumulates in the scutellum, for they are quickly converted to sucrose, directly or via sucrose-6-phosphate.

For a generalized summary of events associated with endosperm reserve hydrolysis, based on the observations for barley grains, see Table 7.2.

Table 7.2. Probable Sequence of Events Involved in the Mobilization of Stored Starch Reserves in the Cereal Endosperm[a]

1. Completion of germination
2. Release (probably after synthesis) of gibberellins from embryo via scutellum
3. Dissolution of intermediate layer and expansion of epithelial layer cells
4. Release of α-amylase and cell-wall-hydrolyzing enzymes (after *de novo* synthesis) from scutellum into endosperm. Start of endosperm cell wall dissolution and starch hydrolysis
5. Release of α-amylase, α-glucosidase, debranching enzymes, and cell-wall-hydrolyzing enzymes (after *de novo* synthesis) from aleurone layer into endosperm. Grand phase of endosperm hydrolysis
6. Activation of β-amylase (concurrent with 4 and 5)
7. Adsorption of glucose by scutellum. Synthesis of sucrose therein for transport to growing axis (concurrent with 4–6)

[a]Note that this scheme is a general guide to events. Variations will occur between grains of different cereal species and cultivars, and timing of the events will vary depending on the conditions (e.g., temperature, moisture level) of germination and growth.

7.3. MOBILIZATION OF STORED CARBOHYDRATE RESERVES IN LEGUMES

Legume seeds have been divided into two groups on the basis of the location of their major stored reserves. The nonendospermic legumes are those in which the endosperm is absorbed during seed development, the cotyledons then becoming the major storage organs (e.g., pea, dwarf and broad bean). Endospermic legumes, which are members of the tribe Trifolieae, retain their endosperm until maturity, and in some seeds (e.g., fenugreek, carob, honey locust, and guar) it is the major storage organ although in others (e.g., soybean) it is only residual and the cotyledons store most of the reserves. Fewer studies have been conducted on the mobilization of legume carbohydrates than on that of cereals; nevertheless, a general pattern of events has emerged.

7.3.1. Nonendospermic Legumes

Early seedling growth in peas draws on reserves of carbohydrate, protein, and fat within the radicle itself, and sucrose, raffinose, and stachyose serve as the primary sources of respirable substrate. Both α- and β-amylase are present in the axes of ungerminated pea seeds, which also contain starch: β-amylase activity increases appreciably after germination, in association with the growth of the epicotyl. After these early events have passed, the further development of the root and shoot depends on a supply of hydrolytic products from the cotyledons. Their stored carbohydrate (and other reserves) are hydrolyzed and transported to the growing axis.

Hydrolysis of reserves in intact legume cotyledons commences after emergence and elongation of the radicle. The depletion of starchy reserves from the cotyledons of peas is biphasic (Fig. 7.5, A), an initial slow rate that lasts for 5–6 days being followed by a phase of rapid decline. The free sugars and dextrins released by starch hydrolysis are rapidly translocated to the growing axis; they do not accumulate to any extent within the cotyledons (Fig. 7.5, A). The initial slow degradation of starch might be by phosphorolysis since starch phosphorylase is the first enzyme to increase in activity in the cotyledons (Fig. 7.5, B). The more rapid degradation at 8 or 9 days probably requires amylolysis and is coincident with an increase in α- and β-amylase activity. It is not known what proportion of the increase in phosphorolytic and amylolytic activity is due to *de novo* synthesis of enzymes, and what proportion is due to inactivation of a latent form. Debranching enzyme (limit dextrinase) activity is high in dry pea seeds and remains so for several days before declining. This decline occurs prior to the major increase in amylolytic activity, and since the limit dextrinase can only debranch products of amylolysis, its role in the overall scheme of

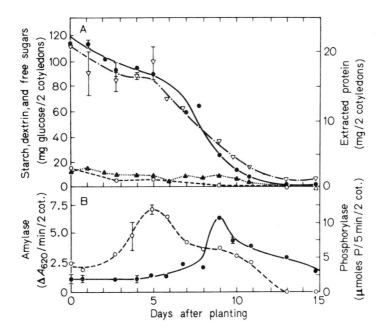

Figure 7.5. (A) Changes in the levels of starch and dextrin (●), dextrin (○), free sugars (▲), and extracted protein (▽) in the cotyledons of Alaska pea. (B) Changes in starch phosphorylase (○) and amylase (●) enzymes in Early Alaska pea. After Juliano and Varner (1969).

starch breakdown is enigmatic. Suffice it to say, at the present time, that debranching must accompany amylolysis, but the appropriate enzyme remains to be identified. In another legume, green gram *(Phaseolus vidissimus)*, α- and β-amylases are found in the cotyledons (at a ratio of 20:1) at the time of starch hydrolysis, and α-glucosidase is present too.

The immediate fate of the products of phosphorolysis and amylolysis is virtually unknown, although eventually they must be made available to the growing axis. Starch phosphorylase activity in the cotyledons of peas produces Glc-1-P; this reacts with UTP to form UDPGlc, which is then converted to sucrose by sucrose (phosphate) synthetase and sucrose phosphatase (see Section 7.1.2). Enzymes for the hydrolysis and transformation of sucrose are absent from the cotyledons but are present in the axis, to which this sugar is presumably transported. Since little α-glucosidase activity has been detected in the cotyledons, maltose might be a major transported form of sugar released by amylolysis, although some glucose will be produced directly by this process and also translocated to the axis. Only low levels of α-glucosidase have been detected in the axis, but it is presumed that this is where maltose is hydrolyzed.

It is possible that glucose released by amylolysis is phosphorylated and utilized in the same way as the products of phosphorolysis, but supportive evidence for this possibility is lacking.

Cotyledons of many legumes (e.g., pea, bean, lupin) contain cell walls that are rich in hemicelluloses. These are mobilized along with other major reserves, but little is known about their composition, their mode of hydrolysis, or resultant products. During the expansion of lupin cotyledons, arabinose and galactose are hydrolyzed from the cell walls and quickly metabolized, leaving a primary wall with a high uronic acid (pectin) content. Dissolution of the hemicellulose components of the cell wall usually accompanies mobilization of the major reserves (e.g., starch and protein) within the cytoplasm. Often, however, the integrity of the primary wall is maintained, in contrast to the complete degradation of the cell walls in the endosperms of cereals and endospermic legumes.

7.3.2. Endospermic Legumes

In many species of the Trifolieae a well-developed endosperm, comprised largely of the storage carbohydrate galactomannan, lies between the seed coat and the cotyledons. In fenugreek *(Trigonella foenum-graecum)* extensive deposition of this polymer on the inside of the primary walls during seed development results in the gradual occlusion of the living contents until in the mature seeds the cells are dead (Section 2.3.2). The outermost layer of the endosperm is the aleurone layer which, unlike the rest of the endosperm, is made up of living cells devoid of galactomannan reserves (Fig. 7.6, A). In some endospermic legumes, e.g., carob, the endosperm cell walls do not completely occlude the cytoplasm, and all cells have living contents.

Under suitable conditions of light and moisture, isolated embryos of fenugreek germinate and grow as well as embryos of intact seeds. Support for this growth comes from the hydrolytic products of proteins and lipids stored in the cotyledons. Hence the endosperm may not be an essential food source for the embryo, although it is difficult to extrapolate from experiments using isolated embryos, germinated and grown under ideal laboratory conditions, to the situation in the natural environment. It is clear, however, that the endosperm has another function besides that of a storage organ. The high affinity of galactomannans for water (when imbibed, many become quite mucilaginous) allows the endosperm to regulate the water balance of the embryo during germination; this may be important to plants in their native habitat, since many members of the tribe Trifolieae appear to have their origins in the dry regions of the eastern Mediterranean. Perhaps it is safe to assume that, under natural conditions, the endosperm has a dual role: to regulate water balance during germination, and to serve as a substrate reserve for the developing seedling after germination.

As in the cereal grains and seeds of nonendospermic legumes, the raffinose-series oligosaccharides (especially raffinose and stachyose) in the embryonic tissues of the fenugreek seed are the first carbohydrates to be utilized, by hydrolytic cleavage of the α-galactosidic link to yield sucrose and galactose. This commences soon after imbibition. After emergence of the radicle, the galactomannan in the endosperm begins to be mobilized. There is a wave of hydrolysis commencing close to the aleurone layer and moving toward the cotyledons (Fig. 7.6, B). This reflects the synthesis and release from this layer of at least three critical enzymes: α-galactosidase, β-mannosidase, and β-mannanase. α-Galactosidase is an exopolysaccharidase that cleaves the α-1,6 link between the unit galactose side chains and the mannose backbone.

α-Galactosidase   Gal   Gal

Man - Man - Man - Man - Man - Man - Man - Man

β-Mannanase   Gal

β-Mannanase is an endoenzyme that hydrolyzes oligomers of mannose (tetramers or larger) to mannobiose or mannotriose, and β-mannosidase then converts these residues to mannose. This latter enzyme might also act as an exomannopolysaccharidase and hydrolyze single mannose residues from the oligomannan chain. Mannan breakdown by phosphorolysis does not occur. A similar sequence of events occurs in other endospermic legume seeds.

Eventually the endosperm reserves become depleted (Fig. 7.6, C), and the released mannose and galactose are absorbed into the cotyledons by passive diffusion, but they do not accumulate there. Instead they are metabolized further, perhaps by initially being phosphorylated (to Gal-1-P and Man-6-P). If not used directly for energy metabolism, they might be transformed to sucrose and then even to starch, which can later be mobilized when the sucrose level in the cotyledons falls after its transport to the axis. A small amount of starch builds up in the cotyledons in the absence of the endosperm (Fig. 7.7), but the massive accumulation of starch occurs only in its presence, showing that it is the products of galactomannan breakdown that are made available for the major starch synthesis. Not surprisingly, an increase in α-amylase activity within the cotyledons coincides with starch hydrolysis (Fig. 7.7). A summary of the events involved in galactomannan breakdown in endospermic legumes is shown in Fig. 7.8. It should be noted, however, that not all the enzymes required for the conversion of mannose to sucrose and starch have been located within the cotyledons, although it is reasonable to assume they are there.

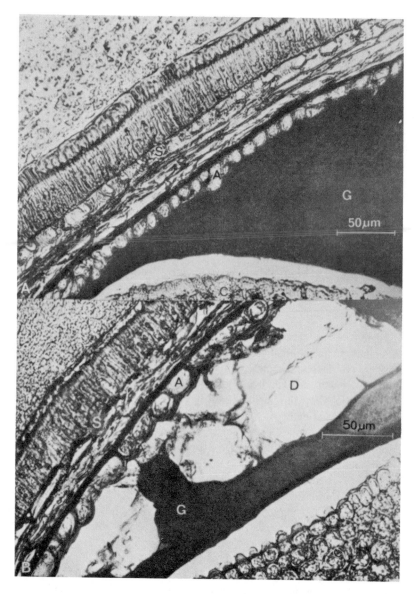

Figure 7.6. Light micrographs of the outer region of the seed of *Trigonella foenum-graecum* (fenugreek). (A) Before mobilization of the seed reserves. The three-layered seed coat (S), a small part of the cotyledon (C), and the endosperm layer (A and G) are shown. The aleurone layer (A) is the outer cell layer of the endosperm, the rest (G) being comprised of large cells with thin primary walls to the inside of which is deposited dark-staining galactomannan that appears completely to fill the cell. (B) During galactomannan breakdown the reserves in the

The similarities between the mobilization of galactomannans in fenugreek and of starch in cereals are quite striking, although the endospermic legume is unique in that the cotyledons do not contain starch until it is synthesized therein as a temporary reserve after the germination is completed. In both cases, reserve hydrolysis is by enzymes released from a living, peripheral aleurone layer, and the products are absorbed and modified by the cotyledons (the scutellum being a reduced cotyledon in cereals) before being passed to the growing axis. In a sense, therefore, more contrast is shown between the patterns of hydrolysis in the endospermic and nonendospermic leguminous seed than between endospermic legumes and cereals.

7.3.3. Mannan-Containing Seeds Other than Legumes

A number of nonleguminous plants also store mannans, although few have received much attention as far as mobilization of their reserves is concerned. Hydrolysis of polysaccharides in the endosperm (89% mannose deposited in the secondary walls) of date *(Phoenix dactylifera)* occurs when a haustorial out-

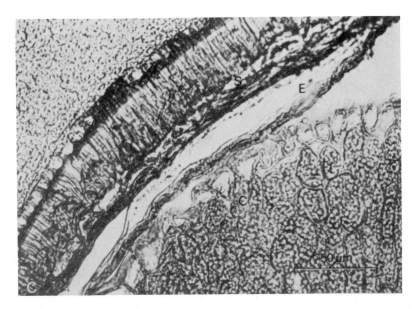

endosperm (G) are being dissolved. The dissolution zone (D) begins at the aleurone layer (A) and spreads toward the cotyledons (C). (C) The endosperm (E) is depleted and only a remnant remains between the seed coat (S) and the cotyledon (C). The aleurone layer has disintegrated. From Reid (1971).

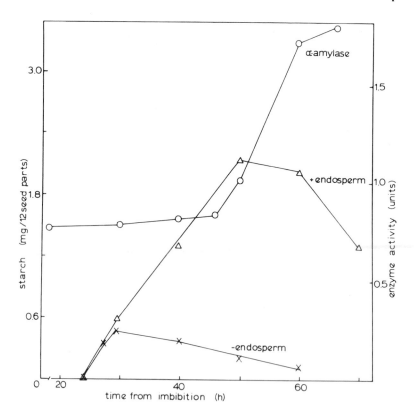

Figure 7.7. Starch synthesis in the cotyledons of germinated fenugreek seeds in the presence (Δ) and absence (X) of the endosperm, and the appearance of α-amylase in whole seeds (o). Data provided by Dr. D. W. M. Leung.

growth from the seedling grows into it releasing exoenzymes. The mannan is hydrolyzed to mannose, which is absorbed by the haustorium and transported to the growing axis, where it is converted to sucrose. Mobilization of mannans from the endosperm cell walls of the lettuce seed endosperm, on the other hand, involves synthesis of mannanase within the endosperm itself. The products of hydrolysis (small oligomannans) are cleaved further by a β-mannosidase located in the cell walls of the cotyledons which absorb and utilize the resultant mannose. No further details of the mobilization of mannanase in lettuce will be presented here, for this will be discussed in Chapter 8 in relation to the control of this process.

Different patterns of mannan hydrolysis between different nonleguminous seeds are to be expected. Since the deposition of mannans within the cell walls of the endosperm and perisperm of hard seeds like date, coffee, and ivory nut results in the destruction of the cell cytoplasm during seed development, a source of enzymes other than the endosperm itself is required. On the other hand, the lettuce endosperm is still comprised of living cells at maturity, and mobilization of their walls is, essentially, by autolysis.

7.4. STORED LIPID CATABOLISM

7.4.1. General Catabolism

As outlined in Chapter 1, triglycerides are the major storage lipids in seeds. Their initial hydrolysis (lipolysis) is by lipases, enzymes that catalyze the three-stage hydrolytic cleavage of the fatty acid ester bonds in triglycerides, ultimately to yield glycerol and free fatty acids (FFA).

$$\text{Triglyceride} \rightarrow \text{Diglyceride} + \text{FFA} \rightarrow \text{Monoglyceride} + 2\text{FFA}$$
$$\downarrow$$
$$\text{Glycerol} + 3\ \text{FFA}$$

Glycerol enters the glycolytic pathway after its phosphorylation and oxidation to the triose phosphates (dihydroxyacetone phosphate $\rightleftharpoons$ glyceralde-

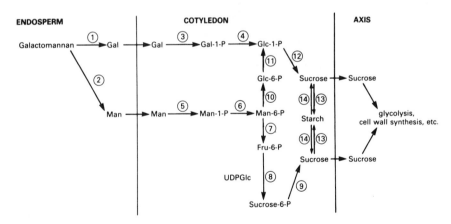

Figure 7.8. Flow diagram to illustrate the potential fate of products of galactomannan mobilization in endospermic legumes. Enzymes: (1) α-galactosidase; (2) β-mannanase and β-mannosidase; (3) galactokinase; (4) hexose phosphate uridyl transferase (a group of three enzymes that convert Gal-1-P + UTP $\rightarrow$ UDPGal $\rightarrow$ UDPGlc $\rightarrow$ Glc-1-P + UTP); (5) mannokinase; (6) phosphomannomutase; (7) phosphomannoisomerase; (8) sucrose-6-P synthetase; (9) sucrose phosphatase; (10) C_2 epimerase; (11) phosphoglucomutase; (12) sucrose synthetase or sucrose-6-P synthetase (see Section 7.1.2); (13) see Section 2.3.1; (14) see Sections 7.1.1 and 7.1.2. Gal, galactose; Man, mannose; Glc: glucose; Fru, fructose.

hyde-3-phosphate), which are then condensed by aldolase in the reversal of glycolysis to yield hexose units (Fig. 7.11, step 21). Alternatively, the triose phosphates may be converted to pyruvate and then oxidized through the citric acid cycle (Fig. 4.11).

The FFA released by lipase may be utilized in oxidation reactions to yield compounds containing fewer carbon atoms. The predominant oxidation pathway is β-oxidation, in which the fatty acid is first "activated" in a reaction requiring ATP and coenzyme A, and then, by a series of reactions involving the sequential removal of two carbon atoms this "active fatty acid" is broken down to acetyl CoA (Fig. 7.11). Saturated fatty acids with an even number of carbon atoms yield only acetyl CoA. Chains containing an odd number of carbon atoms, if completely degraded by β-oxidation, will yield the two-carbon acetyl moieties (acetyl CoA) and one three-carbon propionyl moiety (propionyl CoA, $CH_3CH_2CO\text{-}S\text{-}CoA$). This, in turn, can be degraded in a multistep process to acetyl CoA. The acetyl moiety may be completely oxidized in the citric acid cycle to CO_2 and H_2O or utilized initially via the glyoxylate cycle for carbohydrate synthesis. During seedling growth this latter process is the most important.

α-Oxidation of fatty acids is a process that involves the sequential removal of one carbon at a time from FFA of chain lengths C_{13}–C_{18}. Although it is unlikely that complete oxidation of fatty acids occurs by this pathway, it could serve to shorten odd-chain fatty acids to even lengths and thus facilitate their complete degradation to acetyl CoA by β-oxidation. Alternatively it could convert even-length FFA to odd-length FFA, resulting in increased propionic acid production by β-oxidation; this compound is an important precursor of coenzyme A. α-Oxidation additionally plays a role in catabolism of ricinoleic acid, the fatty acid component of the storage fat in castor bean seeds (Fig. 7.12).

The oxidation of unsaturated fatty acids (e.g., oleic acid; 18:1) is by the same general pathways, although some extra steps are required. The double bonds of naturally occurring unsaturates are in the *cis* configuration, but for step 4 (Fig. 7.11) of β-oxidation to be effected, they must be in the *trans* position. Hence a *cis, trans* isomerase is necessary to convert the fatty acid to its oxidizable form:

$\Delta^{3,4}$ *cis* Enoyl CoA

cis, trans Isomerase

$\Delta^{2,3}$ *trans* Enoyl CoA

Polyunsaturated fatty acids containing two or more double bonds (e.g., linoleic acid, 18:2; linolenic acid, 18:3) cannot be degraded simply by β-oxidation. Alterations to the positions of the double bonds need to be made for this process to be completed: this is achieved by the combined activities of a series of enzymes, including lipoxygenase, a hydroperoxide isomerase, and an epimerase. It is probable that the resultant product can be ultimately degraded to acetyl CoA by β-oxidation, although only incomplete information on the fate of the compounds resulting from lipoxygenase activity in seeds is available.

Directly coupled to the β-oxidation pathway is the glyoxylate cycle, which takes the acetyl CoA and, in a series of enzymatic reactions, links this to the glycolytic pathway, which then operates to produce hexose (Fig. 7.11). The key enzymes for forging this link are malate synthetase and isocitrate lyase. Acetyl CoA is first converted to citrate (in the same manner as precedes its entry into the citric acid cycle: Fig. 7.11, step 7), then to isocitrate, but the decarboxylating steps in the citric acid cycle between isocitrate and succinate are avoided by the action of isocitrate lyase, which cleaves isocitrate directly to succinate and glyoxylate. Another acetyl CoA is incorporated into the cycle (step 10) and is condensed with glyoxylate by malate synthetase to yield malate. With each turn of the cycle one molecule of succinate is released (step 9) and is converted to oxaloacetate by citric acid enzymes in the mitochondria (steps 13–15), and then into the glycolysis pathway as phosphoenolpyruvate (step 16).

Catabolism of the fat (oil) reserves within the storage tissues of germinated seeds involves three distinct organelles found within fat-containing cells. These are (1) the fat-storing oil body, (2) the glyoxysome, and (3) the mitochondrion (Fig. 7.9), and they function as follows: (1) hydrolysis of fats to FFA and glycerol commences within the oil bodies; (2) within the glyoxysome the fatty acids are oxidized, and synthesis of succinate occurs via the glyoxylate cycle; (3) the succinate is converted to malate or oxaloacetate within the mitochondrion. The malate or oxaloacetate is processed further in the cytosol to yield sucrose. We will consider these stages in more detail, and since many of the "classical" studies on fat mobilization in seeds have been carried out on the endosperm of castor bean *(Ricinus communis),* work on this tissue will be emphasized.

7.4.2. Mobilization of Fats from Oil Bodies

As fat reserves are hydrolyzed, the storage vesicles that contain them, i.e., the oil bodies, gradually disappear. In high-fat seeds, such as peanut *(Arachis hypogaea)* and castor bean, the oil bodies change little in the general appearance during mobilization; they gradually decrease in size and disappear as the reserves are depleted.

Since storage fats are present within discrete oil bodies, it is not an unrea-

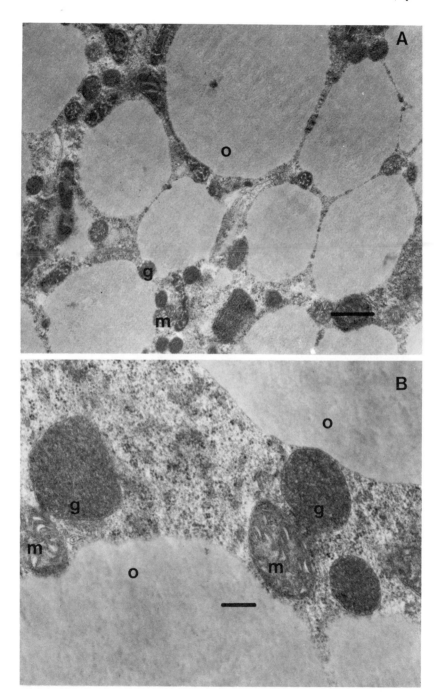

sonable expectation that the enzymes responsible for their hydrolysis should be intimately associated with these structures. In rape *(Brassica napus)* and jojoba *(Simmondsia chinensis)* cotyledons this is indeed the case. There is an alkaline lipase associated with the oil storage bodies that increases in activity coincidentally with the increase in fat mobilization: the enzyme might be *de novo* synthesized, for it is absent from the dry seed. In contrast, in peanut cotyledons, only a very small proportion of the lipase activity is found within the oil bodies. More than 60% of the total activity resides with an enzyme associated with the glyoxysome, and the balance is in a mitochondrial and membrane (ER) fraction. Lipase in this latter fraction can hydrolyse mono-, di-, and triglycerides, but the glyoxysomal lipase can only use the monoglyceride as substrate. Hence we are left with an unclear picture of events in the peanut: somehow the enzyme capable of hydrolyzing triglycerides must come into contact with the stored fats in the oil bodies and effect their breakdown. FFA and monoglycerides diffuse from the oil body to the glyoxysome: FFA can enter the organelle, but the monoglycerides must first be hydrolyzed further by the lipase, which is conveniently located on the glyoxysomal membrane.

Mobilization of stored lipid in the castor bean endosperm begins on about the third day after imbibition, and complete digestion takes a further 4 days (Fig. 7.10). By then the endosperm is liquefied and the contents absorbed by the expanding cotyledons attached to the growing root–shoot axis. Initially an acid lipase (optimum pH 5) is activated, and as it declines the activity of an alkaline lipase (optimum pH 9) increases (Fig. 7.10). The acid lipase is associated with the oil body and is capable of hydrolyzing tri-, di- and monoglycerides. Yet it reaches maximum activity at a time when there is little mobilization of stored fat. In contrast the alkaline lipase is attached to the glyoxysome membrane. This enzyme is at peak activity when the major hydrolysis of stored fats is occurring, and yet its specificity is only for monoglycerides. Presumably, as in peanut cotyledons, the glyoxysomal alkaline lipase plays a role in the degradation of monoglycerides prior to uptake of FFA into the organelle, but how the hydrolysis of the major triglyceride reserves takes place remains a mystery.

Lipases have been reported to be present in the storage organs of the seeds of several species; with the exception of castor bean, they are generally alkaline lipases. Some dry seeds contain lipases, but in most there appears to be an increase in activity after imbibition. Whether this is due to *de novo* synthesis or activation of existing lipases is unknown. A decline in lipase activity with declining triglyceride reserves has been widely reported.

←

Figure 7.9. (A) Electron micrograph of the cells of castor bean endosperm during the period of mobilization of the lipid reserves. Note in (B) the close association of the oil bodies, glyoxysomes, and mitochondria. o, oil (fat) body; g, glyoxysome; m, mitochondrion. Bar in (A) = 1 μm and in (B) = 0.2 μm. Courtesy of Dr. J. S. Greenwood.

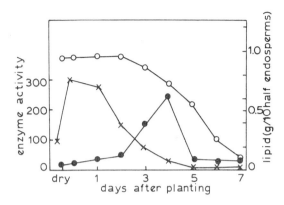

Figure 7.10. Changes in stored lipid content (o), acid lipase—optimum pH 5.0 (X), and alkaline lipase—optimum pH 9.0 (●) in the castor bean endosperm. Both enzyme activities expressed in nmol lipid hydrolyzed $\cdot min^{-1} \cdot$ endosperm^{-1}. After Muto and Beevers (1974).

7.4.3. The Fate of Glycerol and Fatty Acids

The products of lipolysis, glycerol and FFA, usually are rapidly metabolized. In persistent storage organs some of the released glycerol may be reutilized for fat synthesis, some is respired, and the rest is converted to sucrose for transport to the growing axis. As might be expected, in nonpersistent organs the reincorporation of glycerol into lipids is very limited: most of the glycerol is converted to sucrose for export, with some being lost in respiratory processes. The conversion to sucrose requires first that the glycerol be phosphorylated by glycerol kinase in the cytosol, to give α-glycerol phosphate. This is then oxidized in the mitochondria by cytochrome-linked α-glycerol phosphate oxidoreductase, to yield dihydroxyacetone phosphate (Fig. 7.11, step ii). Finally, this is released into the cytosol for conversion to hexose.

Although seeds of most dicots and cereals do not accumulate FFA during lipolysis, the storage tissues of a few species do. For example, in the germinated seeds of the West African oil palm *(Elaeis guineensis)* a specialized structure called a haustorium invades the endosperm, and through its vascular system the products of lipid catabolism are transported to the growing root and shoot. During fat degradation there is a buildup of FFA in the endosperm, and those that are not respired are absorbed directly by the haustorium without prior conversion to sucrose. FFA accumulate also in the haustorium itself and there might be reconverted to triglyceride for temporary storage until required by the growing axis.

7.4.4. Role of the Glyoxysome, Mitochondrion, and Cytosol in Gluconeogenesis

The complete process of degradation of saturated triglycerides and conversion to carbohydrate is summarized in Fig. 7.11. The oil bodies, glyoxysomes, mitochondria, and cytosol are all involved, and the enzymes for each step in this process have been located in their appropriate position in the cell. In view of the cooperative role of the three organelles, it is not surprising that they are found in juxtaposition within the cell (Fig. 7.9). The pathways involved in lipolysis, fatty acid oxidation, and gluconeogenesis have been outlined in Section 7.4.1 and will not be considered in detail again: only a few essential features of the pathways will be mentioned here.

FFA produced by lipolysis enter the glyoxysome, in some tissues after the final hydrolysis of monoglycerides by a lipase attached to the glyoxysome membrane. The fatty acids are activated with CoA within the glyoxysome, and their passage through the β-oxidation pathway and glyoxylate cycle is completed within this organelle too. There are no enzymes present within the glyoxysomes for the reoxidation of NADH, an event that is essential for the sustained operation of the glyoxylate cycle and β-oxidation (Fig. 7.11, steps 5 and 11). Instead the cooperation of the mitochondrion is required, for it is within this organelle that reoxidation occurs. Since NADH itself is not readily transported across glyoxysomal or mitochondrial membranes, a more complex shuttle system involving malate and aspartate operates whereby reducing equivalents are transferred between these two organelles. The involvement of the mitochondrion is also essential for further utilization of succinate released from the glyoxylate cycle by isocitrate lyase (Fig. 7.11, step 9). Either malate or oxaloacetate (steps 14 and 15) is transferred to the cytosol from the mitochondrion, and after conversion to phosphoenolpyruvate (PEP) (step 16) the glycolytic pathway operates in reverse to produce hexose. It has been suggested that the reaction sequence from PEP to sucrose (steps 17–26) occurs within the plastids, from which enzymes of the glycolytic pathway have been isolated. However, the activities of several of these plastid enzymes seem to be too low to account for the *in vivo* rate of gluconeogenesis, whereas levels of all glycolytic pathway enzymes in the cytosol are quite adequate.

One aspect of fat degradation that the scheme in Fig. 7.11 does not take into account is the fact that the storage triglycerides are usually unsaturated and, in the case of ricinoleic acid in castor bean, also hydroxylated. Complete oxidation of unsaturated and polyunsaturated fatty acids requires their modification by processes other than β-oxidation (Section 7.4.1): presumably the appropriate enzymes (e.g., *cis, trans* isomerase, lipoxygenase) are present in all glyoxysomes. For β-oxidation of ricinoleic acid (12-OH 18:1) two additional enzymatic steps are required as follows:

1. Conversion of 12-OH 18:1 to 6-OH 12:1 by β-oxidation leaves a double bond in the *cis* position; a *cis, trans* isomerase must be present to convert 6-OH 12:1 (3 *cis*) to 6-OH 12:1 (2 *trans*).
2. Further β-oxidation of 6-OH 12:1 (2 *trans*) would yield 2-OH 8:0, which cannot then be oxidized. Hence the hydrogen must be removed from

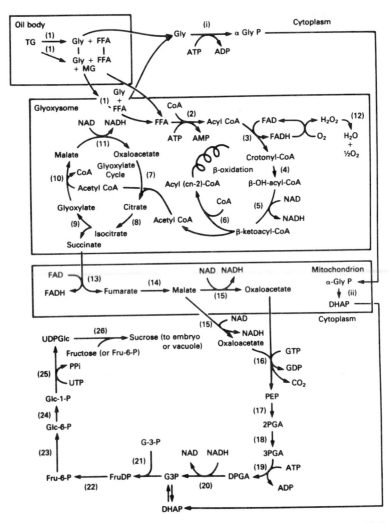

Figure 7.11. Pathways of triglyceride catabolism and hexose assimilation. Enzymes: (1) lipases; (2) fatty acid thiokinase; (3) acyl CoA dehydrogenase; (4) enoyl hydratase (crotonase); (5) β-hydroxyacyl CoA dehydrogenase; (6) β-ketoacyl thiolase; (7) citrate synthetase; (8) aconitase; (9) isocitrate lyase; (10) malate synthetase; (11) malate dehydrogenase; (12) catalase; (13) succinate dehydrogenase; (14) fumarase; (15) malate dehydrogenase; (16) phosphoenolpyruvate carboxykinase; (17) enolase; (18) phosphoglyceromutase; (19) phosphoglycerate kinase; (20) glyceraldehyde-3-phosphate dehydrogenase; (21) aldolase; (22) fructose-1,6-diphosphatase; (23) phosphohexoisomerase; (24) phosphoglucomutase; (25) UDPGlc pyrophosphorylase; (26) sucrose synthetase (or sucrose-6-P synthetase and sucrose phosphatase). (i) Glycerol kinase; (ii) α-glycerol phosphate oxidoreductase. Substrates: TG, triglyceride; MG, monoglyceride; Gly, glycerol; FFA, free fatty acids; PEP, phosphoenolpyruvate; 2PGA, 2-phosphoglyceric acid; 3PGA, 3-phosphoglyceric acid; DPGA, 1,3-diphosphoglyceric acid; G3P, glyceraldehyde-3-phosphate; FruDP, fructose-1,6-diphosphate; Fru-6-P, fructose-6-phosphate; Glc-6-P, glucose-6-phosphate; Glc-1-P, glucose-1-phosphate; UDPGlc, uridine diphosphate glucose; α-Gly P, α-

the hydroxyl to yield an oxygenated (oxo) derivative, which is converted to the 7:0 fatty acid by α-oxidation.

The proposed pathway for the catabolism of ricinoleic acid to acetyl CoA in the glyoxysome is summarized in Fig. 7.12.

7.4.5. Glyoxysome Biosynthesis and Degradation

Regardless of the species or tissue in which they are found, glyoxysomes are remarkably conservative in both size and composition. They have an equilibrium density of 1.25 g/cm^3, are surrounded by a single unit membrane, and, with the exception of the membrane-associated alkaline lipase, have very similar specific activities of individual enzymes. As noted previously, when fat degradation is occurring most actively, glyoxysomes are frequently found in juxtaposition to oil bodies and mitochondria (Fig. 7.9).

We will now follow the synthesis and fate of the glyoxysome in two types of storage organ: (1) that which disintegrates as the major stored reserves are mobilized (exemplified by the castor bean endosperm), and (2) that which persists after reserve mobilization and eventually becomes photosynthetic [using cotton *(Gossypium hirsutum)* and cucumber *(Cucumis sativus)* as the examples]. A survey of the literature suggests that the pattern of events gained from work on these species is typical of most fat-storing seeds.

The appearance and disappearance of the glyoxysomal enzymes in the castor bean endosperm are accompanied by the synthesis and degradation of the glyoxysomal membrane. In the endosperm of ungerminated castor bean seeds, glyoxysomes and mitochondria are barely detectable, but their numbers increase dramatically as the hydrolysis of fat reserves proceeds. The production of these new organelles requires the prior *de novo* synthesis of the protein and phospholipid constituents of the membranes. Hence the castor bean endosperm, whose function is to degrade fatty acids, must concurrently synthesize another group of fatty acids for inclusion into membranes of organelles involved in the catabolic processes. The current view is that after imbibition the plastids present in the dry seeds begin to synthesize FFA, mainly palmitic (16:0) and oleic (18:1). These fatty acids are then transported to the ER where they are activated, combined with glycerol-3-phosphate to form phosphatidic acid, and then converted to all the other membrane phospholipids by

glycerol phosphate; DHAP, dihydroxyacetone phosphate. Coenzymes and energy suppliers: FAD/(H), flavin adenine dinucleotide/(reduced); NAD/(H), nicotinamide adenine dinucleotide/(reduced); GTP, guanosine triphosphate; ATP, adenosine triphosphate; UTP, uridine triphosphate; GDP, guanosine diphosphate; ADP, adenosine diphosphate; AMP, adenosine monophosphate; CoA, coenzyme A.

Ricinoleyl CoA (double bond in *cis* position 9)

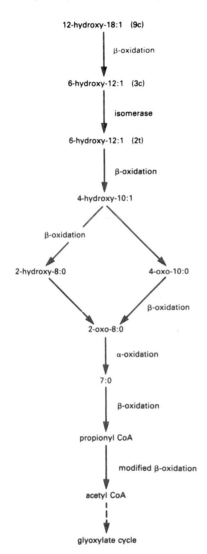

12-hydroxy-18:1 (9c)

β-oxidation

6-hydroxy-12:1 (3c)

isomerase

6-hydroxy-12:1 (2t)

β-oxidation

4-hydroxy-10:1

β-oxidation

2-hydroxy-8:0   4-oxo-10:0

β-oxidation

2-oxo-8:0

α-oxidation

7:0

β-oxidation

propionyl CoA

modified β-oxidation

acetyl CoA

glyoxylate cycle

Figure 7.12. Pathway for the catabolism of ricinoleic acid in the castor bean endosperm. After Hutton and Stumpf (1971).

enzymes associated with the ER. Membrane proteins are inserted into the ER, and then by vesiculation glyoxysomes are formed. The site of synthesis of the enzymes found within the mature glyoxysome of the castor bean endosperm is a subject of considerable controversy. One postulation is that at least some of the enzymes are synthesized on polyribosomes bound to the ER, before they are vectorially discharged across the ER membrane and collected in the central lacuna. Eventually the ER vesiculates to form and excise isolated membrane-bounded organelles in which the enzymes are trapped (Fig. 7.13). But there is equally convincing evidence that at least three key glyoxysomal enzymes, malate synthetase, isocitrate lyase, and catalase, are synthesized on free polyribosomes in the cytosol and then transported inward, by a yet unexplained mechanism, through the membrane of the mature glyoxysome (Fig. 7.13). This second possibility, therefore, involves the formation of the glyoxysome itself from the ER, with its component enzymes being synthesized freely within, and inserted from, the cytoplasm.

Glyoxysomes eventually disappear from the castor bean endosperm as the tissue disintegrates. The integrity of the organelle is maintained until the time of its disappearance. This suggests that instead of enzymes leaking out from the glyoxysome as tissue senescence progresses, the whole organelle is autolyzed at once, perhaps within a cellular compartment containing hydrolytic enzymes.

The castor bean endosperm appears to be exceptional in that there are no glyoxysomes present within the mature dry seed. Most other mature oil-containing seeds contain at least some glyoxysome activity, as determined by assay of the glyoxylate cycle enzyme malate synthetase. Some species, e.g., cotton, contain appreciable levels of this enzyme, as well as others of the glyoxylate cycle and the β-oxidation pathway (Table 7.3). Isocitrate lyase activity is very low, however, and hence glyoxysomes from dry seeds are not completely functional. In contrast to castor bean, then, it appears that in the storage organs of most other seeds the glyoxysome, containing most constituent enzymes, is synthesized during seed development and persists through the desiccation phase,

Table 7.3. The Presence of Enzymes Involved in Fat Catabolism in the Cotyledons of Dry Seeds and at the Time of Fat Mobilization (36–40 h after Planting) in Growing Seedlings of Cotton (*Gossypium hirsutum*).[a]

Enzyme	Fig. 7.11 step	Dry	36–40 h imbibed
Isocitrate lyase	9	0	41
Malate synthetase	10	32	200
Catalase[b]	12	4	59
Malate dehydrogenase	11 and 15	840	1840
Citrate synthetase	7	28	52
Aconitase	8	6	55
Enoyl hydratase	4	110	620
β-Hydroxyacyl CoA dehydrogenase	5	38	208
Fructose-1,6-diphosphatase	22	7	122

[a]Values in nmol/min/cotyledon pair $\times$ 10^{-1}. Data taken from Bortman *et al.* (1981)
[b]Expressed in Lück units.

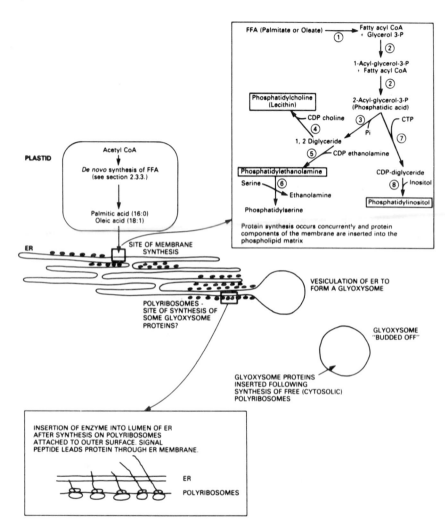

Figure 7.13. Involvement of the plastids and endoplasmic reticulum in the biogenesis of glyoxysomes. Note that the spatial displacement of the site of protein (glyoxysomal enzyme) synthesis and of membrane synthesis is for clarity, and does not imply that the two events cannot be occurring in juxtaposition. Not drawn to scale. Enzymes: (1) thiokinase; (2) fatty acyl transferase; (3) phosphatidic acid phosphatase; (4) CDP-choline-diglyceride transferase; (5) CDP-ethanolamine-diglyceride transferase; (6) phosphatidylethanolamine-serine phosphatidyl transferase; (7) phosphatidic cytidyltransferase; (8) CDP-diglyceride-inositol transferase. The major phospholipids found in the ER of castor bean are boxed.

which marks the end of development. After germination there is an increase in activity as more glyoxysomal enzymes are synthesized (Table 7.3), and in particular isocitrate lyase is added to the preexisting organelle, which increases in size. The synthesis of isocitrate lyase and other enzymes probably occurs on free polyribosomes within the cytosol, with the complete proteins being inserted into glyoxysomes. In addition to this improvement of preexisting organelles by insertion of enzymes, there is a distinct possibility that in many fatty storage organs there is also *de novo* synthesis of complete glyoxysomes.

In cucumber *(Cucumis sativus)* cotyledons there is a marked increase in activity of several key glyoxysomal enzymes (malate synthetase, isocitrate lyase, and catalase) at the time of the major mobilization of lipids—shown for isocitrate lyase in Fig. 7.14, A. *De novo* synthesis of this enzyme (and of malate synthetase) is preceded by, and hence is presumably dependent on, an increase in the appropriate messenger RNA (Fig. 7.14, B). The decline in activity of these enzymes as fat mobilization is completed is also accompanied by a decline in mRNA. This is convincing evidence that the level of glyoxylate cycle enzyme activities is regulated by the availability of translatable mRNA. This could be due to transcription of new mRNA or activation of preexisting messenger precursors.

The cotyledons of those species whose mode of seedling growth is epigeal (Section 3.7) turn green as they emerge from the soil into the light. During greening there is a gradual loss of glyoxysomes and the production of a similarly sized microbody called the peroxisome. There is considerable controversy as to whether the glyoxysomes and peroxisomes appear and disappear as inde-

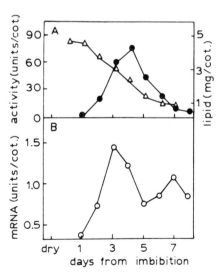

Figure 7.14. Changes in the activity of isocitrate lyase and its mRNA in the cotyledons of dark-grown cucumber seedlings. (A) Isocitrate lyase activity (●) during depletion of lipid reserves (△). (B) Translatable mRNA for isocitrate lyase in total cotyledonary RNA. Total RNA was extracted from the cotyledons at various times after initial imbibition and translated *in vitro*. The isocitrate lyase in the translation products was isolated and quantitated. After Weir *et al.* (1980).

pendent organelles, or whether there is a succession of glyoxysomal and peroxisomal enzymes within the same microbody. Discussion of the alternatives is beyond the scope of this book, for the subject pertains to greening and seedling growth, but the reader is directed to a pertinent review by H. Beevers (1979) on microbodies.

Whatever the ultimate fate of the glyoxysome, at least some of the enzymes contained therein must be destroyed. If the glyoxysome persists as a peroxisome, then fat-catabolizing enzymes such as isocitrate lyase (Fig. 7.14, A), malate synthetase, and the β-oxidation pathway enzymes must be destroyed, while others, e.g., catalase and malate dehydrogenase, are partially or wholly retained. Controlled removal of certain enzymes might involve a special class of inactivating proteins, perhaps proteinases with specificity for particular glyoxysomal enzymes. This mechanism would ensure the destruction of some proteins and conservation of others within the same organelle.

7.4.6. Utilization of the Products of Fat Catabolism

Although the products of triglyceride hydrolysis, FFA and glycerol, may be used for the resynthesis of fats and membrane lipids (particularly in persistent cotyledons), they are to a large extent converted to hexose, and finally to sucrose, by a sequence of reactions outlined in Fig. 7.11. Castor bean endosperms contain high levels of sucrose-6-P synthetase, sucrose phosphatase, and also sucrose synthetase (see Section 7.1.2 for their role in sucrose synthesis). Here the major product of lipid mobilization is sucrose, which is taken up by active transport into the cotyledons. More than 80% of this sucrose is redistributed to the growing axis. Sometimes there is temporary storage of sucrose, in the endosperm, in vacuoles that develop as the storage products are degraded if the embryonic axis is removed. Sucrose uptake by the cotyledons is thereby drastically reduced. Thus, as far as the growing seedling is concerned, removal of the sink (axis) alters replenishment at the source (cotyledons).

The cotyledons of some seeds (e.g., pumpkin, watermelon, sunflower) can utilize acetyl CoA arising from β-oxidation of fatty acids for amino acid synthesis via partial reactions of the glyoxylate and citric acid cycle. The usual products are glycine, serine, glutamic acid, glutamine, and γ-amino butyric acid.

7.5. STORED PROTEIN CATABOLISM

7.5.1. General Catabolism

Hydrolysis of storage proteins (polypeptides) to their constituent amino acids requires a class of enzymes called proteinases, some of which effect total hydrolysis whereas others produce small polypeptides that must be degraded further by peptide hydrolases.

The proteinases can be categorized in relation to the manner in which they hydrolyze their substrates as follows:

1. Endopeptidases: these cleave internal peptide bonds to yield smaller polypeptides.
2. Aminopeptidases: these sequentially cleave the terminal amino acid from the free amino end of the polypeptide chain.
3. Carboxypeptidases: as (2), but single amino acids are sequentially hydrolyzed from the carboxyl end of the chain.

Both (2) and (3) are exopeptidases.

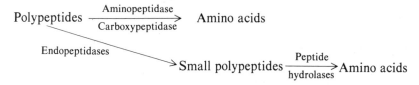

The liberated amino acids may be reutilized for protein synthesis or be deaminated to provide carbon skeletons for respiratory oxidation. Ammonia is produced by deamination, but this is prevented from reaching toxic levels by fixation into glutamine and asparagine, two commonly translocated forms of amino acid.

7.5.2. Protein Mobilization in Cereals

Reserve proteins are stored in two separate regions of the cereal grain: in the aleurone grains of the aleurone layer (up to 30% of total), and in the protein bodies (sometimes disrupted, e.g., wheat, barley) of the starchy endosperm. A minor amount of storage protein is present in the scutellum and the axis, and this may be hydrolyzed to provide amino acids for the growing axis prior to mobilization of the major endosperm reserves.

There appear to be three major sites of proteolytic activity in the germinated cereal grain:

1. Aleurone layer. Proteinases are produced in the aleurone layer, in some species or cultivars under the control of gibberellin (see Chapter 8), and these may be synthesized *de novo*. One or more of these enzymes hydrolyzes the proteins within the aleurone grains, and the resultant amino acids are utilized in the synthesis of proteins (e.g., α-amylase) in this tissue.

2. Starchy endosperm. Proteinases effective in breaking down the major

protein reserves within the starchy endosperm may come from two sources: the aleurone layer or the starchy endosperm itself. In Kristina and Iisvesi cultivars of barley, for example, an endopeptidase released from the aleurone layer may play a central role in endosperm protein hydrolysis. But there are also pre-formed proteinases within the endosperm of dry grains, and these somehow become activated when required for protein mobilization. Besides their action in degrading reserve proteins, proteinases in the starchy endosperm also serve to release and activate bound enzymes (e.g., β-amylase; Section 7.2.2), and to aid in dissolution of the cell walls by hydrolyzing links between glucan and protein components.

   3. Axis and scutellum. Peptide hydrolase activity in the scutellum is responsible for the hydrolysis of peptides being taken up by this structure from the endosperm (see Section 7.5.3). Proteolytic activity is also present in the axis for hydrolysis of the small amount of reserve proteins stored therein. In addition, there are a number of proteolytic enzymes within the embryo that are involved with the normal turnover of proteins associated with growth.

   Only a few of the proteolytic enzymes present in cereal grains have been characterized, but it is likely that a number of exo- and endopeptidases will be identified in the various parts of the grain, each having different substrate specificities. To give but one example of the variety of enzymes involved in protein mobilization: in barley grains at least five endopeptidases, four carboxypeptidases, and four aminopeptidases have been reported, as well as several peptide hydrolases.

   Proteolytic activity within the starchy endosperm results in the production of amino acids, dipeptides, and a number of small oligopeptides (Fig. 7.15). Peptides and amino acids initially accumulate in the endosperm in equal proportions, but with time there is less accumulation of peptides, probably because they are taken up by the scutellum more efficiently than are the free amino acids. Although this uptake of peptides does not appear to involve or require their hydrolysis, they are eventually cleaved by peptide hydrolases within the scutellum, and only free amino acids accumulate to any extent in the growing embryo (Fig. 7.15). Active, i.e., energy-dependent, uptake mechanisms within the scutellum can distinguish between peptides and amino acids, but whereas there are probably several independent uptake systems for amino acids, there is only a single uptake system for peptides, and this is capable of handling di- and oligopeptides. The capacity of the leucine uptake system to transport this amino acid (and probably some others too) into the barley scutellum increases at least 20-fold with time after germination, and it is at its most efficient during the time of the major mobilization of the protein reserves. This, and presumably the other uptake systems, are probably synthesized *de novo,* since the scutellum of dry and early-imbibed grains can take up peptides and amino acids only poorly.

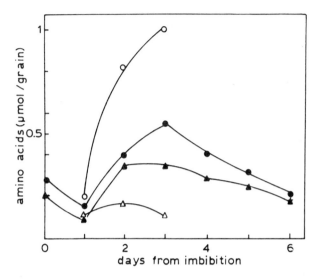

Figure 7.15. Changes in the size of the peptide (△, ▲) and free amino acid pools (○, ●) in the embryo (open symbols) and endosperm (closed symbols) of barley during mobilization of stored reserves. After Higgins and Payne (1981).

7.5.3. Protein Mobilization in Dicots

The hydrolysis of protein reserves has been studied mostly in the seeds of nonendospermic legumes. Cotyledons of the mature dry seed contain little proteinase activity but, as shown in Fig. 7.16 for garden pea *(Pisum sativum)*, there is an increase in this enzyme after several days, associated with the mobilization of the major protein reserves (vicilin and legumin). The hydrolytic process itself generally appears to occur in two stages: (1) modification of the large components (subunits) of the storage polypeptides, followed by (2) their degradation into amino acids and small peptides. Many of the storage proteins (e.g., globulins) in seeds are deposited in an insoluble form, and to make them fully susceptible to hydrolysis they must first be modified. One way in which this can be achieved is by limited endopeptidic attack on the stored protein to yield large, but more soluble, components. An example of this is to be found in pumpkin *(Cucurbita moschata)* cotyledons. Here, it is not possible to extract any high-molecular-weight storage globulins with water (dilute buffer solutions) on the first day after imbibition (Fig. 7.17, A). But with increasing time, as the pumpkin storage globulin begins to be mobilized, there is an increase in high-molecular-weight water-soluble proteins (28–56 kD): this is because limited proteolysis has resulted in solubilization of some globulin components (Fig. 7.17, B–D). These released components can now be hydrolyzed further. This

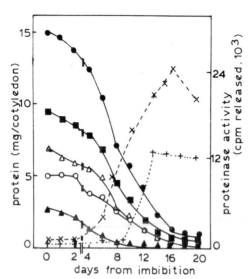

Figure 7.16. Changes in various storage protein fractions and proteinase activity in cotyledons of pea (*Pisum sativum*, cv. Burpeeana). —●—, total protein; —○—, albumins; —■—, globulins; —△—, legumin; —▲—, vicilin; --X--, extractable proteinase activity (pH 5.0) capable of hydrolyzing radioactive legumin; ·+·, extractable proteinase activity (pH 5.0) capable of hydrolyzing radioactive vicilin. Note that the increase in, and peak of, proteinase activity does not coincide with the loss of vicilin and legumin. Enzyme activities and protein changes were not determined in the same experiment: hence temporal variability of events could be due to differences in germination and growth conditions, e.g., between experiments. After Basha and Beevers (1975).

series of events is illustrated in Fig. 7.18. The limited hydrolytic activity (activity I) of the storage globulins is followed by complete hydrolysis (activity II) of the solubilized polypeptides: it remains to be determined whether the enzymes involved in these two stages of hydrolysis are *de novo* synthesized or are present in the dry seed and subsequently activated. Nevertheless, it is likely that the two activities are catalyzed by different enzymes.

Conversion of storage protein from an insoluble to a soluble form might not be activiated in the above manner (as for pumpkin, garden pea, broad bean, and others) in all seeds. An alternative method of solubilization could involve selective hydrolysis near the carboxy- or amino-terminal ends to yield low-molecular-weight oligopeptides, which, in turn, can be hydrolyzed further. In peanut and soybean cotyledons there might be modifications brought about through the removal by deaminases of amido groups from storage proteins prior to their hydrolysis: this activity could provide an early source of ammonia to the developing seedling. Many storage proteins, such as vicilin and other 7 S proteins, are glycosylated, but removal of the sugar components occurs after cleavage of the peptide links, and the glycosyl units are released as complete oligosaccharides. Deglycosylation probably is not an early modification of storage proteins.

It is worth noting, in passing, that different storage proteins within storage tissues are often hydrolyzed at different times after germination, and at different rates. In the garden pea and broad bean *(Vicia faba)*, for example, legumin is degraded earlier and faster than vicilin. Whether this is due to a variation in susceptibility of the storage proteins to proteinases, or due to differences in the times at which enzymes specific for different proteins are produced, or because of some other reason, remains to be determined.

The cellular changes that precede and accompany proteolysis have been studied most thoroughly in the cotyledons of mung beans *(Vigna radiata)*. Here, the major storage protein is vicilin, which comprises some 70–80% of the total, and the enzyme responsible for its hydrolysis is an endopeptidase, vicilin peptidohydrolase—with some participation by a carboxypeptidase. The ER is the site of vicilin peptidohydrolase synthesis, and associated with the synthesis of this enzyme the ER undergoes various modifications. Dry and early-imbibed cotyledons contain tubular ER, which is dismantled some 12–14 h after the

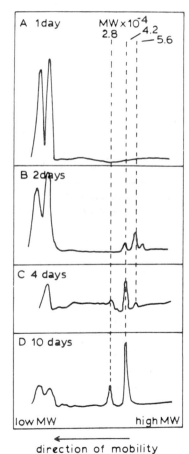

Figure 7.17. (A)–(D) SDS-polyacrylamide gel profiles of extractable water-soluble proteins from cotyledons of pumpkin at different times from the start of imbibition. After Reilly *et al.* (1978).

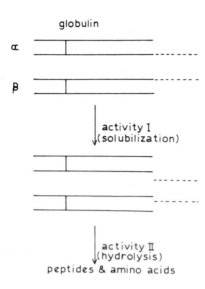

Figure 7.18. Diagrammatic representation of the mode of hydrolysis and initial solubilization of two storage globulins (α and β) from cotyledons of pumpkin. Activity I (solubilization) effects the removal of relatively short peptide chains (----), and the solubilized protein can then be degraded by activity II (proteolysis/hydrolysis). After Hara and Matsubara (1980).

start of imbibition. Although, overall, there is a net loss of membrane, at the same time there is a proliferation of a new type of ER, with ribosomes attached and with obvious cisternae. The vicilin peptidohydrolase is synthesized *de novo* on polyribosomes attached to this new ER, and the enzyme is inserted into the ER lumen and packaged into vesicles. These are transported to the protein bodies, and degradation of the vicilin therein commences only after the peptidohydrolase has been inserted into these organelles. Initially the vicilin is cleaved from 50- to 63-kD components to 20- to 30-kD components, and these are then hydrolyzed more slowly.

Eventually the protein bodies come to contain hydrolytic enzymes other than the peptidohydrolase. During protein hydrolysis a *de novo* synthesized ribonuclease is inserted into the protein body (after synthesis on ER and transport via vesicles), and ultimately these organelles have other hydrolases: α-mannosidase and a glucosaminidase (for the hydrolysis of mannose and glucosamine residues from vicilin, a glycoprotein), acid phosphatase, phosphodiesterase, and phospholipase D (for digestion of membrane phospholipids). As protein digestion proceeds, the emptying storage vesicles fuse to form a large vacuole containing an array of hydrolytic enzymes, analogous in function to the lysosome. Digestion of cell contents by the enzymes of the vacuole is achieved when vesicles are internalized by an autophagic process in which a portion of the cytoplasm is engulfed and sealed off by the protein body membrane (Fig. 7.19). Hence disintegration of the cotyledons can be achieved by cellular autolysis. Figure 7.20 summarizes the sequence of events in mung bean

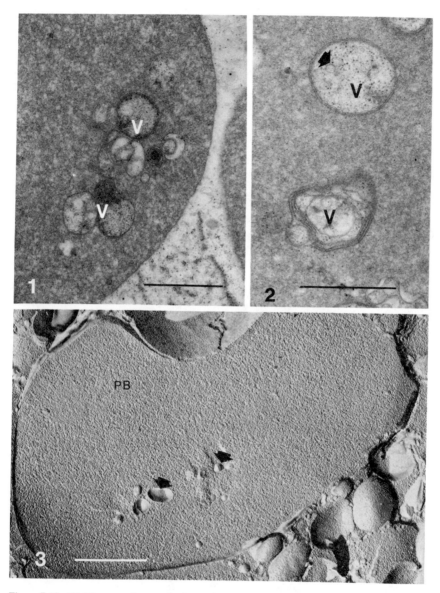

Figure 7.19. (1) Electron micrograph of autophagic vesicles (V) in a protein body of a cotyledon cell of a 3-day-old seedling of mung bean. The vesicles contain sequestered membrane elements and numerous ribosomes. Bar: 1 μm. (2) Portion of protein body with two autophagic vesicles (V). The top vesicle contains free ribosomes and membranous vesicles (arrow). The bottom vesicle contains several sequestered membranes. Bar: 1 μm. (3) Freeze–fracture replica of a protein body (PB) from a 3-day-old seedling cotyledon cell showing small vesicles (arrows). Bar: 1 μm. Photographs courtesy of Dr. M. J. Chrispeels. See also Chrispeels and Jones (1980/81) and Herman *et al.* (1981).

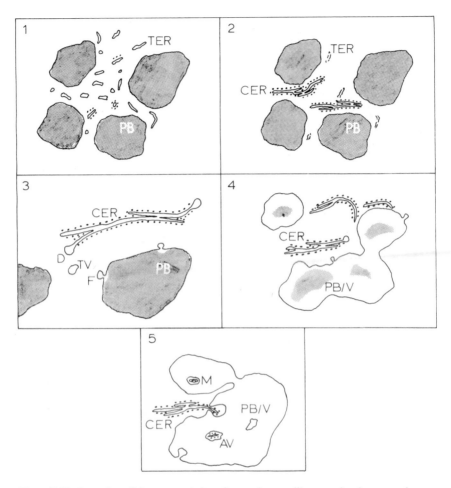

Figure 7.20. Part of a cell from a cotyledon of mung bean to illustrate the changes undergone by the protein body and ER during reserve hydrolysis and cell autolysis. (1) Dry state and up to 12 h from the start of imbibition. Intact protein bodies and tubular endoplasmic reticulum (TER) with few ribosomes attached. (2) Starting 12–24 h from imbibition. Dismantling of TER and synthesis of cisternal endoplasmic reticulum (CER) with ribosomes attached. No change to protein bodies. (3) Three to five days from start of imbibition. (Expanded scale to show one protein body.) Vicilin peptidohydrolase synthesized on polyribosomes attached to newly formed CER and inserted into CER lumen. Dilations (D) of the cisternae form containing the enzyme; these break off as transport vesicles (TV) and carry the peptidohydrolase to the protein bodies, with which they fuse (F). The proteinase commences hydrolysis of the vicilin. Other enzymes, e.g., ribonuclease, start to be inserted into the protein body. (4) As proteins are hydrolyzed and the protein bodies coalesce to form larger vacuoles (PB/V) and other hydrolytic enzymes are inserted. (5) Autophagic vacuoles (AV) form engulfing cell contents such as the CER and mitochondria (M). More protein bodies fuse to form a large central vacuole, with autolytic enzymes. Note that this is an illustrative scheme of events and is not drawn to scale. Moreover organelles and cell structures other than the ER and protein bodies are deliberately omitted for clarity. Based on the studies of Chrispeels and co-workers.

cotyledons associated with protein, and ultimately cell, hydrolysis. It is likely that this scheme of events has some features that are to be found in many dicotyledon storage organs.

7.5.4. Utilization of the Liberated Amino Acids

The major transported forms of amino acids from the storage organs into and throughout the growing seedlings are the amides, i.e., asparagine and glutamine. Hence the amino acids liberated from storage proteins must be further metabolized, including the conversion of amino nitrogen to amido nitrogen. The extent to which the carbon skeletons of liberated amino acids undergo interconversions can be seen by comparing the amino acid composition of the cotyledon exudate of mung beans (i.e., the translocated amino acids) with that of the major storage protein, vicilin (Table 7.4). Of particular interest is the increase in aspartyl amino acids in the exudate, with the predominance of asparagine as the transported amino acid being evident. The synthesis of aspar-

Table 7.4. Comparison of the Amino Acid Composition of the Cotyledon Exudate from Cotyledons after 4–5 Days from the Start of Imbibition, and the Composition of the Major Storage Glycoprotein Fraction (Vicilin)[a]

| | Amino acid composition (mole %) | |
	Cotyledon exudate	Vicilin
Asp	4.2	[b] 13.4
AspN	31.3	
Thr	4	3
Ser	3.8	7.3
Glu	3.8	
GluN	2.0	[b] 19.9
Pro	5.9	2.9
Gly	0.2	5.4
Ala	2.1	5.6
Val	8.9	6.6
Ile	4.8	4.5
Leu	6.7	9.4
Tyr	2	1.9
Phe	5.7	6.1
Lys	4.8	6
His	4.6	2
Arg	4.8	5.5

[a]Data taken from Ericson and Chrispeels (1973) and Kern and Chrispeels (1978).
[b]Amino acid and amide not determined separately in the vicilin fraction.

agine involves the donation of an amino group from glutamine in a reaction catalyzed by the ATP-dependent enzyme asparagine synthetase (AS) as follows:

$$\text{Aspartate} + \text{glutamine} + \text{ATP} \xrightarrow{\text{AS}} \text{Asparagine}$$
$$+ \text{glutamate} + \text{AMP} + \text{PPi}$$

Glutamine itself is formed from glutamate as follows:

$$\text{Glutamate} + \text{NH}_3 + \text{ATP} \xrightarrow[\text{(GS)}]{\text{Glutamine synthetase}} \text{Glutamine} + \text{ADP} + \text{Pi}$$

and glutamine can donate its amino group to form glutamate from α-ketoglutarate using the enzyme glutamate: NAD(P) oxidoreductase (transaminating), or GOGAT.

$$\text{Glutamine} + \alpha\text{-ketoglutarate}$$
$$+ \text{NAD(P)H} \xrightarrow{\text{GOGAT}} 2 \text{ Glutamate} + \text{NAD(P)}$$

Alternatively, glutamate:NAD(P) oxidoreductase (deaminating), also known as glutamate dehydrogenase (GDH), can add ammonia to α-ketoglutarate to yield glutamate in the presence of NAD(P)H.

The various fates of the amino acids released from storage proteins in relation to their transport and subsequent utilization in the growing seedling are detailed in Fig. 7.21.

In cotyledons of various legumes seeds there is, not surprisingly, an increase in the activity of the aforementioned enzymes involved in glutamate, glutamine, and aspargine synthesis at a time when the major protein reserves are being hydrolyzed. Hence, as amino acids are being liberated from the stored form they undergo the appropriate conversions to the readily transportable asparagine. GDH, GS, and AS are present only in low amounts in the cotyledons of mature dry cotton seed *(Gossypium hirsutum)* (Fig. 7.22) but increase appreciably during the first 2 days from the start of imbibition (perhaps by *de novo* synthesis), peaking approximately at the time of, or just prior to the commencement of, protein mobilization. In this seed, as in the legumes, the major transport form of amino acid is asparagine.

The pattern of amino acid metabolism in pea *(Pisum sativum)* cotyledons is unusual in that a major transported form of amino acid, and the one that accumulates in the cotyledon, is homoserine (Table 7.5). Glutamine is another transport form of amino acid, but not asparagine to any great extent. Homo-

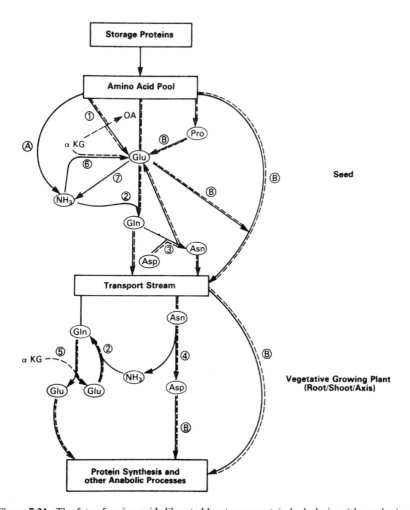

Figure 7.21. The fate of amino acids liberated by storage protein hydrolysis, with emphasis on the fact that glutamine and asparagine are the major transport amino acids. Enzymes: (1) amino-transferase; (2) glutamine synthetase; (3) asparagine synthetase; (4) asparaginase; (5) GOGAT; (6) glutamate dehydrogenase, (7) deaminase. Reactions: (A) specific deaminations; (B) direct interconversions of amino acid skeletons, or direct transfer without interconversion. Compounds: Glu, glutamic acid; Gln, glutamine; Asp, aspartic acid; Asn, asparagine; NH_3, ammonia; Pro, proline (high in amino acid pool of cereals when storage prolamins are broken down); OA, oxalo-acetic acid; αKG, α-ketoglutaric acid. Solid lines show the path of N, and dashed lines the path of C. Based on Miflin *et al.* (1981) and P. J. Lea (personal communication).

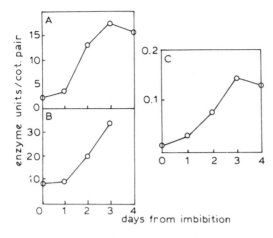

Figure 7.22. Activities of (A) glutamate dehydrogenase, (B) glutamine synthetase, and (C) asparagine synthetase in the cotyledons of cotton. After Dilworth and Dure (1978).

serine is an amino acid that is not incorporated into protein, nor is it readily converted in the seedling to other amino acids. Hence the effectiveness of homoserine as a nitrogen donor can be questioned, and we may wonder why the amino nitrogen stored in the cotyledons is converted into a compound that is not readily utilizable within the growing seedling. The answer is not readily apparent. Presumably a sufficient supply of amino acids for seedling growth is provided through translocation of glutamine and other amino acids from the cotyledons.

In castor bean seeds the site of protein storage is the endosperm, and the predominant form of transported nitrogen from this region is glutamine. Some of the amino acids released from the storage protein by hydrolysis, e.g., aspartate, glutamate, alanine, glycine, and serine, can be converted to sucrose and transported as the sugar. The amide nitrogen derived from the deamination of these gluconeogenic amino acids is probably used in the production of gluta-

Table 7.5. Changes in the Content of Glutamate/Glutamine, Aspartate/Asparagine, and Homoserine in the Cotyledons of Light-Grown Alaska Peas[a]

	Cotyledons			Shoots		
Time after sowing (days)	0	2.5	10	3	7	10
Glu/GluN	0.7	5.8	1.9	0.5	0.9	3.6
Asp/AspN	0.1	0.5	1.3	0.1	5	11.2
Homoserine		2.7	0.6	3.4	28.5	24.2

[a]Values in μmoles amino acid/seedling. Based on Larson and Beevers (1965).

mine. By comparison, amino acids that are not gluconeogenic are probably transported unchanged to the growing seedling; some might undergo modifications of their carbon skeleton to form glutamate.

The major transported amino acid from the endosperm of castor bean, glutamine, is taken up into the cotyledons by an active process, even against a concentration gradient. In species where the cotyledons are the storage organs, these are connected to the growing regions by a network of vascular bundles, and the products of hydrolysis must be translocated to the axis largely within the phloem. Loading of amino acids into the translocation stream might be aided in some species, e.g., broad bean *(Vicia faba),* by the presence of transfer cells that border the xylem and phloem; these are specialized cells with an increased surface area to aid the transport of solutes over short distances. In other species, e.g., mung bean *(Vigna radiata),* there are no transfer cells, but the parenchyma cells adjacent to the phloem have extensive evaginations of the plasmalemma to form fine tubules (plasmalemmasomes): these also serve to increase the surface area for transport.

7.5.5. Proteinase Inhibitors

Within both monocot and dicot seeds are proteins that specifically inhibit the action of proteinases in animals and, to a lesser extent, in plants. The function of these inhibitors is incompletely understood, but it could include one or more of the following:

1. Storage. Proteinase (trypsin) inhibitors constitute some 5–10% of the water-soluble proteins of some cereal grains, but in legumes the amounts are much smaller, varying from 0.25–3.6 g/kg seed.

2. Control of endogenous proteins. There is still a considerable controversy over the role of inhibitors in controlling the activity of endogenous proteinases. Some workers believe that in early-imbibed seeds proteolytic enzymes are kept inactive by the inhibitors and then become active as the inhibitor content declines. The evidence is not entirely convincing, however. Vicilin peptidohydrolase activity in mung bean cotyledons increases as the concentration of two inhibitors declines (Fig. 7.23). But these two phenomena are not causally related, for fractionation of the cells shows that the peptidohydrolase is associated exclusively with the protein bodies, and the inhibitors are only in the cytosol. Perhaps the inhibitors function to protect the cytoplasm should the proteinase-containing bodies accidentally rupture.

3. Protection or dissuasion. Proteinase inhibitors might inhibit proteolytic digestive enzymes of invading insects or the secreted proteinases of invading microorganisms.

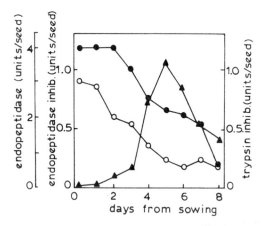

Figure 7.23. Time course of increase of endopeptidase (vicilin peptidohydrolase activity) (▲) and decline of two proteinase inhibitors, large endopeptidase inhibitor (○) and trypsin inhibitor (●) in cotyledons of mung bean. After Baumgartner and Chrispeels (1976).

7.6. STORED PHOSPHATE CATABOLISM

7.6.1. General Catabolism

Phytic acid (*myo*-inositol hexaphosphate) is the major phosphate reserve in many seeds, and since, in its storage form, it is a mixed salt with such elements as K^+, Mg^{2+} and Ca^{2+} (and as such is called phytin or phytate), it is also a major source of these macronutrient elements in the seed.

Phytase hydrolyzes the phytin to release phosphate, its associated cations, and *myo*-inositol. Breakdown of the phytin is rapid and complete, for *myo*-inositol phosphate esters with fewer than six phosphate groups do not accumulate within seeds. The released *myo*-inositol may be used by the growing seedling for cell wall synthesis, since this compound is a known precursor of pentosyl and uronosyl sugar units normally associated with pectin and certain other cell wall polysaccharides.

Lipid phosphate, protein phosphate, and nucleic acid phosphate occur in smaller amounts in seeds. Phospholipids and phosphoproteins are probably dephosphorylated during their hydrolysis (acid phosphatases may play a role here); the free phosphate is translocated to the growing axis, but the lipid and protein moieties are catabolized *in situ* as outlined in previous sections.

7.6.2. Phosphate Catabolism in Seeds

This has been studied in most detail in oat *(Avena sativa)* grains. In Fig. 7.24, the changes in various phosphate fractions over the first 8 days from the start of imbibition are shown. Essentially the results indicate that the most abundant source of phosphate in the dry grain is phytin (Fig. 7.24, acid soluble-P) in the endosperm (actually the aleurone layer: see later). This, along with other phosphate-containing fractions, declines in the endosperm after about 2 days, an event that is accompanied by a rise in various phosphate fractions within the growing axis (Fig. 7.24). This is suggestive of a release of phosphate from its storage form in the endosperm, its transport to the axis, and its reutilization there for the synthesis of essential phosphate-containing compounds. Phytin does not accumulate in the axis at any time.

The largest store of phytin in the cereal grain is found within the aleurone layer, where it is present as globoids in the aleurone grains. No phytin is associated with the protein bodies of the starchy endosperm, although a little is present in protein bodies in the scutellum. Generally, phytase activity is quite high in the mature dry cereal grain and appears to be closely associated with the protein bodies of the starchy endosperm, although a little is present in pro-

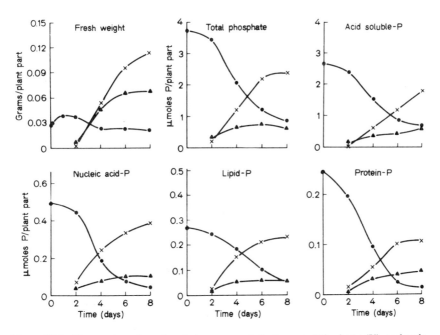

Figure 7.24. Changes in various phosphate components in the roots (△), shoots (X), and endosperm (●) of germinated oat grains. The phosphate-containing region of the endosperm is the aleurone layer. Acid soluble phosphate in the endosperm is mostly phytic acid, but in the shoot and root it is inorganic phosphate and some acid soluble organic phosphate, but not phytic acid. After Hall and Hodges (1966).

tein bodies in the scutellum. Generally, phytase activity is quite high in the mature dry cereal grain and appears to be closely associated with the aleurone grains. The enzyme increases in activity after germination, although it has been calculated that sufficient enzyme is present in the dry aleurone layer to hydrolyze all of the stored phytate: hence the significance of the increase is not known.

Changes in the phosphate fractions in the cotyledons and axes of the legumes exhibit a pattern similar to that shown already for cereals. That is, there is a decline in total phosphate (comprised largely of phytin, phospholipids, phosphoproteins, and nucleic acids) in the cotyledons after germination, coincident with an increase in phytase and acid phosphatase activity. After the catabolites of the phosphate-containing compounds are transported to the axis, they are used in the synthesis of cell components such as phospholipids and nucleotides. There is little accumulation of free inorganic phosphate within the cotyledons during the degradation of phosphate-containing compounds, but this released phosphate can account for up to 50% of the phosphate present in the growing axis. As with cereals, phytin is not synthesized within the axis. Unlike in cereals, phytase is not present in the dry seed, but its activity increases in the cotyledons after imbibition, possibly by *de novo* synthesis.

7.6.3. Mobilization of Nucleic Acids from the Storage Regions

Changes in the nucleic acid phosphate fraction in the endosperm and growing regions of the germinated oat grain are shown in Fig. 7.24. Here, as in other cereals, nucleic acids account for less than 10% of the stored phosphate: most is present in the aleurone layer, with little in the starchy endosperm. Neither the endosperm (including the aleurone layer) nor scutellum can supply enough of the nucleotide precursors required for RNA and DNA synthesis within the axis. These are provided by *de novo* synthesis, the nitrogen source being the amino acids present in, or transported to, the embryo.

In legumes seeds also there is a depletion of RNA and DNA in the cotyledons, particularly during the final disintegration of this tissue. Ribonuclease (and presumably deoxyribonuclease) activities increase from a low level and, as mentioned in Section 7.5.3, may be deposited in the protein bodies as they assume their hydrolytic function. The nucleotide products of ribonuclease activity do not accumulate in the cotyledons but are translocated to the growing axis where they may support nucleic acid synthesis. But, as with cereals, the increase in axis RNA is greater than the decline in cotyledonary RNA, indicating net nucleotide synthesis using an alternative nitrogen source.

USEFUL LITERATURE REFERENCES

GENERAL REFERENCES

Ap Rees, T., 1974, in: *Plant Biochemistry,* Biochemistry Series One, Volume 11 (H. L. Kornberg and D. C. Phillips, eds.,) Butterworths, London, pp. 89–127 (carbohydrate catabolism pathways).

Ashton, F. M., 1976, *Ann. Rev. Plant Physiol.* **27**:95–117 (storage protein mobilization).

Beevers, H., 1979, *Ann. Rev. Plant Physiol.* **30**:159–193 (microbodies).

Briggs, D. E., 1973, in: *Biosynthesis and Its Control in Plants* (B. V. Milborrow, ed.), Academic Press, London, New York, pp. 219–277 (hormones and carbohydrate metabolism).

Brown, H. T., and Morris, G. H., 1890, *J. Chem. Soc.* **57**:458–528 (classic studies on cereal reserve mobilization).

Chrispeels, M. J., and Jones, R. L., 1980/81, *Isr. J. Bot.* **29**:225–245 (endoplasmic reticulum and reserve hydrolysis).

Macleod, A. M., 1969, *Sci. Prog. Oxf.* **57**:99–112 (cereal reserve catabolism).

Manners, D. J., 1973, in: *Plant Carbohydrate Biochemistry* (J. B. Pridham, ed.), Academic Press, London, pp. 109–125 (starch catabolism).

Pernollet, J. C., 1978, *Phytochemistry* **17**:1473–1480 (protein bodies).

Richardson, M., 1977, *Phytochemistry* **16**:159–169 (proteinase inhibitors).

SECTIONS 7.1 AND 7.2

Adams, C. A., and Novellie, L., 1975, *Plant Physiol.* **55**:7–11 (mobilization in sorghum).

Ballance, G. M., and Manners, D. J., 1978, *Carbohydrate Res.* **61**:107–118 (endosperm cell wall hydrolysis).

Edelman, J., Shibko, S. I., and Keys, A. J., 1959, *J. Exp. Bot.* **10**:178–189 (sucrose transport and scutellum).

Fincher, G. B., and Stone, B. A., 1974, *Aust. J. Plant Physiol.* **1**:297–311 (hydrolysis in wheat endosperm).

Gibbons, G. C., 1979, *Carlsberg Res. Comm.* **44**:353–366. (amylase production in barley).

Glennie, C. W., Harris, J., and Liebenberg, N. V. D. W., 1983, *Cereal Chem.* **60**:27–31 (endosperm digestion in sorghum).

Humphreys, T., and Echeverria, E., 1980, *Phytochemistry* **19**:189–193 (sugar uptake by scutellum).

James, A. L., 1940, *New Phytol.* **39**:133–144 (barley axis metabolism).

Miyata, S., and Akazawa, T., 1983, *J. Cell Biol.* **96**:802–806 (amylase synthesis in rice).

Palmer, G. H., 1982, *J. Inst. Brew.* **88**:145–153 (scutellar α-amylase production).

Yamada, J., 1981, *Agric. Biol. Chem.* **45**:747–750 (rice debranching enzymes).

SECTION 7.3

Davis, B. D., 1977, *Plant Physiol.* **60**:513–517 (amylase in pea axis).

Juliano, B. O., and Varner, J. E., 1969, *Plant Physiol.* **44**:886–892 (starch catabolism in pea cotyledons).

Keusch, L., 1968, *Planta* **78**:321–350 (mannans in date).
Matheson, N. K., and Saini, H. S., 1977, *Phytochemistry* **16**:59–66 (lupin cotyledons).
McCleary, B. V., and Matheson, N. K., 1976, *Phytochemistry* **15**:43–47 (galactomannan mobilization).
Reid, J. S. G., 1971, *Planta* **100**:131–142 (fenugreek endosperm hydrolysis).
Reid, J. S. G., and Bewley, J. D., 1979, *Planta* **147**:145–150 (roles of fenugreek endosperm).
Yamasaki, Y., and Suzuki, Y., 1979, *Plant Cell Physiol.* **20**:553–562 (amylase in bean).
Yellowlees, D., 1980, *Carbohyd. Res.* **83**:109–118 (debranching enzyme of pea).

SECTION 7.4

Bortman, S. J., Trelease, R. N., and Miernyk, J. A., 1981, *Plant Physiol.* **68**:82–87 (glyoxysomes in cotton).
Breidenbach, R. W., and Beevers, H., 1967, *Biochem. Biophys. Res. Comm.* **27**:462–469 (discovery of glyoxysome).
Huang, A. H. C., 1975, *Plant Physiol.* **55**:555–558 (glycerol metabolism).
Huang, A. H. C., and Morgan, R. A., 1978, *Planta* **141**:111–116 (lipases in oil seeds).
Hutton, D., and Stumpf, P. K., 1971, *Arch. Biochem. Biophys.* **142**:48–60 (ricinoleic acid catabolism).
Khan, F. R., Saleemuddin, M., Siddiqi, M., and McFadden, B. A., 1979, *J. Biol. Chem.* **254**:6938–6944 (destruction of isocitrate lyase).
Kindl, H., Köller, W., and Frevert, J., 1980, *Hoppe-Seyler's Z. Physiol. Chem.* **361**:465–467 (cytosolic synthesis of glyoxysome enzymes).
Lord, J. M., and Bowden, L., 1978, *Plant Physiol.* **61**:266–270 (glyoxysome biogenesis).
Mettler, I. J., and Beevers, H., 1980, *Plant Physiol.* **66**:555–560 (glyoxysomal NADH reoxidation).
Moore, T. S., Jr., Lord, J. M., Kagawa, T., and Beevers, H., 1973, *Plant Physiol.* **52**:50–53 (membrane phospholipid biosynthesis).
Muto, S., and Beevers, H., 1974, *Plant Physiol.* **54**:23–28 (lipid hydrolysis).
Nishimura, M., and Beevers, H., 1979, *Plant Physiol.* **64**:31–37 (gluconeogenesis).
Opute, F. I., 1975, *Ann. Bot.* **39**:1057–1061 (oil palm lipid catabolism).
Rosnitschek, I., and Theimer, R. R., 1980, *Planta* **148**:193–198 (rapeseed lipase).
Vick, B., and Beevers, H., 1978, *Plant Physiol.* **62**:173–178 (plastid phospholipid synthesis).
Weir, E. M., Riezman, H., Grienenberger, J.-M., Becker, W. M., and Leaver, C. J., 1980, *Eur. J. Biochem.* **112**:469–477 (cucumber glyoxysomal enzyme synthesis).

SECTION 7.5

Basha, S. M. M., and Beevers, L., 1975, *Planta* **124**:77–87 (proteolysis in pea cotyledons).
Baumgartner, B., and Chrispeels, M. J., 1976, *Plant Physiol.* **58**:1–6 (mung bean proteinase inhibitors).
Dilworth, M. F., and Dure, L., III, 1978, *Plant Physiol.* **61**:698–702 (asparagine and glutamine synthesis).
Ericson, M. C., and Chrispeels, M. J., 1973, *Plant Physiol.* **52**:98–104 (mung bean storage proteins).
Gilkes, N. R., and Chrispeels, M. J., 1980, *Planta* **149**:361–369 (ER synthesis).

Hara, I., and Matsubara, H., 1980, *Plant Cell Physiol.* **21**:219–232 (pumpkin seed globulins).

Harris, N., 1981, *Plant Cell Environ.* **4**:169–175 (plasmalemmasomes).

Herman, E. M., Baumgartner, B., and Chrispeels, M. J., 1981, *Eur. J. Cell Biol.* **24**:226–235 (autophagic vacuoles).

Higgins, C. F., and Payne, J. W., 1981, *Plant Physiol.* **67**:785–792 (peptide uptake by scutellum).

Kern, R., and Chrispeels, M. J., 1978, *Plant Physiol.* **62**:815–819 (amides in mung bean).

Khavin, E. E., Misharin, S. I., Markov, Y. Y., and Peshkova, A. A., 1978, *Planta* **143**:11–20 (maize embryo proteins).

Larson, L. A., and Beevers, H., 1965, *Plant Physiol.* **40**:424–432 (homoserine in peas).

Lichtenfeld, C., Manteuffel, R., Müntz, K., Neumann, D., Scholz, G., and Weber, E., 1979, *Biochem. Physiol. Pflanzen.* **174**:255–274 (proteolysis in *Vicia faba*).

Miflin, B. J., Wallsgrove, R. M., and Lea, P. J., 1981, in: *Current Topics in Cellular Regulation,* Volume 20, Academic Press, New York, pp. 1–43 (glutamine metabolism in plants).

Moureaux, T., 1979, *Phytochemistry* **18**:1113–1117 (proteolysis in maize endosperm).

Reilly, C. C., O'Kennedy, B. T., Titus, J. S., and Splittstoesser, W. E., 1978, *Plant Cell Physiol.* **19**:1235–1246 (pumpkin globulin solubilization).

Sopanen, T., Uuskallio, M., Nyman, S. and Mikola, J., 1980, *Plant Physiol.* **65**:249–253 (scutellar leucine uptake).

Stewart, C. R., and Beevers, H., 1967, *Plant Physiol.* **42**:1587–1595 (castor bean amino acids).

Sundblom, N.-O., and Mikola, J., 1972, *Physiol. Plant.* **27**:281–284 (proteinases in barley aleurone).

Van Der Wilden, W., Gilkes, N. R., and Chrispeels, M. J., 1980, *Plant Physiol.* **66**:390–394 (vicilin peptidohydrolase synthesis).

Van Der Wilden, W., Herman, E. M., and Chrispeels, M. J., 1980, *Proc. Natl. Acad. Sci. USA* **77**:428–432 (autophagic vacuoles).

SECTION 7.6

Hall, J. R., and Hodges, T. K., 1966, *Plant Physiol.* **41**:1459–1464 (P metabolism in oats).

Maiti, I. B., and Loewus, F. A., 1978, *Planta* **142**:55–60 (*myo*-inositol metabolism).

Sugiura, M., and Sunobe, Y., 1962, *Bot. Mag. (Tokyo)* **75**:63–71 (dicot P metabolism).

Walker, K. A., 1974, *Planta* **116**:91–98 (phytin in development and germination).

Chapter 8

Control of the Mobilization of Stored Reserves

The majority of work on the control of mobilization of stored reserves has been carried out on cereal grains, although in recent years there has been a resurgence of interest in the control mechanisms in dicot seeds. For an account of mobilization *per se,* the reader is directed to Chapter 7.

8.1. CONTROL OF RESERVE MOBILIZATION IN CEREALS

That degradation of the barley endosperm is markedly slowed down or even prevented by removal of the embryo has been known since the end of the last century. But only about 1960 was it appreciated that the role of the embryo is to release a diffusible promotive factor that stimulates hydrolytic events within the endosperm. This factor has been identified as a plant hormone, a gibberellin. Not surprisingly, then, embryoless grains can be stimulated to commence hydrolytic events by application of gibberellins, with liquefaction of the endosperm occurring virtually as in the intact kernel. The control of reserve mobilization in barley has thus been established: the embryo produces gibberellin, which induces the living cells on the periphery of the endosperm, the aleurone layer, to produce and secrete food-mobilizing enzymes. These then hydrolyze the reserves stored within the nonliving cells of the starchy endosperm. Although this pattern of events will now be elaborated upon, we must add a cautionary note that not all cultivars and harvests exhibit such a clear-cut response to gibberellin. In fact, the aleurone layers of some may synthesize α-amylase without any apparent requirement for this hormone.

8.1.1. Gibberellin and the Induction of α-Amylase Synthesis in Barley

Starch is the major stored reserve found within the barley endosperm. Not unexpectedly, then, the enzyme that has been most studied in relation to control of reserve mobilization is the one largely responsible for starch hydrolysis,

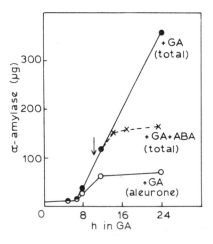

Figure 8.1. Time course of α-amylase synthesis and release by aleurone layers of barley cv. Himalaya isolated from 3-day-old water-imbibed embryoless barley grains and incubated with 1 μM GA₃ and 5 μM ABA. Enzyme activity measured in the medium surrounding the aleurone layers and in the supernatant of a 0.2 M NaCl extract of the aleurone layer (aleurone). Total (for ABA, and GA + ABA) refers to the sum of these two activities. Arrow points to the time of ABA application to the GA₃-treated aleurone layers. After Chrispeels and Varner (1967a, 1967b).

i.e., α-amylase. Much of our understanding of the mode of action of gibberellin has come from work using isolated aleurone layers, i.e., those that have been dissected out of the imbibed grain and then incubated in a sterile medium, to which appropriate additions (e.g., gibberellic acid, GA₃) have been made. Isolated aleurone layers incubated in GA₃ solution begin to secrete α-amylase about 8 h after being introduced to the hormone; this release proceeds linearly over the next 16 h (Fig. 8.1). The aleurone layer itself accumulates only a little enzyme and this might be external to the protoplasts, within the cell walls. ABA, an antagonist of gibberellin action in many plant systems, is inhibitory to α-amylase production (Fig. 8.1).

There is now overwhelming evidence that the increase in α-amylase activity is a consequence of *de novo* synthesis of this enzyme. However, there is no accumulation of an inactive precursor or zymogen of α-amylase during its synthesis—the mature, active enzyme is synthesized directly.

Because of the considerable interest in the action of gibberellins at the molecular level many studies have been conducted to determine how GA₃ induces the synthesis of α-amylase. Consequently, the events of the lag period between GA₃ addition and the start of α-amylase production (about 8 h; Fig. 8.1) have received much attention. Changes occur in the internal organization of the aleurone layer cells during this period: e.g., at least 2 h before the lag period is completed the protein-storing aleurone grains lose their spherical appearance and undergo distinct volume and shape changes. This is probably because of the hydrolysis of the proteins within these grains to provide amino acids for the *de novo* synthesis of α-amylase and other enzymes.

An attractive hypothesis, which was popular for some time, was that GA₃ induces an increase in the protein-synthesizing capacity of the aleurone layer cells, and that α-amylase is made on newly formed membrane-bound poly-somes. Evidence seemed to suggest that GA₃ induced an increase in polysome levels during the lag phase, as well as a proliferation of membranes—the polysomes then associated with the membranes, resulting in an increase in rough endoplasmic reticulum. It is now evident, however, that GA₃ does not enhance the total lipid, total lipid phosphorus, or membrane phospholipid levels in aleurone layer cells up to and beyond the time of α-amylase synthesis, nor does it induce an increase in polysomes. The original claims unfortunately were based on results obtained from inadequate techniques.

The addition of GA₃ to isolated aleurone layers causes a qualitative rather than a quantitative shift in the pattern of protein synthesis. After 10 h the major protein that is synthesized, and secreted (Fig. 8.2), is α-amylase, and later this enzyme accounts for up to 70% of total protein synthesized *de novo* by the aleurone layer. An important question in relation to gibberellin action is, therefore, whether this hormone acts at the nuclear level to induce transcription of the gene for α-amylase. Is the lag period, in fact, the time required

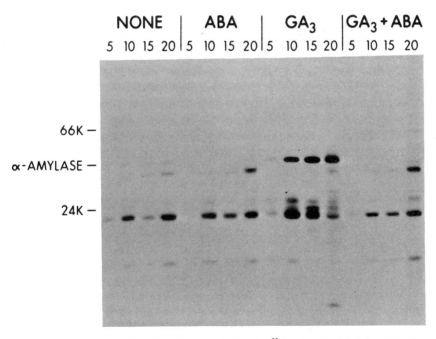

Figure 8.2. Fluorogram of a polyacrylamide gel of *in vivo* ³⁵S-methionine-labeled proteins (i.e., *de novo* synthesized proteins) secreted into the medium of aleurone layers incubated for the indicated number of hours with no hormones, ABA, or GA₃. The locations of two protein standards of mol. wt. 66 and 24 kD and of purified α-amylase are indicated on the left. Note that α-amylase is produced and secreted only in GA₃-treated aleurone layers. After Mozer (1980).

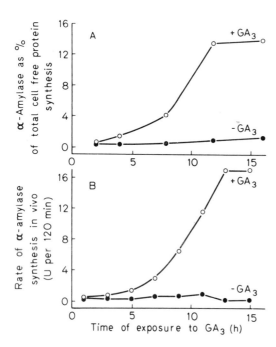

Figure 8.3. The increase with time in (A) the level of translatable messenger RNA for α-amylase, and (B) the increase in the rate of synthesis of the enzyme *in vivo* in response to GA₃ treatment. U, units of α-amylase activity. For (A) the poly(A) RNA was extracted from aleurone layers treated with GA₃ for different time periods and used to support α-amylase synthesis *in vitro*. After Higgins *et al.* (1976).

for the synthesis of the mRNA for this enzyme? Recent studies have taken advantage of the observation that many mRNAs in eukaryotic cells contain a covalently linked poly(adenylic acid) segment—poly(A). This allows RNA fractions containing such a segment to be isolated by their affinity for an inert matrix containing their complementary nucleotides: oligo(deoxythymidylic acid)—oligo(dT), or poly(uridylic acid)—poly(U). It is now accepted that GA₃ induces an increase in the poly(A)-containing RNA within the isolated aleurone layers and that this RNA when placed in an *in vitro* protein-synthesizing system catalyzes the synthesis of several proteins, including α-amylase (Fig. 8.3, A), although mRNAs for other proteins are made also. This increase in *in vitro* activity of the mRNA coincides well with the *in vivo* production of α-amylase, which occurs over the same time period (Fig. 8.3, B). These observations suggest that α-amylase messages start to be transcribed within 3–4 h of the additon of GA₃; these are translated over a subsequent 10-h period. The newly synthesized α-amylase in barley (and wheat) has a slightly higher molecular weight than the secreted enzyme. The larger α-amylase molecule is not a zymogen, but a precursor, which is present transiently until the signal peptide (see Section 2.3.4) is cleaved from it.

We saw in Fig. 8.1 how the addition of ABA prevents the increase in GA₃-induced α-amylase: this is shown in Fig. 8.2 where it can be seen that there is virtually no enzyme secreted (nor is it synthesized) in the presence of this inhibitor. Yet, surprisingly perhaps, there is still an increase in mRNAs for α-amylase in aleurone layers incubated in GA_3 + ABA, just as in layers incubated in GA_3 alone—although not to the same extent. Hence, it appears that GA_3 acts at the transcriptional level to increase the mRNAs for α-amylase, and ABA somehow modulates the translation mechanism to reduce the synthesis of this enzyme by limiting utilization of the mRNA. At the present time this point of view is by no means the only one, and some workers maintain with equal conviction that both GA_3 and ABA act at the transcriptional level to provide or retard the production of mRNAs for α-amylase.

Assuming that at least GA_3 can activate the gene for α-amylase, the next question to be resolved is, of course—how? Even the increase in mRNAs for the enzyme occurs only after 2–4 h (Fig. 8.3), and hence there is still a lag phase to be explained. Perhaps the GA_3 has to bind to a receptor site and the GA_3-receptor complex must then initiate one or more primary events that eventually lead to transcription of the α-amylase gene. That time is required for the posttranscription processing of mRNAs (i.e., adenylation at the 3′ end, or "capping" of the 5′ end) appears to have been ruled out. In mammalian systems, the activation of adenyl cyclase, an enzyme that catalyzes cyclic 3′,5′-adenosine monophosphate (cyclic AMP) synthesis—the "second messenger"—constitutes the initial molecular event in the target cells of several hormones. The nucleotide then initiates a series of events leading to a final response characteristic of the hormone. Although this is still a controversial topic, the weight of evidence is strongly against the mediation of GA_3 action via cyclic AMP.

After synthesis, the α-amylase must be secreted from the aleurone layer cells (Figs. 8.1 and 8.2). Two modes of secretion are possible: either the enzyme is packaged into vesicles prior to its export from the cells, or else it is released without prior packaging. Current evidence favors the former possibility. The ER in GA_3-treated aleurone cells may undergo some modifications to form transport vesicles in which the *de novo* synthesized α-amylase is sequestered. That the enzyme is closely associated with the endoplasmic reticulum (probably in membrane-bound vesicles) is shown in Fig. 8.4, for separation of this cellular fraction by sucrose gradient centrifugation shows a coincidence between its position on the gradient (NADH cytochrome c reductase is a marker enzyme for ER) and that of α-amylase, even when the density of the ER is changed by varying the Mg^{2+} concentration in the gradient. Some electron microscope studies of the barley aleurone layer cells seem to show an aggregation of discrete vesicles along the outer cell membrane, particularly toward the basal end of the cell (i.e., toward the starchy endosperm) (Fig. 8.5). Presumably, to effect secretion, the transport vesicles fuse with the plasmalemma, and in doing so the α-amylase and other vesicular contents are expelled from the cell (Fig. 8.5).

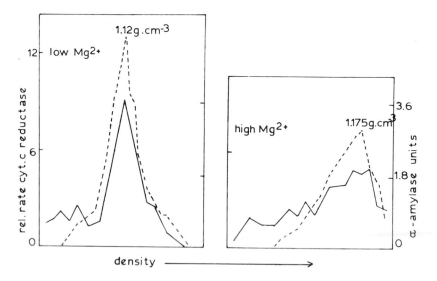

Figure 8.4. Distribution of NADH–cytochrome c reductase (----) (an endoplasmic reticulum-associated enzyme) and α-amylase activity (——) on isopycnic sucrose gradients in an organelle fraction from GA_3-treated aleurone layers. Aleurone layers were incubated for 18.5 h in GA_3, a time when synthesis and secretion of α-amylase are high. Note that when the density of the ER-rich membrane fraction is changed due to the presence of different Mg^{2+} concentrations in the gradient (peak density given in grams per cubic centimeter), then so is the density at which α-amylase sediments. This shows that the enzyme is closely associated with the membranes and may be sequestered within the vesicles. After Chrispeels and Jones (1980/1981).

8.1.2. Regulation of α-Amylase Production within the Intact Grain

In intact barley grains subjected to malting conditions (steeped in excess water under partially anaerobic conditions at 14°C) the content of gibberellin-like material (i.e., material giving a positive response in a bioassay for gibberellins) rises to a peak by the second day after imbibition starts, a time when α-amylase synthesis commences (Fig. 8.6, A). Production of the enzyme continues even as the content of gibberellin-like material declines; the rate of enzyme synthesis itself declines about a day later. In barley germinated on moist filter paper at 25°C the increase in gibberellin-like compounds follows the increase in α-amylase synthesis, which starts to decline when the synthesis of gibberellin-like material commences (Fig. 8.6, B). In order to explain this apparent anomaly, it could be argued that the low amount of gibberellin present in the germinated grains over the first 1–2 days after imbibition starts is sufficient to trigger α-amylase synthesis, although this does not explain the subsequent and massive synthesis (or release) of gibberellins. Perhaps this latter event is concerned with embryo growth rather than with reserve mobilization.

As noted already in Fig. 8.6, A and B, the synthesis of α-amylase within the intact grain eventually diminishes. One possible reason for this is that production by the embryo of the stimulus (i.e., gibberellins) for enzyme synthesis decreases, although current evidence is against this being the mechanism whereby enzyme synthesis is switched off. An alternative, but still very tenta-

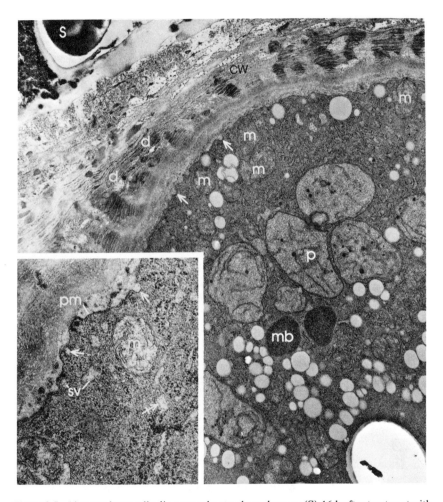

Figure 8.5. Aleurone layer cell adjacent to the starchy endosperm (S) 16 h after treatment with GA₃. The cell wall (CW) shows signs of hydrolysis in dissolution zones (d). The plasmalemma (pm) appears to be undulated (arrows), possibly owing to the fusion with small secretory vesicles (SV), some of which lie just below the membrane (see insert). Numerous mitochondria (m) are close to the plasmalemma, and plastids (p) and microbodies (mb) are present in the cytoplasm. The rough endoplasmic reticulum has areas where it is devoid of ribosomes (crossed arrow in inset). After Vigil and Ruddat (1973).

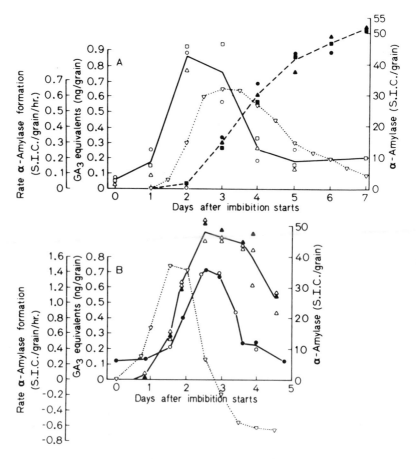

Figure 8.6. Changes in the levels of α-amylase and gibberellin-like materials in barley (*Hordeum distichon* cv. Proctor) (A) malted at 14°C or (B) germinated on moistened filter paper at 25°C. Symbols in (A) (represent values from three separate experiments): □-○-△, gibberellin-like material; ■-●-▲, α-amylase; ▽---▽, rate of formation of α-amylase. Symbols in (B) (represent values from two experiments): ○—●, gibberellin-like material; △---▲, α-amylase; ▽----▽, rate of formation of α-amylase. After Groat and Briggs (1969).

tive possibility is that gibberellins released from the embryo are enzymically degraded or converted to conjugates of low biological activity within the grain before they can become effective in the aleurone layer. But factors other than hormone availability might restrict α-amylase production. An attractive hypothesis is that inhibition is caused by the accumulation of the products of starch hydrolysis (glucose and maltose) which reach osmotic concentrations as high as 400–500 milliosmolar by the third day after imbibition starts. This probably imposes an osmotic stress on the aleurone layer cells and causes a general reduction in their metabolic activity, including protein synthesis. Certainly, α-amylase synthesis by isolated aleurone layers is increasingly inhibited when they are incubated in 0.1–0.4 M glucose or maltose (Fig. 8.7) and in osmotic agents that are not products of starch hydrolysis, e.g., polyethylene glycol and mannitol.

Thus, the control mechanism for α-amylase synthesis in the aleurone layer of intact barley grain might operate as follows (details can be followed in Fig. 8.8). Following imbibition, gibberellins are synthesized within the embryo (A); they then diffuse across the starchy endosperm and into the aleurone layer (B). In the meantime, the aleurone layer cells must undergo some metabolic changes to make them receptive to stimulation by gibberellin, which then promotes the synthesis of various hydrolases, but particularly α-amylase. This is secreted into the starchy endosperm (C), and there hydrolysis of the reserve starch commences (D). Maltose and glucose, released by amylolysis may be converted to sucrose by the aleurone cells and transported as such to the growing embryo, but most are absorbed directly through the scutellum where sucrose is formed (F). If the production of these monosaccharide sugars exceeds their rate of transport to, and utilization by, the growing embryo, they accumulate in the endosperm. There they act as an effective switch to stop production of further α-amylase (E), an enzyme whose activity in the endosperm already exceeds the requirement for the products of its hydrolytic activity (i.e., sugars).

In closing this section, it is worth noting that the initial amylolytic activity that is released from the scutellum, prior to α-amylase production by the aleu-

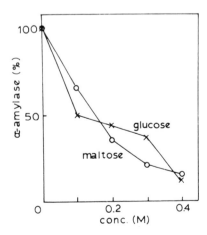

Figure 8.7. Reduction of α-amylase activity achieved by incubating isolated GA$_3$-treated aleurone layers in glucose and maltose solutions of different concentrations for 24 h. After Jones and Armstrong (1971).

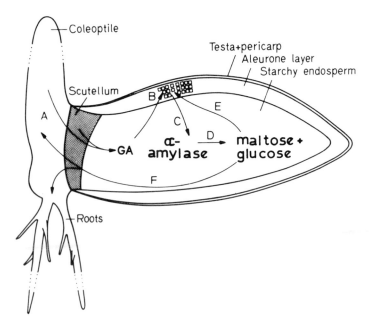

Figure 8.8. Summary diagram of the control mechanism for α-amylase production in intact barley grains. See text for details. After Jones and Armstrong (1971).

rone layer (Section 7.2.2), appears not to be under the control of gibberellins: the promotive mechanism for scutellar α-amylase synthesis remains to be determined.

8.1.3. Regulation of Other Hydrolases in the Barley Aleurone Layer

α-Amylase is not the only hydrolase whose synthesis in, or release from, the aleurone layer is controlled by gibberellin. There is, for example, GA₃-induced *de novo* synthesis of proteinase. This enzyme exhibits the same time course of synthesis and release from the aleurone layer as does α-amylase; moreover, the GA_3 dose-response curves for the two enzymes are very similar. The role of this proteinase in reserve mobilization has not been elucidated; some of its activity might be important within the aleurone layer cells themselves to provide amino acids for *de novo* enzyme synthesis by hydrolysis of the protein bodies (aleurone grains) therein. Proteinase secreted into the starchy endosperm might be involved in hydrolysis of the protein reserves there, and in the activation of bound enzymes, e.g., β-amylase (Section 7.5.2).

Release of α-amylase from the aleurone layer is facilitated by digestion of the aleurone-layer cell wall in the basal area, toward the starchy endosperm (Fig. 8.5). The cell wall has a high content of arabinoxylan (85%), which has a linear β-1,4-xylan backbone, and only a little cellulose (8%). Three pentosanases capable of degrading this polymer, viz., β-1,4-endoxylanase, β-xylanopyranosidase, and α-arabinofuranosidase, increase in activity in germinated barley grains. They are also stimulated by GA_3 in isolated aleurone layers, resulting in an increase in released pentose residues (60% xylose, 40% arabinose) into the incubation medium (Fig. 8.9). The pentosanases may also play a role in degrading the walls of the starchy endosperm, thus rendering the starch and other reserves more accessible for enzymic attack. β-1,3-Glucanase is synthesized by the aleurone layers in the absence of GA_3, but it is only released in the presence of this hormone. The glucanase is also involved in the degradation of the cell walls of the starchy endosperm cells.

Both α-amylase and acid phosphatase increase in amount in the walls of the aleurone layer prior to their release. Isolated aleurone layers incubated in the absence of GA_3 accumulate acid phosphatase (Fig. 8.10) in the inner region of the wall, indicating that the synthesis and release of this enzyme from the cytoplasm are not gibberellin dependent. But release from the cell walls (and into the surrounding medium, in the case of isolated layers) is GA_3 dependent (Fig. 8.10). Release of acid phosphatase commences 6–12 h after the introduction of the hormone and occurs via areas of the cell wall that under the microscope show obvious signs of digestion. It is likely, then, that the release of this enzyme (and also of β-1,3-glucanase) is only indirectly attributable to the action of GA_3: first, synthesis of the cell-wall-hydrolyzing enzymes, the pentosanases, is induced and these digest channels in the aleu-

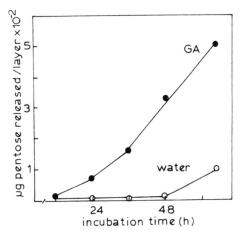

Figure 8.9. Release of pentose residues by isolated aleurone layers into the incubation medium in the presence (●) or absence (○) of 1 μM GA_3. The extent of pentose release is proportionate to the level of pentosanases induced by the hormone. After Dashek and Chrispeels (1977).

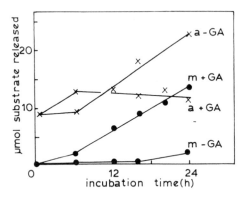

Figure 8.10. Changes in acid phosphatase activity in the incubation medium (m, ●) and in isolated aleurone layers (a, X) incubated in the presence or absence of 1 μM GA$_3$. After Ashford and Jacobsen (1974).

rone-layer cell wall to liberate enzymes that are trapped (e.g., acid phosphatase, β-1,3-glucanase) or enzymes whose release is impeded by the cell wall (e.g., α-amylase). It is likely that the cell wall channels eventually provide a common path for the release of all enzymes from the aleurone layer.

Two other enzymes that are intimately involved in starch hydrolysis are synthesized *de novo* in aleurone layer cells in response to GA$_3$; these are limit dextrinase and α-glucosidase. β-Amylase is not synthesized *de novo* in the presence or absence of GA$_3$ but is carried over in an inactive form from the developing grain. It may be activated indirectly by gibberellin through induction of a proteinase that releases it from its latent, bound form.

Besides stimulating the production and release of hydrolases, GA$_3$ can also retard enzyme synthesis. Consistent with the observation that GA$_3$ induces the degradation of aleurone layer cell walls is the observation that it inhibits the synthesis of pentosan (xylose and arabinose) components of the wall, in part owing to the reduced activity of a membrane-bound arabinosyl transferase and perhaps also of xylosyl transferase. Changes in such membrane-bound enzymes could reflect a shift in the role of membranes from the production of cell wall components to the synthesis and packaging of hydrolases.

In summary, then, gibberellin released from the embryo induces the *de novo* synthesis of α-amylase, limit dextrinase, and α-glucosidase in the aleurone layer, which are released into the starchy endosperm to effect starch hydrolysis. This release is aided by the action of gibberellin-induced pentosanases, which are produced by the aleurone-layer cells and are secreted into their walls. These pentosanases hydrolyze the walls of the aleurone layer and, along with β-1,3-glucanases, also degrade the walls of the starch-laden endosperm cells. The synthesis of α-amylase (and perhaps other carbohydrases) is reduced by the low-molecular-weight products of starch hydrolysis. The "feedback" signal presumably signifies that more sugars are available to the growing seedling than can be transported there, and that sufficient enzymes have been produced to complete hydrolysis of the stored starch.

8.1.4. Gibberellin Induction of Hydrolytic Enzymes in Other Cereals

The mechanism of control of starch hydrolysis has been studied in less detail in wild and domestic oats (*Avena fatua* and *A. sativa*), rice *(Oryza sativa)*, wheat *(Triticum aestivum)*, and maize *(Zea mays)* than in barley. But in these cereals too it is recognized that gibberellin is a controlling factor in that it induces the aleurone layer to synthesize a variety of hydrolytic enzymes. In wheat, gibberellin-stimulated α-amylase synthesis might be enhanced by cytokinin released from the endosperm. In both wild and domestic oats more α-amylase can be detected when GA_3 is supplemented by a mixture of amino acids.

Some cultivars or lines of cereals (including barley) do not appear to require the presence of the embryo for α-amylase to be produced; furthermore, de-embryonated kernels of some maize lines or hybrids do not require added GA_3 for enzyme induction. Here enzyme synthesis might be stimulated by gibberellins deposited within the aleurone cells and/or starchy endosperm during grain development. Alternatively, hydrolases might be preformed in the endosperm during maturation and released, rather than synthesized during and after germination.

In germinated rice grains the epithelial layer of the scutellum is a rich source of both α- and β-amylase and is of comparable importance to the aleurone layer. It is unclear whether gibberellin plays a role in the synthesis or release of these amylases. In maize, too, the scutellum is an important early source of α-amylase.

Maltase also increases in activity in dormant wild oat grains treated with GA_3. Imbibed dormant grains contain high α- and β-amylase activity, but starch is not broken down appreciably. Thus, maltase could be important in removing the products of starch hydrolysis by amylases, which might otherwise accumulate to concentrations that inhibit the continuation of amylolysis. There is some tentative, indirect evidence that maltase plays such a role, since the hydrolysis of raw starch from wild oat grains by amylase (salivary—not native enzyme) is enhanced *in vitro* by the addition of maltase.

8.2. CONTROL PROCESSES IN OTHER SEEDS

In cereals and some other grasses almost all the mobilizing enzymes are produced in a tissue separate from the location of the reserves. Some, probably the minor quantity, of these enzymes are formed in the embryo, but most come from a digestive tissue, the aleurone layer, which is activated by hormonal factors—gibberellins—from the germinated embryo. No strictly comparable system occurs in other seeds, however. In most species, the mobilizing enzymes are generally produced at the same tissue site as the reserves, i.e., in the endosperm or cotyledons. And even in the relatively few instances where there is a discrete enzyme-secreting tissue, such as the aleurone layer of fenugreek (which makes enzymes to attack the endosperm galactomannans), there is no evidence that factors from the embryo participate in a regulatory fashion. But mobilization of storage reserves must be associated with germination and embryo growth just as it is in cereals. If this were not the case some, possibly substantial, breakdown of reserves could occur in an imbibed, ungerminated seed (e.g., a dormant seed), with deleterious consequences. What are the controls that operate to secure production and activity of mobilizing enzymes at the right time and to link these with the embryo and subsequent seedling formation? These are the questions we will attempt to answer in the following sections.

8.2.1. When Are the Mobilizing Enzymes Produced?

Little or no activity of the major hydrolytic enzymes—proteinases, lipases, and amylases—can be detected in dry or newly imbibed seeds, but in most cases activity increases gradually as the seeds hydrate (Fig. 8.11). It seems likely that enzyme synthesis accounts for the rise in extractable activity, a conclusion that arises largely from the observation that inhibitors of protein synthesis and DNA-dependent RNA synthesis prevent the increase in certain mobilizing enzymes. Examples of enzymes that can be inhibited in this way are isocitrate lyase in squash cotyledons, lipase in castor bean endosperm, and

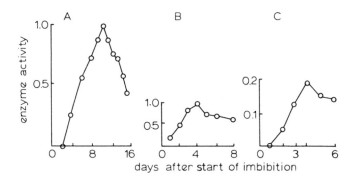

Figure 8.11. Changes in activity of some mobilizing enzymes. All seeds were allowed to germinate in darkness. (A) α-Amylase in dwarf bean cotyledons. After Van Onckelen *et al.* (1977). (See also Fig. 7.5.) (B) Lipase in cucumber cotyledons. After Davies *et al.* (1981). (See also Fig. 7.10.) (C) Proteinase (carboxypeptidase) in castor bean endosperm. After Yamamoto *et al.* (1982). (See also Figs. 7.16, 7.23, and 8.12, B.)

α-amylase in dwarf bean cotyledons. Better evidence that the enzymes are newly synthesized is rare; however, when hydrated cotyledons of watermelon and mung bean are supplied with D_2O the isocitrate lyase of the former and the endopeptidase of the latter become density labeled, showing that their synthesis occurred in the presence of the label. A few apparent exceptions have been reported, for example, a lipase which is present in newly hydrated castor bean endosperm and which is likely to have been in the dry seed (Fig. 7.10). And in lettuce, α-galactosidase and β-mannosidase, which participate in the final stages of hydrolysis of the galactomannans of the endosperm cell walls, occur in the cotyledons of the dry seeds.

It is not clear why activity of the food-mobilizing enzymes diminishes after a time (see Fig. 8.11). The fall does not always accompany the completion of mobilization of the stored reserves but often precedes it. One suggestion is that products of reserve hydrolysis (e.g., amino acids from protein) act by feedback mechanisms to arrest the synthesis of further enzyme molecules. Although the application of such products does reduce enzyme formation in some cases (e.g., proteinase in pea cotyledons), it fails to do so in others (e.g., endopeptidase in mung bean cotyledons), and so wide experimental proof for the concept of feedback inhibition is lacking. There is some evidence that synthesis of an enzyme closely associated with protein mobilization in mung beans—asparagine synthetase—is arrested when an enzyme inhibitor develops, but it is not known if such a mechanism is widespread.

8.2.2. Regulation of Mobilization

What regulates the rise in mobilizing activity that follows the germination of a seed? This question really inquires about two processes, (1) enzyme formation, and (2) enzyme activity, since regulation can be achieved by controlling either or both of these. Unfortunately, it is not always possible to determine which is responsible simply by measuring the overall rates of mobilization of the reserves. Further, measurements of the activity of extracted enzymes do not necessarily reflect their *in vivo* activity, and storage tissues that have the same extractable enzyme activity may be mobilizing reserves to quite different extents. Therefore, an understanding of the regulation of mobilization should include knowledge about rates of reserve utilization, rates of enzyme formation, and *in vitro* and *in vivo* enzyme activity.

It is clear in many cases that mobilization is controlled by the embryonic axis, i.e., the radicle/hypocotyl and plumule, or, where the endosperm is the storage tissue, by the embryo as a whole. This point has been established by means of surgical experiments in which the axis is excised at different times during germination and seedling growth. An illustration of the role of the axis

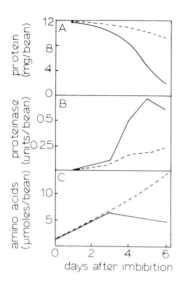

Figure 8.12. The effect of the axis on protein mobilization in mung bean cotyledons. The embryonic axis was carefully removed from dry seeds. The isolated cotyledons (----) were placed on moist sand and the intact seeds (——) on wet vermiculite in darkness (where their germination occurred). Storage protein (A), proteinase (vicilin peptidohydrolase) (B), and free amino acids (C) were measured at daily intervals over 6 days. After Kern and Chrispeels (1978).

is provided by the example shown in Fig. 8.12—protein mobilization in the cotyledons of mung bean. In the 6 days following the start of imbibition, the seed germinates and seedling growth ensues. During this time, the amount of storage protein in the cotyledons falls by about 75%, a change that is accompanied by a rise in activity of extractable proteinase (vicilin peptidohydrolase) (Fig. 8.12, A and B). The enzymatic breakdown of stored protein leads at first to an accumulation of free amino acids in the cotyledons, but these decrease after about 3 days as they, or their products, are transported into the growing axis (Fig. 8.12 C). The pattern is quite different, however, in isolated, hydrated cotyledons, i.e., when the axis is previously carefully removed from the dry seeds. Here, the rates of protein hydrolysis and the increase in peptidohydrolase activity are reduced by about 75%, and the amino acids that arise from the limited protein breakdown accumulate in the cotyledons (Fig. 8.12, A–C). Hence, the mung bean axis appears to regulate the breakdown of the proteins stored in the cotyledons, at least partially by controlling the formation of peptidohydrolase.

A similar effect of the axis is found in relation to the food reserves of several species, and in many the formation and activity of mobilizing enzymes depend upon the presence of the axis (Table 8.1). But there are also many cases in which removal of the axis has little or no effect on enzyme activity (Table 8.1), and so the tissues containing the enzyme seem to be autonomous. We should be clear, however, that the increase in *extractable* enzyme activity in the absence of the axis does not necessarily reflect the change in activity within the storage tissue itself. Some enzymes are sensitive to product inhibition, and their activity ceases when the products of even limited reserve hydrolysis accumulate, a situation that would occur if exit of materials were impeded by the absence of the growing axis. Thus, the enzyme would be inactive *in vivo* but active when extracted.

An interesting case is the cucumber seed, which illustrates the important regulatory role of the tissues enclosing its storage organs, the cotyledons. After germination of an intact seed, the fat stored in the cotyledons rapidly becomes depleted, but if the axis is removed from a newly imbibed seed, the level of stored fat hardly falls (Fig. 8.13, curve A). Removal of the testa and the inner membrane around the axis-free cotyledons allows substantial fat breakdown to proceed, however, indicating that these enclosing tissues normally inhibit the process (Fig. 8.13, curve B). The inhibition is probably due to a limitation of oxygen entry, thereby affecting both enzyme synthesis and the oxidation of fatty acids, for which molecular oxygen is needed. The coat is apparently important in seeds which have had no "surgical" treatments and which have germinated normally. Fat breakdown in the cotyledons of these seeds does not begin when the axis elongates but only when the testa is pushed off as a result of its being wedged against a peg of tissue on the elongating hypocotyl (Fig.

Table 8.1. Effect of the Attached Axis on Extractable Enzyme Activity

Enzyme	Species	Effect of axis[a]
Peptidohydrolase	Mung bean	+
Proteinase	Pea	+
	Squash	+
	Cucumber	−
	Castor bean	+
α-Amylase	Pea	+
Amylolytic activity	Dwarf bean	+
	Groundnut	−
	Mung bean	−
Lipase	Cotton	+
	Cucumber	−
Isocitrate lyase	Cucumber	−
	Squash	+
	Groundnut	+
	Castor bean	−
Glutamine synthetase	Mung bean	−
Asparagine synthetase	Mung bean	−
β-Mannanase	Lettuce	+
α-Galactosidase	Lettuce	+

[a] +, extractable enzyme activity promoted by the attached axis; −, no promotive effect of axis.

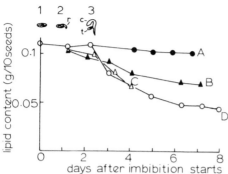

Figure 8.13. Fat mobilization in cucumber cotyledons. Total lipid was extracted and measured at daily intervals from cotyledons of seeds treated in the following ways: (A) Radicle/hypocotyl axis removed from the dry seed; testa still around the cotyledons. (B) As A but testa and membrane also removed. (C) Testa removed from newly imbibed seed; radicle/hypocotyl axis intact. (D) Fully intact seed; axis present and testa intact at the start of imbibition. The appearance of the intact seed/seedling over the first 2½ days is shown at top of diagram. Stage 1, dry seed; stage 2, radicle emergence; stage 3, testa displacement. The testa becomes displaced at day 2; lipid breakdown in the cotyledons then begins. c, cotyledons; t, testa; r, radicle. Adapted from Slack *et al.* (1977).

8.13, curve D). Moreover, the start of fat mobilization is advanced when the testa is experimentally removed from newly germinated seeds (Fig. 8.13, curve C). In cucumber, therefore, both the axis and the testa have important regulatory influences in the mobilization of fats by the cotyledons.

In conclusion, some of the processes taking place in reserve mobilization are regulated by the axis, which is needed for the formation of enzymes in some species (e.g., vicilin peptidohydrolase in mung bean) and/or for the maintenance of enzyme activity in others. The testa plays an important, additional role in cucumber, probably by affecting entry of oxygen, and it seems likely that this kind of regulation might also occur in other species.

8.2.3. Mode of Regulation by the Axis

Two possibilities have been considered to explain regulation by the axis: (1) Specific regulatory substances move from the axis to the storage organs or tissues where enzyme formation occurs, i.e., a hormonal mechanism. This explanation invokes a system similar to the one in certain cereal grains, where the embryo regulates enzyme production in the aleurone layer through the action of gibberellins that it secretes. (2) The axis is a sink, drawing off the products of reserve mobilization in the cotyledons or endosperm, which would otherwise arrest continued enzymic activity by feedback inhibition mechanisms. What is the evidence for these two possibilities?

8.2.3.1. Hormonal Control by the Axis

Most of the support for the possibility that hormones control the development of activity of the mobilizing enzymes comes from testing the effects of adding growth regulators to isolated storage tissue. This approach attempts to answer the question, can these chemicals replace the influence of the axis? In many cases, applied growth regulators induce greater breakdown of reserves in isolated tissue and/or increase the activity of enzymes concerned with mobilization. Cytokinins, for example, cause increases in amylolytic activity of isolated dwarf bean cotyledons (Fig. 8.14), and in the levels of certain proteolytic enzymes and protein hydrolysis in excised squash cotyledons, and enhance the activities of isocitrate lyase (in the glyoxylate cycle—Section 7.4) in watermelon and sunflower cotyledons. Activities of enzymes of the glyoxylate cycle in hazel cotyledons, for β-oxidation of fatty acids and for hydrolysis of stored protein reserves in castor bean endosperm, are increased by gibberellin, which also promotes α-amylase activity in excised pea cotyledons. It is important to note, however, that the effects of applied growth regulators are generally relatively small, even though sometimes comparable with the action of the axis (see Fig. 8.14), and nowhere match the dramatic control of enzyme production that is achieved by gibberellin in cereal aleurone cells.

Direct evidence for the involvement of hormones is tenuous. There are some instances, such as fructose-1,6-diphosphatase in castor bean endosperm and isocitrate lyase in megagametophytic tissue of Ponderosa pine, where diffusates or extracts of embryos stimulate enzyme activity in isolated storage tissue. But in no case has it been shown convincingly that a known hormone, e.g., a cytokinin or gibberellin, moves from the embryo or axis to the storage tissue and thereby regulates mobilizing activity. Evidence from detailed investigations on lettuce seeds does, nevertheless, support the concept of hormonal control. Although the endosperm of lettuce makes a quantitatively minor contribution to the overall reserve economy, it does appear to furnish usable food material in the earliest stages of seedling growth, before utilization of the cotyledonary reserves begins. The mannans (polymers of mannose, with galactose

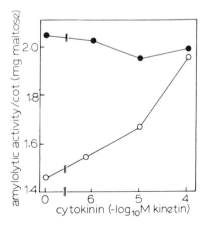

Figure 8.14. Effect of cytokinin on amylolytic activity in dwarf bean cotyledons. Intact seeds (●) were germinated: isolated cotyledons (○) were allowed to hydrate. Different concentrations of the cytokinin kinetin were applied to both. Extracted amylolytic activity was determined after 4 days. After Ilan and Gepstein (1980/81).

side groups) of the endosperm cell walls are hydrolyzed by endo-β-mannanase, produced by the endosperm itself, and the products—oligomannans—move into the cotyledons, to be attacked by β-mannosidase and α-galactosidase. The β-mannosidase is present in the dry seed, and its formation does not depend on axial influences after germination. But endo-β-mannanase forms in the endosperm of the intact seed only when the the axis has been stimulated to germinate, e.g., by light. Also, the level of α-galactosidase activity in the cotyledons increases in response to factors coming from the axis, within 2–3 h after illumination. As far as these two enzymes are concerned, the effect of the axis can

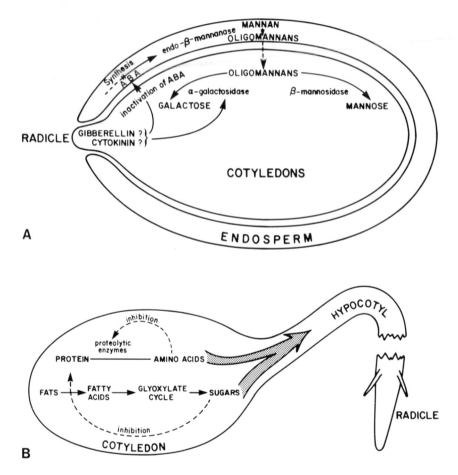

Figure 8.15. Hormonal and feedback regulation of mobilization. (A) Lettuce. After Bewley and Halmer (1980/81) and from data of Bewley *et al.* (1983). (B) Cucumber. From data of Davies and Chapman (1980) and Slack *et al.* (1977).

be replaced by a combination of added gibberellin and cytokinin. The presence of axes, stimulated to germinate by light, in the same incubation medium as isolated cotyledons, promotes an increase in α-galactosidase in these organs. There is good evidence that an inhibitor in the endosperm, probably ABA, normally stops the formation of endo-β-mannanase, and that the promotive factor(s) from the axis are needed to overcome the inhibition. The lettuce seed is therefore a case where mobilizing enzymes present at two sites—the endosperm and the cotyledons—are controlled by promotive influences coming from the axis, which may be cytokinins and gibberellins (Fig. 8.15, A).

8.2.3.2. The Axis as a Sink

Many enzymes are inhibited by the products of the reactions they catalyze. The effect can involve repression of enzyme synthesis and inhibition of the activity of already existing enzyme molecules. Such feedback inhibition may be important in the regulation of activity of the enzymes for reserve mobilization. Growth of the axis uses the products of reserve breakdown, which therefore do not accumulate in the storage tissues. The continual withdrawal of these products by the axis could account for the rise and maintenance of activity of the mobilizing enzymes; i.e., the axis regulates enzyme activity simply by virtue of its action as a sink.

This mechanism seems, in several cases, to account for the beneficial effect of the axis on mobilization. In cucumber, for example, although removal of the axis does not hinder development of several lipolytic enzymes in the cotyledons, fat breakdown itself is much reduced; hence, the activity of the enzymes apparently stops in the absence of the axis. Reducing sugars and sucrose accumulate in the excised cotyledons as lipolytic activity slows down. Moreover, addition of sucrose to isolated cotyledons leads to an even greater inhibition of lipolysis. The fact that fat breakdown proceeds in isolated cotyledons when the testa is removed might be thought to argue against regulation by a sink, since the normal sink—the growing axis—is missing. However, such isolated cotyledons enlarge, making additional cell wall material (e.g., cellulose) as they do so, presumably from the sugars coming from fat utilization: and, in addition to that of cellulose, synthesis of starch also occurs. So even though the axial sink is absent, two other sinks—cellulose and starch synthesis—serve to drain off the products of fat mobilization and, by preventing their accumulation, permit the activity of the lipolytic enzymes to continue. The activity of extracted proteolytic enzymes from the cotyledons is similarly unaffected by removal of the axis, but within the cotyledons protein hydrolysis itself is minimal. Accumulated amino acids, especially leucine and phenylalanine, as well as the dipeptide tryptophylphenylalanine, inhibit the activity of aminopeptidase in the cotyledons thus preventing protein degradation.

The cucumber seed is probably the best known case in which sink effects adequately account for the influence of the axis in certain aspects of reserve mobilization (Fig. 8.15, B), but there is evidence to suggest that a similar mechanism operates in respect to isocitrate lyase in castor bean endosperm and squash cotyledons, and to proteolytic activity in pea cotyledons. Detailed examination of other systems is needed before we can tell how widespread this type of regulatory control is.

USEFUL LITERATURE REFERENCES

SECTION 8.1 (BARLEY)

Ashford, A. E., and Jacobsen, J. V., 1974, *Planta* **120**:81–105 (release of acid phosphatase).

Bernal-Lugo, I., Beachy, R. N., and Varner, J. E., 1981, *Biochem. Biophys. Res. Comm.* **102**:617–623 (transcription of α-amylase mRNA).

Chrispeels, M. J., and Jones, R. L., 1980–81, *Isr. J. Bot.* **29**:225–245 (ER and α-amylase secretion).

Chrispeels, M. J., and Varner, J. E., 1967a, *Plant Physiol.* **42**:398–406 (α-amylase synthesis and release).

Chrispeels, M. J., and Varner, J. E., 1967b, *Plant Physiol.* **42**:1008–1016 (GA_3 and ABA action).

Dashek,, W. V., and Chrispeels, M. J., 1977, *Planta* **134**:251–256 (cell-wall-hydrolyzing enzymes).

Filner, P., and Varner, J. E., 1967, *Proc. Natl. Acad. Sci. USA* **58**:1520–1526 (*de novo* α-amylase synthesis).

Firn, R. D., and Kende, H., 1974, *Plant Physiol.* **54**:911–915 (GA_3 and membrane synthesis).

Groat, J. I., and Briggs, D. E., 1969, *Phytochemistry* **8**:1615–1627 (gibberellin and α-amylase in whole grains).

Hardie, D. G., 1975, *Phytochemistry* **14**:1719–1722 (*de novo* synthesis of several hydrolases).

Higgins, T. J. V., Zwar, J. A., and Jacobsen, J. V., 1976, *Nature* **260**:166–169 (GA_3 and mRNA synthesis).

Ho, T.-H. D., and Varner, J. E., 1978, *Arch. Biochem. Biophys.* **187**:441–446 (no pre-α-amylase synthesis).

Jacobsen, J. V., and Knox, R. B., 1974, *Planta* **115**:193–206 (GA_3 and enzyme release).

Jacobsen, J. V., and Varner, J. E., 1967, *Plant Physiol.* **42**:1596–1600 (proteinase induction by GA_3).

Johnson, K. D., and Chrispeels, M. J., 1973, *Planta* **111**:353–364 (regulation of pentosan biosynthesis).

Jones, R. L., 1971, *Plant Physiol.* **47**:412–416 (β-1,3-glucanase release).

Jones, R. L., and Armstrong, J. E., 1971, *Plant Physiol.* **48**:137–142 (osmotic regulation of α-amylase).

Locy, R., and Kende, H., 1978, *Planta* **143**:89–99 (secretion of α-amylase).

Mozer, T. J., 1980, *Cell* **20**:479–485 (transcriptional and translational control by GA_3 and ABA).

Paleg, L. G., 1960, *Plant Physiol.* **35**:293–299 (discovery of GA_3 induction of α-amylase).

Smith, M. T., and Briggs, D. E., 1980, *Phytochemistry* **19**:1025–1033 (control in intact grains).

Vigil, E. L., and Ruddat, M., 1973, *Plant Physiol.* **51**:549–558 (E. M. study of secretory vesicles).

SECTION 8.1 (OTHER CEREALS)

Boston, R. S., Miller, R. J., Mertz, J. E., and Burgess, R. R., 1982, *Plant Physiol.* **69:**150–154 (α-amylase signal peptide in wheat).

Colborne, A. J., Morris, G., and Laidman, D. L., 1976, *J. Exp. Bot.* **27:**759–767 (ER formation in wheat).

Eastwood, D., Tavener, R. J. A., and Laidman, D. L., 1969, *Nature* **221:**1267 (GA_3 and cytokinin activity in wheat).

Gibson, R. A., and Paleg, L. G., 1975, *Aust. J. Plant Physiol.* **2:**41–49 (secretion of α-amylase in wheat).

Harvey, B. M. R., and Oaks, A., 1974, *Planta* **121:**67–74 (GA_3 and enzyme induction in maize).

Naylor, J. M., 1966, *Can. J. Bot.* **44:**19–32 (α-amylase induction in wild oats).

Palmiano, E. P., and Juliano, B. O., 1972, *Plant Physiol.* **49:**751–756 (hydrolytic enzyme induction in rice).

Simpson, G. M., and Naylor, J. M., 1962, *Can. J. Bot.* **40:**1659–1673 (maltase, GA_3, and wild-oat α-amylase synthesis).

Varty, K. J., and Laidman, D. L., 1976, *J. Exp. Bot.* **27:**748–758 (phospholipid metabolism in wheat).

SECTION 8.2

Bewley, J. D., and Halmer, P., 1980/81, *Isr. J. Bot.* **29:**118–132 (embryo/endosperm interactions).

Bewley, J. D., Leung, D. W. M., and Ouellette, F. B., 1983, in: *Mobilization of Reserves in Germination* (C. Nozzolillo, P. J. Lea, and F. A. Loewus, eds.), Recent Advances in Phytochemistry, Volume 13, Plenum Press, New York, pp. 137–152 (regulation of enzymes for endosperm and oligomannans utilization in lettuce).

Davies, H. V., and Chapman, J. M., 1980, *Planta* **149:**288–291 (protein mobilization and feedback inhibition in cucumber).

Davies, H. V., and Slack, P. T., 1981, *New Phytol.* **88:**41–51 (review on regulation of reserve mobilization).

Davies, H. V., Gaba, V., Black, M., and Chapman, J. M., 1981, *Planta* **152:**70–73 (lipid mobilization in cucumber seeds).

Ilan, I., and Gepstein, S., 1980/81, *Isr. J. Bot.* **29:**193–206 (hormones and mobilization).

Kern, R., and Chrispeels, M. J., 1978, *Plant Physiol.* **62:**815–819 (protein mobilization in mung beans).

Slack, P. T., Black, M., and Chapman, J. M., 1977, *J. Exp. Bot.* **28:**569–577 (testa and lipid mobilization in cucumber).

Van Onckelen, H. A., Caubergs, R., and DeGreef, J., 1977, *Plant Cell Physiol.* **18:**1029–1040 (hormonal control of amylase in dwarf bean).

Yamamoto, T., Shimoda, T., and Funatsu, G., 1982, *Sci. Bull. Fac. Agr., Kyushu Univ.* **36:**71–78 (proteinase in castor bean).

Chapter 9

Seeds and Germination
Some Agricultural and Industrial Aspects

9.1. INTRODUCTION

The fundamental importance of germination physiology to agriculture and horticulture is so obvious that it need hardly be stated, for almost all of man's reliance on plants depends on the germinability of their seeds. The most straightforward dependence is when seeds are the starting materials for crops; in this case, we require that the seeds have high viability, that their level of germination is high (and therefore that they have no dormancy, at least under the conditions experienced during cultivation), and that germination is uniform so as to produce vigorous plants closely similar in their stage of growth. All these requirements concern aspects of seed physiology and biochemistry that have been covered in previous chapters. Also critical, especially when seeds are used directly as human or animal food, are the events occurring during seed development and maturation when the seeds' storage reserves are deposited, for the processes taking place then govern the quality and amount of materials that are nutritionally important (Section 2.2).

For this final chapter, we have chosen some aspects of industry or agriculture in which seed physiology is equally important, but perhaps in less obvious ways. These examples place events such as viability, the processes of germination, and facets of reserve mobilization in an applied context. It is hoped that the reader will obtain from them some further appreciation of how the physiology and biochemistry of seeds are of fundamental relevance to man's affairs.

9.2. MALTING

This section is not intended as a comprehensive account of malting (see the reference list for other sources) but instead it highlights the germination physiology upon which the process rests.

Malt is a cereal grain that has been allowed to germinate under controlled conditions, to make a limited amount of seedling growth, and then been dried and lightly cooked. Wheat, oats, rye, millet, sorghum, and triticale can be malted, but by far the most commonly used cereal is barley. In some countries, barley is grown almost exclusively for malting, and in others, such as the United Kingdom, the malting industry consumes a substantial proportion (ca. 25%) of the barley crop. There are many uses for malt—malt extracts, syrups, malt flour—but most of the malt is used to provide the fermentable materials during the manufacture of beers, spirits such as whisky, gin, and vodka, and malt vinegar. It is estimated that the annual world production of barley exceeds 180 million metric tons, of which about 35% is destined for malting. This industry, then, depends on germination under controlled conditions on a massive scale.

The function of malting is to achieve the production of enzymes which, at a later stage, hydrolyze the starch reserves of the endosperm to make sugars available for fermentation. Amylolytic enzymes are of prime importance, but other enzymes also play important roles. Other essential processes occur, for example the production of flavor compounds, all of which contribute to the quality of the malt. The enzymes are derived principally from the aleurone layer and to a lesser extent from the embryo. The physiological and biochemical basis of malting therefore centers on the control of reserve mobilization in cereals, which was described in Chapter 8. It is necessary, however, that mobilization does not proceed to completion, when the reserves are utilized by the growing seedling: thus, an important feature of malting is the minimization of reserve mobilization and growth so as to avoid loss of potentially fermentable material to the seedling. To this extent, then, malting involves the imposition of certain constraints on some of the normal germinative and postgerminative events.

Malting begins with imbibition by the barley grains during the process of steeping (Fig. 9.1), when large quantities of grain are submerged in water in steep tanks. The aim of steeping is to hydrate the grain to a water content sufficient to support germination, enzyme formation, and the movement of the mobilizing enzymes into the starchy endosperm, but not to encourage excessive growth. Steeping may continue for 50–70 h, but unless certain measures are adopted the quality of the grain can be severely impaired. For example, prolonged submergence in water restricts germination because of the so-called water sensitivity of the grain. This results from the anaerobic conditions imposed by excess water and leads to considerable delays in the germination of individual grains. Germination is then uneven, with a consequent reduction in extract yield of the malt. Nonuniform germination is avoided by interpolating periods of "air resting," when the water is drained off and the grain is exposed to air, say for 16 h. However, this treatment can itself have some undesirable effects, because the rootlets of the grains begin to grow ("chitting"), a

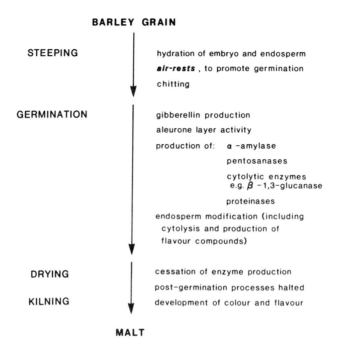

Figure 9.1. Stages in the malting of barley.

process that utilizes some of the endosperm reserves and thus lowers the eventual yield of malt. Chitting and continued rootlet elongation may be checked by the application, for short time periods, of various chemicals, such as calcium hydroxide, sodium carbonate, or potassium bromate, a chemical that also decreases proteolysis. In some malting procedures, steeped grain is heated to 45°C in order to inhibit rootlet growth.

The overall water content of the grain is generally allowed to reach about 43% by the end of steeping, though slower-malting or high-nitrogen cultivars may be permitted higher moisture levels (ca. 48%). The early phase of steeping takes the grain to about 35% water content, the minimum level needed for subsequent embryo growth; there then follows the air rest to encourage even germination; and, finally, overall moisture contents of 35–45% hydrate the endosperm sufficiently to permit enzyme production, migration, and some activity. Several steps may be taken to accelerate water uptake, including warm water or abrasion of the grain. It is essential, however, that the water content should not rise above the quoted levels, otherwise seedling growth will be excessive, at the expense of the storage reserves of the endosperm and with a reduction in malt yield.

   The conditions during steeping can also differentially affect enzymes that are important in malting. Higher temperatures (e.g., 20°C) depress the formation of β-1,3-glucanase, a cytolytic enzyme that hydrolyzes components of the endosperm cell walls (β-1,3-, β-1,4-glucans), to expose the starch grains and make them more accessible for amylolytic attack. Steep temperatures of 13–15°C are consequently preferred, to encourage the production of this enzyme.

   By careful practice, therefore, grains emerge from steeping with a water content set within narrow limits, with a small, even amount of germinative growth, with some cytolytic enzymes produced, but still without the substantial quantities of the amylolytic enzymes necessary for modification of the endosperm. Production of these enzymes, especially α-amylase, occurs largely after the steeped grain has been transferred to so-called germination beds, when growth of the germinated embryo continues to some extent, but more importantly, when the production of enzymes proceeds apace (Fig. 9.1). This stage generally takes place in huge beds accommodating as much as 50 metric tons of grain, which is turned and aerated periodically. The time spent in the germination bed varies with the malting practice but is generally from 4 to 8 days. During this time, the water content and the temperature are carefully regulated so as to encourage efficient enzyme production and satisfactory endosperm modification, but not utilization of the products of hydrolysis by the seedling which would lead to losses of malt yield (see Fig. 9.2). Addition of chemicals to the malting barley accelerates endosperm modification (gibberellic acid) and reduces rootlet growth (potassium bromate).

   The production of gibberellins is considered to be the crucial role of the germinated embryo in malting. These hormones diffuse from the scutellum to the aleurone layer, to promote enzyme synthesis and secretion into the starchy endosperm (Section 8.1). As a consequence of enzyme activity, modification of the endosperm occurs with the pattern shown in Fig. 9.3. There is some debate, however, concerning the role of embryo-derived α-amylase in this process, and when the embryonic contribution is significant, the modification pattern may be altered accordingly. Addition of gibberellic acid to malting grains is now a fairly common, perhaps routine procedure in some countries. This growth reg-

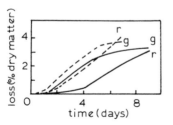

Figure 9.2. Effect of temperature on losses occurring during malting. Temperatures during germination and growth: ----, 19–22°C; ——, 13–17°C. r, Losses due to respiration. g, Losses due to growth of rootlets (chitting). After Briggs (1978).

Figure 9.3. Endosperm modification in a malted barley grain. wm, Well-modified endosperm; mm, moderately modified endosperm; pm, partly modified endosperm; um, undermodified endosperm; al, aleurone layer; c, coleoptile; h, husk; pt, pericarp/testa; r, rootlets; s, scutellum; se, scutellar epithelium; v, vascular tissue. For a comparison with the pattern of α-amylase production under nonmalting conditions, see Fig. 7.3. After Palmer (1980).

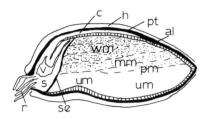

ulator accelerates malting (thus cutting costs) and makes endosperm modification more uniform. Even greater rates and uniformity of malting are achieved by abrading grains before the addition of GA_3. The entry of the growth regulator over the whole of the grain surface, to reach the underlying aleurone layer, is hence greatly enhanced. α-Amylase is the gibberellin-induced enzyme whose activity is central to endosperm modification during malting (Section 8.1.1). But the activities of other enzymes such as pentosanases, β-glucanases, and proteinases are also of great importance. The pentosanases (which are also gibberellin induced) act on the aleurone-cell walls and thus assist in the outward passage of other enzymes, especially α-amylase, into the starchy endosperm. As we noted previously, β-1,3-glucanase is a cytolytic enzyme responsible for the breakdown of the cell walls of the starchy endosperm (which consist of about 25% β-1,3-glucan) thus facilitating the accessibility of the enclosed starch grains for α-amylase, especially when the malt is later mixed with water ("mashing") to make fermentable wort. The secretion, but not the synthesis, of this cytolytic enzyme is promoted by gibberellin. Proteolytic attack by the gibberellin-induced proteinases also assists in rendering starch grains more available for amylolysis by removing the surrounding protein bodies.

By judicious regulation of the conditions, as outlined previously, satisfactory malt is thus obtained. When the malting process is terminated, the malted grain consists of a friable endosperm containing high levels of enzymes, including α-amylase (which will later participate in the hydrolysis of the starch to produce sugars for fermentation), altered protein, and various other chemicals that are important for the quality of the final product.

Since germination is clearly central to this industrial process, nongerminable grains cannot be malted! Hence, the viability and dormancy of barley selected for malting are of critical importance. Viabilities lower than 98% are unacceptable, as is dormancy of more than 3–4% of the grain. Dormancy can be a serious problem in freshly harvested grain, especially in some seasons and geographical locations. Such grain has to be stored, sometimes at elevated temperatures so as to accelerate afterripening, which increases the costs incurred in malting. On the other hand, if the grain does not have some dormancy, especially while still on the ear, it may be subject to preharvest sprouting in wet, cool weather (see Section 9.3).

In conclusion, we can see how the important industrial process of malt production involves so many different aspects of seed physiology and biochemistry, including water uptake, embryo growth, hormone production, enzyme production, some of the processes of reserve mobilization, dormancy, and viability. Successful, economically efficient malting requires an understanding of the basic seed biology in order to control and manipulate all these events.

9.3. PREHARVEST SPROUTING

The cereals are the major direct or indirect food source for the majority of the world's human population. The annual production of the major cereals—maize, wheat, and rice—is estimated at nearly 1 billion tons, and other cereals—barley, oats, sorghum, rye, and millets—also make highly important contributions to our food economy. Rice is usually consumed as the grains themselves, whereas maize and wheat generally are not. Wheat is first converted into flour, and much of the maize is fed to animals.

It is essential, in most cases, that changes associated with germination do not occur in the grains before they are harvested, as these could have serious deleterious effects on quality and usefulness. Cereal grains, especially wheat, maize, rice, and barley, are prone to germinate while still on the ear of the mother plant. This sometimes occurs just after harvest, when the stooked sheaves are wetted by rain, but in parts of the world where mechanical harvesting and threshing are practiced, germination at this stage is avoided. In all cases, however, sprouting can take place on plants still standing in the field: this is called preharvest sprouting and is a phenomenon that in some years is responsible for large losses to the agricultural industry. In this section, we will concentrate on its occurrence in wheat, an advanced state of which is shown in Figure 9.4.

9.3.1. Conditions under Which Preharvest Sprouting Occurs

Preharvest sprouting occurs in wet or humid conditions in many regions of the world, including northwest Europe, North and South America, Australia, and New Zealand. The extent of the problem is illustrated by a few examples. In Australia, where white-grained cultivars predominate, 1.8 million metric tons of grain was spoiled in northern New South Wales in 1969 owing

Figure 9.4. Sprouting in the wheat ear. The ear (B) shows a very advanced case of sprouting, induced in the laboratory. This ear would also be prone to sprouting in the field but germination would not be as extensive. The ear on the far right (C) shows sprouting as it is often found in the field. Two nonsprouted ears (A) are shown for comparison. Photograph by courtesy of Dr. M. Gale.

to germination in the ear. Though most of the sprouting damage in the United States occurs in the white-grained wheat-producing areas of the northwestern seaboard and Idaho (as a result of which Japan ceased importing these wheats for a period in 1968), preharvest sprouting sometimes also appears in the plains states from Texas to North Dakota. In Nebraska in 1977, for example, about 12% of the red-grained wheat and 19% of the durum wheat was affected. Preharvest sprouting is considered to be the most serious problem facing wheat growers in the northern region of Brazil, and it causes great losses to the producer. In Europe, the United Kingdom experienced a particularly bad year in 1977, especially in southeast England, where an average sprouting level of about 19% was recorded. Consequently, very little of the wheat crop could be used for milling, and farmers suffered a cash loss of £10–15 million. To redress the loss of homegrown material, extra wheat had to be imported, leading to a balance of payments deficit of £50–70 million.

The extent of sprouting varies according to cultivar, provenance, and growing season. Many countries routinely screen wheat cultivars and grade them for sprouting susceptibility under local conditions. Wheat in regions that are warm and dry during grain development and maturation is not likely to suffer from preharvest sprouting, but that in cooler, wet localities is at risk. Obviously, since meteorological conditions vary from year to year, crops of some seasons may completely escape sprouting damage, whereas those in other years are badly afflicted.

The period over which sprouting can occur is wide, extending from 2–3 wk after fertilization until harvest; hence, sprouting is not confined to grains that have been dried but also happens when moisture contents are still quite high. This is seen in the 1977 United Kingdom crop (Fig. 9.5), when high sprouting levels were observed in so-called dough-ripe grains—of about 45% (FW) water content. In this figure, two modes of sprouting are expressed— visible and nonvisible. The former category includes grains in which some root-let and coleoptile growth occur, i.e., a typical, germinated grain. In nonvisible sprouting, however, there is no radicle or rootlet growth, and though the coleoptile elongates, it fails to emerge from beneath the testa/pericarp, extending, instead, underneath these enclosing tissues. Hence, it is not immediately apparent that the grains have germinated, and only close inspection reveals that they have done so.

Sprouted grain is obviously of imperfect appearance and will not be purchased for milling, though it can be diverted into animal feed. The major objection to sprouted grain is not aesthetic, however, but is because of its relatively high α-amylase content. Enzymic hydrolysis of starch by this enzyme proceeds in the sprouted grain, but since the enzyme persists in the flour made from the

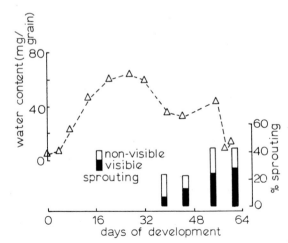

Figure 9.5. Time course of preharvest sprouting. Sprouting on the ear of wheat grains of different ages, southeast England, 1977. Water content (% FW) is also shown ($\triangle$---$\triangle$). See text for explanation of visible and nonvisible sprouting. After Mitchell *et al.* (1980).

grain, it also continues during the baking process, producing undesirable quantities of sugars. The bread-making quality of the flour is thus seriously impaired, for the loaf has a sticky crumb and a darkly pigmented crust caused by caramelization of the sugars. The production line for mass-produced, sliced loaves is halted when the slicing blades cease to cope with the gummy texture of the bread. It can be appreciated, therefore, why millers reject even slightly sprouted grain. Certain cultivars, however, have high α-amylase levels even in the absence of sprouting, a feature that can be detected in the routine testing of incoming batches of grain being offered for purchase.

9.3.2. The Physiology of Preharvest Sprouting in Wheat

Preharvest sprouting occurs because grain is too germinable! During seed development, a period of rest normally intervenes between embryo development and germination itself. The controlling factors enforcing rest are poorly understood, but they involve constraints imposed by maternal tissues, and perhaps inhibitors such as ABA (see Section 2.5.1.). These controls, which should operate in the early phases of seed development, obviously are ineffectual in those cases where sprouting occurs in very young grain. It is not known why, but the absence of the inhibitor ABA does not seem to be the explanation. In older grain, embryo development has long been completed and drying has begun. Such grain will sprout when wetted as long as there is no dormancy: hence, preharvest sprouting is intimately dependent on the presence or absence of dormancy. We should recall that dormancy in wheat is strongly expressed at germination temperatures above about 13°C (see Fig. 5.10)—a condition referred to as relative dormancy. Sprouting will therefore take place even in grains with relative dormancy when the temperature is low. This is one reason why preharvest sprouting is common when cool, wet weather prevails at certain times during maturation.

For a further appreciation of preharvest sprouting, and of how this occurs at different times on the ear, we clearly need to know something about the time course of dormancy in the developing and maturing grain. The degree of dormancy is dependent on the environmental factors operating during seed development, particularly the temperature. Grains of wheat, barley, and the grass wild oats are more dormant at maturity when development has taken place at relatively low temperatures (e.g., 10–20°C) than when it has been at higher temperatures (e.g., 20–28°C). Especially important seem to be the temperatures occurring later in grain maturation (Fig. 9.6). This explains why sprouting susceptibility of *mature* grain (i.e., at or close to harvest dryness) is less when temperatures during maturation have been low.

When wheat grains are examined at intervals during their development,

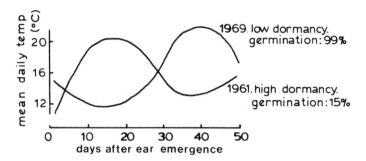

Figure 9.6. Temperature during grain development and dormancy of mature barley grain. Mean daily temperatures experienced over the course of grain development and maturation are shown for two crops of barley. Germinability was measured 3 wk after harvest. After Reiner and Loch (1975).

we can see how the pattern of dormancy varies with the temperature (Fig. 9.7). The first point to note is how the low dormancy of mature grains is achieved if higher temperatures (20°C) are obtained during development. Such grains actually enter a period of dormancy (at 30–40 days, when the germination tested at 20°C is low), but they rapidly pass out of it as they afterripen on the ear. Afterripening occurs because the water content of 40-day-old grain is low (ca. 30–35%), and it is also enhanced by the higher temperature. On the other hand, grains developing and ripening at 10°C have a low germinability when young (ca. 25 days old) and show only marginal afterripening; so even at 60 days old they are still substantially dormant and do not respond expeditiously even to a germination temperature of 10°C.

Let us consider briefly what might be the fate of these two batches of grain—developing at 10°C and 25°/20°C, respectively—if subjected to rain at different times. Grain developing and maturing at 25°/20°C has a fairly low dormancy at ages up to about 25 days and is therefore highly susceptible to sprouting when wetted, whether the temperature remains at about 20°C or falls to 10°C for a few days. Thereafter—up to age 40 days—grains would have a propensity for sprouting only if cool, wet conditions arise, because at a germination temperature of 20°C they express dormancy. After about 40 days, cool, wet weather encourages sprouting, but since the grains are afterripening, heavy rainfall even at warmer temperatures increasingly induces grain germination. At maturity, when afterripening has removed dormancy, grains are highly susceptible to sprouting in both cool and warm wet weather. Grains that are in the trough of dormancy (30–40 days old) are sensitive to a day or two of low temperatures (e.g., 10°C), which promotes subsequent germination, probably because of the low-temperature breaking of dormancy. But when subjected to cool conditions throughout their entire developmental and maturation period, grains remain dormant and resist sprouting even when severely wetted. In nature, such extreme temperature differences during development as those quoted here will not occur; notwithstanding this, temperature-induced variation in dormancy and germination will arise to some extent or other.

We can see, therefore, that preharvest sprouting in wheat depends substantially on the dormancy characteristics of the grain, on the temperature dependence of dormancy expression, and on factors involved in the relief of dormancy (low temperatures and afterripening) which are effective even on the ear.

Some steps can be taken to reduce the incidence of preharvest sprouting. Taking the weather conditions into account, sprouting susceptiblity can sometimes be predicted, and late sprouting can thus be circumvented by harvesting the crop earlier than usual. Further, breeding programs can produce wheat cultivars with deeper grain dormancy. Though protecting against sprouting, this can have undesirable consequences, for it can interfere with seed testing after harvest or delay germination in the field, which is particularly disadvantageous for winter wheats sown in the fall. It could be argued that real progress in dealing with the sprouting problem may have to await the time when we

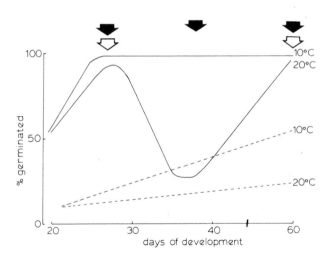

Figure 9.7. Diagram showing patterns of dormancy in developing wheat grains (cv. Sappo). Plants bearing developing grains were transferred to two different temperature regimes: ——, 16 h day at 25°C, 8 h night at 20°C; ----, 16 h day at 10°C, 8 h night at 10°C. At intervals throughout development and maturation grains were removed from the ears and tested for germinability (i.e., dormancy) at two temperatures, 10°C and 20°C. Arrows indicate sprouting susceptibility: black arrows, at a germination temperature of 10°C; white arrows, at a germination temperature of 20°C. By R. Butler and Black (unpublished observations).

have a fuller knowledge of the biochemical basis of dormancy, since only when we understand the mechanism of dormancy can be begin to control it.

9.4. GENETIC CONSERVATION—SEED GENE BANKS

To meet the demands of increasing population and industrialization, society has developed an array of plants for consumption by the people themselves or by their domesticated animals. Yet only about 15 species actually feed the world; these include five cereals (rice, wheat, maize, barley, and sorghum), two sugar plants (sugar cane and sugar beet), three subterranean crops (potato, sweet potato, and cassava), three legumes (bean, soybean, and peanut), and two tree crops (coconut and banana). Of these, just three species, wheat, rice, and maize, produce nearly 70% of the world's seed crop. Thus the fate of hundreds of millions of lives hangs on the precarious balance of the genetic systems of these three crops, their diseases and pests, and their interactions with their environment. Unfortunately, the current trend to develop a few cultivars of high-yielding crops is also increasing genetic uniformity, at the expense of genetic diversity or variability, setting up the potential for global disaster should such crops become susceptible to specific diseases or to a changing environment. For example, losses on a relatively major scale occurred in the United States in 1970, when overuse of a single genetic strain of maize allowed southern corn leaf blight to become rampant. Losses reached 50% in some states and 15% nationally. Genetic resources are being lost also by cultivation of undisturbed lands where wild progenitors might be found, by abandonment of old farming systems, and by the destruction or deterioration of cultivars no longer in use. This inestimable loss of potentially valuable genes necessary for future plant improvement, particularly in relation to pest, disease, and stress resistance, has reached alarming proportions. The gradual loss of genetic variability in the form of germ plasm is known as genetic erosion. Slowly, necessary steps are being taken to assemble germ plasm resources of our cultivated plants and their relatives in order to preserve them in germ plasm banks—gene banks.

Ideally, we should conserve as complete as possible a range of genetic diversity of all species that have actual or potential economic importance. Clearly this is impossible, for even ignoring the problem of deciding the potential value of species not yet investigated we would need to sample a known species right through its range of genetic diversity. This requires at least 10–20×10^3 samples per single crop species, and probably more for the major world crop cereals. Moreover, each sample should contain about 3000 seeds to encompass the normal genetic variation within a population. Thus there are definite limits to the number of samples that can be handled effectively in programs for the conservation and utilization of crop genetic resources. These limits are imposed by financial and personnel considerations, and since governments and world bodies have been generally reluctant to commit themselves to genetic conservation programs, gene banks are few and, on a global basis, distressingly inadequate.

Several stages involved in the establishment of a gene bank are outlined in the following flow chart.

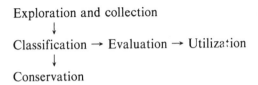

For most species there is one major limiting factor in these processes which determines what the sample size will be. For example, in the case of the primitive land races of the temperate cereals in the Mediterranean basin, which are rapidly facing extinction, adequate personnel resources are not available to collect the material before it is lost forever. On the other hand, in the case of the hexaploid weed relatives of cultivated oats there is an abundance of material present in the field, and the major limiting factor is the breeders' capacity to evaluate and utilize the collected material.

It is not appropriate here to discuss the economical, political, and logistical difficulties associated with exploring for and collecting populations of seeds, nor with the evaluation by breeders of the potential usefulness, in relation to certain desirable genetic traits. Instead, let us turn briefly to the critical features germane to the establishment of the seed gene bank itself. But we should also recognize that for some species, particularly those whose seeds have low storability, the formation of seed gene banks is inappropriate, and other methods, e.g., meristem banks, seedling banks, might have to be found.

There is little point in collecting the maximum amount of genetic diversity in a species unless the material can be adequately conserved to prevent subsequent loss of this diversity. To this end, the International Board for Plant Genetic Resources (IBPGR) has been involved, for the last decade, in the establishment and development of a world network of crop genetic conservation activities. It is on their guidelines and recommendations for conservation of material in gene banks that some of the following section is based.

After collection seeds must be dried under carefully controlled conditions to avoid damage, then cleaned and sorted for storage in suitable containers, e.g., hermetically sealed cans or laminated foil packages. Prior to storage, germination tests must be carried out to determine initial viability. Since germination is essentially a destructive process as far as storage is concerned, tests must use the minimum number of seeds to yield statistically significant results and, at the same time, not deplete the stored stocks. A general rule of thumb is that for heterogeneous material 12,000 seeds are required for storage, and for homogenous material 4000 seeds. These should make up a "base" collection, i.e., one that is undisturbed and specifically laid down for long-term conservation. A "working" collection of the same seeds is also desirable, and these seeds can be used for medium-term storage, regeneration, evaluation, and distribution to other users. If a seed collection is initially unsuitable for storage, or its viability deteriorates to an unacceptably low level during storage, then facilities must be available for the propagation and regeneration of the species with the same genetic composition of the original seed.

Gene banks vary in their size and design. In the National Seed Storage Laboratory in Fort Collins, Colorado, U.S.A., for example, different storage rooms are maintained at between 4°C and 12°C and up to 35% relative humidity. The accessions are arranged in numbered steel trays, placed in numbered steel racks (Fig. 9.8); each room has a capacity of about 180,000 pint cans. Liquid nitrogen storage is a newly developing technique and potentially can confer infinite longevity on seeds. An advantage of this method is that the need for germination tests is removed. Certainly, the lower the storage temperature,

Figure 9.8. Many thousands of seed containers, each labeled as to serial number, kind of seeds, and storage location, are housed in the National Seed Storage Laboratory in Fort Collins, Colorado, U.S.A. Photograph courtesy of Dr. P. C. Stanwood (pictured).

even below $-12°C$, the less frequently will monitoring of viability be necessary.

The problems of short-lived seeds, particularly those which cannot withstand drying or cold treatment, still have to be surmounted. For further details, see Section 3.2.5.

9.5. ENHANCEMENT OF SEED GERMINATION BY OSMOTIC STRESS AND DEHYDRATION TREATMENTS

9.5.1. Osmotic Stress

The imposition of water stress upon a germinating seed generally leads to a delay in radicle emergence, and, provided that the stress is of sufficient magnitude, this terminal event of germination can be completely suppressed. From this observation has developed the horticultural practice of "priming" or "osmopriming," where seeds are initially imbibed in an osmoticum at a concentration that allows for water uptake and permits all the processes of germination to be completed, except for the final event of radicle emergence. Hence, primed seeds are poised at a stage just prior to the completion of germination. Upon subsequent introduction of the seeds to water, germination is rapidly completed, the radicle elongates after only a short delay, and thus seedling establishment is hastened (Section 5.5.2.4).

Although not used widely on a commercial scale, priming to hasten germination has several potential advantages. For example, seeds sown in cold soils in temperate climates seem to germinate and become established faster after priming, and even the final yield may be higher. This means that the seeds do not remain so long in the soil where they are prone to pest and disease attacks. Moreover, by emerging early, seedlings can compete more effectively with weeds, and, if necessary, post-emergence herbicides can be applied before weeds grow large and resistant. Under controlled-environment conditions, such as in greenhouses, faster germination (Table 9.1) and establishment of bedding plants helps reduce the time that the facilities are committed to a particular crop, thus enhancing turnover.

The priming treatments required to optimize faster germination are quite variable and must be determined empirically for each species—perhaps even for particular cultivars and harvests. Some examples of the species in which the time lag between imbibition and germination has been successfully reduced are shown in Table 9.1. The commonly used priming agent is polyethylene glycol (PEG), a nontoxic, high-molecular-weight compound that does not penetrate the cell walls. Low-molecular-weight compounds such as certain metabolically inert sugars (e.g., mannitol), and salts ($MgSO_4$, KNO_3), have also

Chapter 9

Table 9.1. The Hastening of Germination by Priming Conditions[a]

Species	Percent germinated[b]		Days to 50% germination[b]	
	P	NP	P	NP
Antirrhinum	88	92	1	4
Cyclamen	65	61	3	15
Freesia	88	92	10	22
Polyanthus	81	64	5	10
Primula	68	72	4	9
Geranium	79	67	1	3

[a]Data taken from Heydecker (1975a and 1975b).
[b]P, primed; NP, unprimed (sown without any prior treatment).

been used, although there is no obvious advantage to doing so, and their accumulation to deleterious concentrations within cells is an undesirable risk.

Two general rules of priming appear to apply: (1) for any given osmotic potential, a longer priming treatment is required at lower temperatures, and (2) at any given temperature, increasing the concentration of the osmoticum (up to a certain limit) shortens the priming time. For most seeds the efficacious osmotic potential is in the range of -5 to -20 bars, in a temperature range of $10-20°C$. Presumably, at more negative osmotic potentials and lower temperatures than these, normal germination processes do not take place, and hence primed seeds will be arrested at a germination phase prior to the stage immediately before radicle protrusion, which is of no advantage as far as the speeding up of germination is concerned. Other precautions need also be taken during priming treatments. For example, the seed must be provided with an ample supply of oxygen, presumably to allow metabolic processes essential for germination to proceed. Also the imbibed seed must be protected against microbial attack or the proliferation of seed-borne pathogens. For some seeds, the addition of growth regulators (e.g., cytokinins, gibberellins, or fusicoccin) to the priming solution may be of advantage.

Osmotic priming is not used on a large scale, and it is questionable if it will ever be used to prime the enormous quantities of crop seeds planted outdoors in cool or temperate climates, where priming is likely to be seen to best advantage. Economically, it is probably not feasible to build the facilities needed to prime large seeds sown in huge amounts—the cereals or legumes for example. Moreover, since seeds lose the advantages of priming if subsequently dried, conventional sowing techniques cannot be used. The application of priming might be to small-seeded ornamentals and vegetables grown on a relatively minor scale for greenhouses or market-garden plots.

The physiological and biochemical responses of seeds to priming are not understood, although it would appear that radicle elongation is more sensitive to stress than the other processes involved in germination. We have seen previously (Fig. 3.19) that incipient radicle elongation is particularly sensitive to water stress, such that considerably greater stresses are required to prevent the start of elongation than to interfere with subsequent radicle growth.

9.5.2. Dehydration–Rehydration

We saw in Chapter 2 (Section 2.4) that seeds develop a tolerance of desiccation during development, and that the drying process itself might be important in switching seeds from a developmental mode of metabolism to a germination mode. During germination, seeds remain tolerant of desiccation, but at some stage after axis elongation there is generally a loss of this tolerance. It is claimed that drying of seeds during germination, especially at times just prior to radicle elongation, or at the stage of incipient seedling formation, confers certain advantages on the subsequent vegetative plant. In particular, it is reported that the hydration–dehydration treatment results in increased frost and drought resistance—this treatment is thus called "presowing drought hardening."

Essentially, the treatment involves allowing seeds or grains to take up moisture at 10–25°C for several hours before drying them in a stream of air. For some seeds one wetting and drying cycle is sufficient, but for others two or three cycles apparently yield the best results. The time of drying in relation to germination and seedling growth appears to be critical; as these events proceed, the seeds become less desiccation tolerant, and yet the more advanced the seed at the time of drying, the greater the degree of hardening conferred. Thus, the optimum time for the drying treatment(s) must be a compromise between the two opposing tendencies.

The advantages of presowing drought hardening are somewhat enigmatic. Some scientists claim both an increase in resistance of the vegetative plant to drought and cold conditions and substantial increases in final yield—by as much as 100% in some cases, although increases in the 5–25% range are more comon. Some examples are given in Table 9.2. Other scientists, on the other hand, have failed to find any substantial beneficial effects resulting from wetting and drying cycles, on either hardiness or yield. This could be because optimal conditions for presowing drought hardening were not achieved or because, indeed, there is no consistent benefit to be derived from this treatment. The practice of wetting and drying to harden seeds prior to planting is not commonly used in agriculture or horticulture, presumably because it is considered that the means do not justify the end.

Table 9.2. Presowing Drought-Hardening Treatment Effects on Crop Plant Yields, as Shown in Production Tests[a]

Crop	Productivity in 100 lbs/acre		
	Nonhardened	Hardened	Percent increase
Barley	9.3	10.4	11
Maize	75.4	93.2	24
Sugar beet	22.6	28	24
Table beet	55.6	59.6	7
Carrot	27	30.1	11
Tomato	94.9	174	83

[a]Recalculated from Henckel *et al.* (1964).

USEFUL LITERATURE REFERENCES

SECTION 9.2

Briggs, D. E., 1978, *Barley,* Chapman and Hall, London (barley utilization, including malting).

Macleod, A. M., 1979, in: *Brewing Science,* Volume 1 (J. R. A. Pollock, ed.), Academic Press, London, pp. 145–232 (physiology of malting).

Palmer, G. H., 1980, in: *Cereals for Food and Beverages* (G. E. Inglett and L. Munck, eds.), Academic Press, London, pp. 301–338 (morphology and physiology of malting barleys).

Palmer, G. H., and Bathgate, G. N., 1976, in: *Advances in Cereal Science and Technology* (Y. Pomeranz, ed.), American Association of Cereal Chemists, St. Paul, Minnesota, pp. 237–324 (malting and brewing).

SECTION 9.3

Mitchell, B., Armstrong, C., Black, M., and Chapman, J., 1980, in: *Seed Production* (P. D. Hebblethwaite, ed.), Butterworths, London, pp. 339–356 (physiology of preharvest sprouting).

Reiner, L., and Loch, U., 1975, *Cereal Res. Commun.* **4:**107–110 (temperature and barley dormancy).

Sawhney, R., and Naylor, J. M., 1979, *Can. J. Bot.* **57:**59–63 (temperature and dormancy in wild oats).

SECTION 9.4

Bass, L. N., 1979, in: *The Plant Seed. Development, Preservation and Germination* (I. Rubenstein, R. L. Phillips, C. E. Green, and B. G. Gengenbach, eds.), Academic Press, New York, pp. 145–170 (aspects of seed preservation).

Frankel, O. H., and Hawkes, J. D. (eds.), 1975, *Crop Genetic Resources for Today and Tomorrow,* Cambridge University Press, Cambridge, U.K. (review of genetic conservation).

Mooney, P. R., 1980, *Seeds of the Earth. A Private or Public Resource?* Inter Pares, Ottawa (controversial look at seed resource ownership).

Muhammed, A., Aksel, R., and Von Borstel, R. C., 1976, *Genetic Diversity in Plants,* Plenum Press, New York.

Withers, L. A., and Williams, J. T., 1982, *Crop Genetic Resources. The Conservation of Difficult Material.* I.U.B.S. Publications, R. Royer, Paris (storage of recalcitrant seeds).

SECTION 9.5

Bodsworth, S., and Bewley, J. D., 1981, *Can. J. Bot.* **59:**672–676 (osmotic priming advantages at low temperatures).

Evenari, M., 1964, *Nature (London)* **204:**1010–1011 (presowing drought hardening).

Henckel, P. A., Martyanova, K. L., and Zubova, L. S., 1964, *Soviet Pl. Physiol.* **11:**457–461 (presowing drought hardening).

Heydecker, W., 1975a, *Grower,* Sept. 27, 1975 (osmotic priming advantages).

Heydecker, W., 1975b. *Commercial Grower,* Oct. 17, 1975 (osmotic priming advantages).

Heydecker, W., and Wainwright, H., 1976, *Sci. Hort.* **5:**183–189 (osmotic priming advantages, with *Cyclamen*).

Rennick, G. A., and Tiernan, P. I., 1978, *Seed Sci. Technol.* **6:**695–700 (osmotic priming advantages, with celery).

Siminovitch, D., and Cloutier, Y., 1982, *Plant Physiol.* **69:**250–258 (desiccation and hardening).

Szafirowska, A., Khan, A. A., and Peck, N. H., 1981, *Agron. J.* **73:**845–848 (osmotic priming of carrot).

Index

The number in boldface is the first page of several in which the subject receives extensive reference.